Miner ▷ P9-CCM-470

Mineral
Fiber

A. Knop, L. A. Pilato

Phenolic Resins

Chemistry,
Applications and Performance

Future Directions

With 109 Figures and 114 Tables

Springer-Verlag
Berlin Heidelberg New York Tokyo

Dr. Andre Knop
Rütgerswerke AG, Frankfurt, FRG

Dr. Louis A. Pilato
Temecon Group International Inc., Bound Brook, NJ, USA

With participation of Volker Böhmer (Chapter 4)

This volume continues the monograph "Chemistry and Application
of Phenolic Resins" by A. Knop and W. Scheib

ISBN 3-540-15039-0 Springer-Verlag Berlin Heidelberg New York Tokyo
ISBN 0-387-15039-0 Springer-Verlag New York Heidelberg Berlin Tokyo

Library of Congress Cataloging in Publication Data.
Knop, A. (Andre), 1941– Phenolic resins. Includes bibliographies. 1. Phenolic resins. I. Pilato, L. (Louis), 1934–.
II. Title. TP1180.P39K562 1985   668.4'222   85-14708

Typesetting: Brühlsche Universitätsdruckerei, Giessen; Offsetprinting: Heenemann, Berlin
Bookbinding: Lüderitz & Bauer, Berlin
2154/3020-543210

Preface

In 1983/84 American Chemical Society, N.Y., Kammer der Technik, Berlin/GDR, and chemical companies closely associated with phenolic resins underscored the embryonic efforts of Leo H. Baekeland and celebrated the 75th anniversary of the first wholly synthetic plastic material during symposia held in Washington, D.C., August 1983, and Berlin, September 1984, respectively.

Since their introduction in 1910, the highly versatile family of phenolic resins has demonstrated an important role in the continuing development of the electrical, automotive, construction and appliance industries. In the 80's the wave of high technology has fostered their active participation in "high tech" areas ranging from electronics, computers, communication, outer space/aerospace, biomaterials, biotechnology and advanced composites. Many phenolic resin systems are actively involved in the "leading edge" of these innovative technologies. Thus, they demonstrate an uncanny versatility to be adaptable to prevailing times as today's society is transforming from an industrial to an information/communication society.

The excellent participation at the recent scientific symposia and the acceptance of the early edition "Chemistry and Application of Phenolic Resins" by A. Knop and W. Scheib – including the Japanese and Russian translation – by the industrial and chemical community demonstrated a high level of interest in the broad subject of phenolic resins and has provided the stimuli of this present publication.

This volume covers fundamentals, the chemical and technological progress, and new applications including the literature generally up to July 1984. Special emphasis was assigned to advanced instrumental and analytical techniques and environmental aspects.

We would like to express our gratitude to all colleagues engaged in the phenolic discipline, in particular to those who have assisted us with advice and suggestions.

Frankfurt and Bound Brook, July 1985   A. Knop
                                                                        L. Pilato

Table of Contents

Abbreviations XIII

Units XV

1   **Introduction** 1
1.1   Origin 1
1.2   Commercial Development of Phenolic Resins 3
1.3   References 4

2   **Raw Materials** 5
2.1   Phenols 5
2.1.1 Physical Properties of Phenol 5
2.1.2 Supply and Use of Phenol 6
2.1.3 Phenol Production Processes 7
2.1.4 Cresols and Xylenols – Synthesis Methods 9
2.1.5 Alkylphenols 10
2.1.6 Phenols from Coal and Petroleum 11
2.1.7 Other Phenolic Compounds 13
2.2   Aldehydes 14
2.2.1 Formaldehyde, Properties, and Processing 14
2.2.2 Paraformaldehyde 18
2.2.3 Trioxane and Cyclic Formals 19
2.2.4 Hexamethylenetetramine, HMTA 20
2.2.5 Furfural 20
2.2.6 Other Aldehydes 21
2.3   References 22

3   **Reaction Mechanisms** 24
3.1   Molecular Structure and Reactivity of Phenols 25
3.2   Formaldehyde-Water and Formaldehyde-Alcohol Equilibria 29
3.3   Phenol-Formaldehyde Reaction under Alkaline Conditions 31
3.3.1 Inorganic Catalysts and Tertiary Amines 31
3.3.2 Ammonia, HMTA, and Amine-Catalyzed Reactions 34
3.3.3 Reaction Kinetics of the Base-Catalyzed Hydroxymethylation 36
3.3.4 Prepolymer Formation 40
3.3.5 Resol Crosslinking Reactions. Quinone Methides 42
3.4   Phenol-Formaldehyde Reactions under Acidic Conditions 46
3.4.1 Strong Acids 46
3.4.2 Reaction Kinetics in Acidic Medium 48

3.4.3   Reaction under Weak Acidic Conditions. "High-*Ortho*" Novolak Resins   49
3.5   Novolak Crosslinking Reaction with HMTA   52
3.6   Reaction with Epoxide Resins   54
3.7   Reaction with Diisocyanates   55
3.8   Reaction with Urea, Melamine, and UF/MF-Resins   56
3.9   Reaction with Imide Precursors   57
3.10   Statistical Approach and Computer Simulation   58
3.11   References   58

4   Structurally Uniform Oligomers (by Volker Böhmer)   62
4.1   Synthesis   63
4.1.1   Principles, Protective Groups   63
4.1.2   Linear Oligomers   65
4.1.3   Cyclic Compounds, Calixarenes   67
4.1.4   Chemical Modifications   69
4.2   Properties   73
4.2.1   Physical Properties   73
4.2.2   Acidity   74
4.2.3   Complexation of Cations   76
4.2.4   Kinetic Studies   77
4.2.5   Crystal Structure   80
4.2.6   Conformation in Solution   85
4.3   References   88

5   Resin Production   91
5.1   Novolak Resins   93
5.2   Resols   95
5.3   High-*Ortho* Resins   97
5.3.1   Novolak   97
5.3.2   Liquid Resol   98
5.4   In Situ Dispersions   98
5.4.1   Aqueous Dispersions   99
5.4.2   Solid Resols   99
5.5   Dust Explosions   100
5.6   Spray Dried Resins   101
5.7   References   101

6   Toxicology and Environmental Protection   103
6.1   Toxicological Behavior of Phenols   104
6.2   Toxicological Behavior of Formaldehyde   104
6.3   Environmental Protection and Regulation   106
6.4   Waste Water and Exhaust Air Treatment Processes   108
6.4.1   Microbial Transformation and Degradation   108
6.4.2   Chemical Oxidation and Resinification Reactions   109
6.4.3   Thermal and Catalytic Incineration   110
6.4.4   Extraction Processes and Recovery   111
6.4.5   Activated Carbon Process   112
6.4.6   Gas Scrubbing Processes   112

6.5   Smoke and Gas Evolution from the Combustion of Phenolics   113
6.6   References   116

7   Analytical Methods   118
7.1   Monomers   118
7.1.1   Phenols   118
7.1.2   Formaldehyde.   119
7.2   Polymers   119
7.2.1   Infrared Spectroscopy   119
7.2.2   Nuclear Magnetic Resonance Spectroscopy   120
7.2.3   Electron Spectroscopy   123
7.2.4   Mass Spectrometry   124
7.2.5   High Performance Liquid Chromatography   124
7.2.6   Paper and Thin Layer Chromatography   128
7.2.7   Gas Chromatography   128
7.2.8   Gel Permeation Chromatography   130
7.2.9   Thermogravimetric Analysis.   130
7.2.10 Differential Thermal Analysis   130
7.2.11 Differential Scanning Calorimetry   130
7.2.12 Dynamic Mechanical Analysis   131
7.2.13 Torsional Braid Analysis   131
7.3   X-Ray Diffraction Analysis   132
7.4   Other Characterization   132
7.4.1   Thermodynamic Properties   132
7.4.2   Solution Properties   133
7.4.3   Dipole Moments   133
7.5   Resin Macrostructure   134
7.5.1   Nitrogen and Water   134
7.5.2   Resin Properties and Quality Control   135
7.6   Structural Analysis of Cured Resins   137
7.7   References   137

**8   Degradation of Phenolic Resins by Heat, Oxygen, and High-Energy
        Radiation**   140
8.1   Thermal Degradation   140
8.2   Oxidation Reactions   143
8.3   Degradation by High-Energy Radiation   144
8.4   References   146

9   Modified and Thermal-Resistant Resins   147
9.1   Etherification Reactions   148
9.2   Esterification Reactions   149
9.2.1   Boron-Modified Resins   149
9.2.2   Silicon-Modified Resins   150
9.2.3   Phosphorus-Modified Resins   151
9.3   Heavy Metal-Modified Resins   151
9.4   Nitrogen-Modified Resins   152
9.5   Sulfur-Modified Resins   152

9.6   Others 153
9.7   References 154

10   High Technology and New Applications 156
10.1   Carbon and Graphite Materials 156
10.2   Phenolic Resin – Fiber Composites 160
10.3   Reaction Injection Molding 164
10.4   Phenolic Fibers 165
10.5   Photo Resists 166
10.6   Carbonless Paper 168
10.7   Ion Exchange Resins 169
10.8   Interpenetrating Polymer Networks and Polymer Blends 171
10.9   Enhanced Oil Recovery 172
10.10   References 172

11   Composite Wood Materials 175
11.1   Wood 175
11.2   Adhesives and Wood Gluing 176
11.3   Physical Properties of Composite Wood Materials 177
11.4   Particle Boards 177
11.4.1 Wood Chips, Resins, and Additives 179
11.4.2 Production of Particle Boards 182
11.4.3 Properties of Particle Boards 183
11.5   Wafer Board and Oriented Strand Board 185
11.6   Plywood 186
11.6.1 Resins, Additives, and Formulations 187
11.6.2 Production of Plywood 188
11.6.3 Compressed Laminated Wood 189
11.7   Fiber Boards 190
11.7.1 Wood Fibers, Resins, and Additives 190
11.7.2 Production of Fiber Boards 191
11.8   Structural Wood Gluing 192
11.8.1 Resorcinol Adhesives 192
11.9   References 193

12   Molding Compounds 196
12.1   Standardization and Minimum Properties 199
12.2   Composition of Molding Compounds 200
12.2.1 Resins 200
12.2.2 Fillers, Reinforcements, and Additives 201
12.3   Production of Molding Compounds 204
12.4   Thermoset Flow 205
12.5   Manufacturing of Molded Parts 207
12.6   Selected Properties 209
12.7   References 211

13   Heat and Sound Insulation Materials 213
13.1   Inorganic Fiber Insulating Materials 213
13.1.1 Inorganic Fibers and Fiber Production 215

13.1.2 Resins and Formulation 217
13.1.3 Properties of Fiber Mats 218
13.2   Phenolic Resin Foam 219
13.2.1 Resins 220
13.2.2 Foaming Equipment 221
13.2.3 Foam Properties 222
13.3   Bonded Textile Felts 226
13.4   References 228

14   Industrial Laminates and Paper Impregnation 230
14.1   Electrical Laminates 230
14.1.1 Materials 233
14.1.2 Production of Electrical Laminates 235
14.2   Laminated Tubes and Rods 237
14.3   Cotton Fabric Reinforced Laminates 238
14.4   Decorative Laminates 238
14.5   Filters 241
14.6   Battery Separators 242
14.7   References 243

15   Coatings 244
15.1   Automotive Coatings 245
15.1.1 Water-Borne Coatings and Electrodeposition 246
15.2   Coatings for Metal Containers 247
15.3   Marine Paints 249
15.3.1 Shop Primers 250
15.3.2 Wash Primers 250
15.3.3 Oil-Modified Phenolic Resin Paints 251
15.4   Printing Inks 251
15.4.1 Rosin-Modified Phenolic Resins 252
15.5   Other Applications 253
15.6   References 254

16   Foundry Resins 256
16.1   Mold- and Core-Making Processes 256
16.1.1 Inorganic Binders 256
16.1.2 Organic Binders 257
16.1.3 Requirements of Foundry Sands 257
16.2   Shell Molding Process 259
16.3   Hot-Box Process 261
16.4   No-Bake Process 263
16.5   Cold-Box Process 265
16.6 SO_2-Process 266
16.7   Ingot Mold Hot Tops 266
16.8   References 267

17   Abrasive Materials 269
17.1   Grinding Wheels 269
17.1.1 Composition of Grinding Wheels 270

17.1.2 Manufacturing of Grinding Wheels 273
17.2   Coated Abrasives 275
17.2.1 Composition of Coated Abrasives 276
17.2.2 Coating Process 277
17.2.3 Abrasive Papers 278
17.2.4 Abrasive Tissues 279
17.2.5 Vulcanized Fiber Abrasives 279
17.3   References 280

18   Friction Materials 281
18.1   Formulation of Friction Materials 282
18.2   References 286

19   Phenolic Resins in Rubbers and Adhesives 288
19.1   Mechanisms of Rubber Vulcanization with Phenolic Resins 288
19.2   Thermosetting Alloy Adhesives 290
19.2.1 Vinyl-Phenolic Structural Adhesives 290
19.2.2 Nitrile-Phenolic Structural Adhesives 291
19.3   Phenolic Resins in Contact Adhesives 292
19.3.1 Chloroprene-Phenolic Contact Adhesives 292
19.3.2 Nitrile-Phenolic Contact Adhesives 295
19.4   Phenolic Resins in Pressure-Sensitive Adhesives 295
19.5   Rubber-Reinforcing Resins 296
19.6   Resorcinol-Formaldehyde Latex Systems 297
19.7   References 297

20   Phenolic Antioxidants 299
20.1   References 301

21   Other Applications 303
21.1   Refractory Linings and Taphole Mixes 303
21.2   Phenolics for Chemical Equipment 303
21.3   Socket Putties 304
21.4   Brush Putties 304
21.5   Synthanes 304
21.6   Concrete Additives 305
21.7   Casting Resins 305
21.8   References 306

Subject Index 307

Abbreviations

ASTM	American Society of Testing and Materials
BHMP	*Bis*(hydroxymethyl)phenol, dimethylphenol
BP	Boiling point
BR	Butyl rubber
CHP	Cumene hydroperoxide
COD	Chemical oxygen demand
CP	Cross polarization
CR	Chloroprene rubber
DIN	Deutsche Industrie Normen
DMA	Dynamic mechanical analysis
DMF	Dimethylformamide
DPM	Diphenylmethane
DSC	Differential scanning calorimetry
DTA	Differential thermal analysis
EPA	Environmental Protection Agency
ESCA	Electron spectroscopy for chemical analysis
F	Formaldehyde
FA	Furfuryl alcohol
FAO	Food and Agriculture Organization
FDA	Food and Drug Administration
FDMS	Field desorption mass spectrometry
FTIR	Fourier transform infrared spectroscopy
GC	Gas chromatography
GPC	Gel permeability chromatography
HPL	High-pressure laminate
HPLC	High-performance liquid chromatography
HMP	Hydroxymethylphenol
HMTA	Hexamethylenetetramine
IB	Internal bond (=tensile strength vertical to the surface)
IE	Ion exchange resin
IPN	Interpenetrating polymer network
IR	Polyisobutylene rubber
LC_{50}	Medium lethal concentration
MAK	Maximum workplace exposure, 8 h
MAS	Magic angle spinning
MF	Melamine-formaldehyde resin
MOE	Modulus of elasticity

MP	Melting point
MW	Molecular weight
MWD	Molecular weight distribution
NBR	Nitrile butadiene rubber
NMR	Nuclear magnetic resonance spectroscopy
OSHA	Occupational Health and Safety Administration
P	Phenol
PB	Particle board
pbw	Parts by weight
PF	Phenol-formaldehyde resin
PPO	Polyphenylene oxide
PS	Patent specification
PVAc	Polyvinyl acetate
PVB	Polyvinyl butyral
PVF	Polyvinyl formal
Py	Pyridine
QM	Quinone methide
RH	Relative humidity
SEM	Scanning electron microscopy
SG	Specific gravity
TBA	Torsional braid analysis
TBBA	Tetrabromobisphenol A
TEM	Transmission electron microscopy
TGA	Thermogravimetric analysis
THMP	*Tris*(hydroxymethyl)phenol, Trimethylolphenol
TKN	Total Kjeldahl nitrogen
TLC	Thin layer chromatography
TTT	Time temperature transformation
UF	Urea-formaldehyde resin
UMP	Urea-melamine-phenol resin
VC	Volatiles content
XPS	X-ray photoelectron spectroscopy

Units

Force	1 kp	$= 9.80665\,\text{N} \approx 10\,\text{N}$
Mechanical tension	$1\,\text{kp/cm}^2$	$= 0.0981\,\text{N/mm}^2 \approx 0.1\,\text{N/mm}^2$
	$1\,\text{N/mm}^2$	$= 145\,\text{psi}$
Pressure	1 at	$= 1\,\text{kp/cm}^2 = 0.980665\,\text{bar} \approx 1\,\text{bar}$
	1 Pa	$= 10^{-5}\,\text{bar}$
Temperature	°F	$= °C \cdot 1.8 + 32$
Dynamic viscosity	1 cP	$= 1\,\text{mPa} \cdot \text{s}$
Heat quantity	1 kcal	$= 4.187\,\text{kJ}$
Thermal conductivity	1 W/Km	$= 0.86\,\text{kcal/m h °C}$
	1 W/Km	$= 0.579\,\text{BTU/ft h °F}$
	1 W/Km	$= 6.95\quad\text{BTU in/ft}^2\,\text{h}$
Length	1 mm	$= 0.0394\,\text{in}$
	1 m	$= 3.2808\,\text{ft}$
Area	$1\,\text{mm}^2$	$= 0.0016\,\text{sq in}$
	$1\,\text{m}^2$	$= 10.764\,\text{sq ft}$
Mass	1 kg	$= 2.2046\,\text{lb}$
Density	$1\,\text{g/cm}^3$	$= 62.41\,\text{lb/ft}^3$

1 Introduction

1.1 Origin

Synthetic polymers are now quite ubiquitous in modern life, yet the polymer industry is barely 75 years old tracing its origin to phenolic resins and the early efforts of Leo H. Baekeland. Resinous products based on phenol and formaldehyde were obtained as early as 1872 by A. von Baeyer, but the reddish-brown intractible mass was not of any technical or commercial interest.

It was to Leo H. Baekeland's credit that he was able to develop in 1907 an economical method to convert these resins to moldable formulations which were transformed by heat and pressure to hard and resistant molded parts[1]. At the same time he disclosed in numerous patents[2] a multitude of applications for this new material.

The first commercial phenolic resin plant, Bakelite GmbH, was started on May 25, 1910 by Rütgerswerke AG at Erkner near Berlin. This was the first company in the world to produce wholly synthetic resins. On October 10, 1910 Baekeland founded the General Bakelite Company in the US.

Fig. 1.1. Leo Hendrik Baekeland

Fig. 1.2. L. H. Baekeland Laboratory

Fig. 1.3. Early applications of phenolic resins. (Photo: ZEITmagazin/Jo Röttger)

In the ensuing years, many famous scientists worldwide, from Europe, US, Japan, and USSR have made many important technical contributions to the understanding of phenolic resin chemistry. Excellent review articles and books by Hultzsch[3], Martin[4], Megson[5] and others have provided a summary of chemical research and existing phenolics technology during that particular period of time.

During phenolics initial period of existence and concurrent with the rapidly emerging electrical industry a plethora of advantages were readily recognized for this novel synthetic material. Phenolic resins possessed excellent insulation properties and were considered a convenient replacement material for natural resins like shellac and gutta percha which were heretofore quite scarce and expensive.

The remarkable ease of fabrication and lightweight features of phenolic materials as compared to prevailing materials (metals, wood) allowed industrial designers unparalleled versatility in transforming these synthetic compositions into simple yet functional designs. All these circumstances contributed to the rapid development of phenolic resins as an important synthetic polymeric material in the first half of the 20th century.

Some early applications of phenolic resins in home appliances are shown in Fig. 1.3. Similarly to radio and telephone housings, they are nostalgic in appearance and popular collector items today.

Since their introduction in 1910, the relatively inexpensive and highly versatile family of phenolic resins has played a vital role in construction, automotive, electrical, and appliance industries. In some of these applications, they were later substituted by less expensive and more easily fabricated thermoplastics. On the other hand, many new applications were developed.

Phenolics today are indeed irreplaceable materials for selective high technology applications offering high reliability under severe circumstances. New products and applications continue to emerge and demonstrate the versatility and uniqueness of phenolics and its potential to cope and adjust to the ever changing requirements of our industrial society.

1.2 Commercial Development of Phenolic Resins

The most important market segments of phenolic resins are those that relate to the wood working industry, thermal insulation, and molding compounds. About 75% of all phenolic resins are consumed in these three market areas. Quite remarkably nearly all early applications initially established by Baekeland continue to utilize phenolic resins and maintain significant volumes (Table 1.1).

In the USA the strong position of plywood products identifies this industry as the major consumer of phenolic resins. According to annual SPI figures (USA) as well as West Europe quantities in Table 1.1, reported volumes relate to resins in delivered condition, i.e. include water, solvents, and fillers. In 1983 the wood working industry used 60% of the total US-production followed by fiber insulation (15%) and molding compounds (9%); these industries represent almost 85% of the total U.S. phenolic volume.

As phenolic resin production is segmented according to various nations, the USA leads by a clear margin ahead of the USSR, Japan, and West-Germany. By volume

Table 1.1. Use of phenolic resins in West Europe in 1983 (per cent)

	%
Wood working industry	32
Thermal insulation	17
Molding compounds	17
Decorative laminates	6
Foundry	5
Coatings	4
Electrical laminates	4
Filters and separators	3
Abrasives	3
Felt bonding	3
Friction	2
Rubber and adhesives	2
Others	3

Tabelle 1.2. Phenolic resin production [6-10] in 1983 (captive production excluded)

	Plastics total 1,000 t	Phenolics 1,000 t	Phenolics share %
USA	13,739	1,124	8.2
Japan	5,600	306	5.5
West-Germany	6,274	165	2.7
France[a]	3,124	72	2.3
Great Britain	2,406	63	2.6
Italy[a]	2,400	85	3.5
USSR[a]	3,028	357	11.7

[a] 1982

phenolic resins are about 4% of world-wide plastics production today (including cel-lulosics, excluding elastomers).

In total plastics volume phenolic resins are comparable to polyester or polyurethanes in volume. In most applications phenolics are combined with reinforcing fillers or fibers to function as the adherent or the critical binder of the composition. These phe-nolic resin bonded materials, i. e. particle boards, molding materials, fiber insulation products, foundry cores, grinding wheels, friction elements, represent the largest volume consumption of phenolics and compare in volume with thermoplastics.

The vital nature of phenolic resins in a diversity of applications attests to its present day importance within a spectrum of industries. Key properties such as high tempe-rature resistance, infusibility and flame retardance are recognizable features that con-tribute to further market growth. Considering the multitude of application areas, it is not too surprising that phenolic resin consumption is closely related to the prevailing gross national product.

Finally, the question arises whether phenolic resins would be competitive with other synthetic resins in the future. Even under protracted long term petrochemical supply, phenolic resins benefit by an extraordinarily well secured raw material supply scenario coupled to relative price stability as compared to other synthetic resins. The actual pe-riod of time when petroleum reserves will finally be exhausted will not be discussed herein, however, it is without question that the chemical industry is rapidly approa-ching that era when basic chemicals derived from coal are able to compete. If crude oil continues to escalate in price or diminishes significantly, the leading basic organic chemicals of today, ethylene and propylene, must be produced by relatively expensive processes (e. g. ethanol dehydration, Fischer-Tropsch synthesis). However, the basic raw materials of phenolic resin production, phenol, cresols and methanol (formalde-hyde) – as primary products of modern coal gasification processes – will be available under relatively economical conditions (see also Chap. 2).

1.3 References

1. Baekeland, L. H.: US-PS 942 699 (July 13, 1907); DE-PS 233 803
2. Knop, A., Scheib, W.: Chemistry and Application of Phenolic Resins, Heidelberg, N.Y., Springer-Verlag, 1979
3. Hultzsch, K.: Chemie der Phenolharze, Berlin, Göttingen, Heidelberg, Springer, 1950
4. Martin, R. W.: The Chemistry of Phenolic Resins, New York, J. Wiley, 1956
5. Megson, N. J. L.: Phenolic Resin Chemistry, London, Butterworths, 1958
6. N. N.: Chemical & Engineering News, June 11, 1984, P. 39
7. Wendenburg, L.: Chem. Ind. 36, 80–81 (1984)
8. Verband Kunststofferzeugende Industrie, Frankfurt, Geschäftsbericht 1983
9. N. N.: Kunststoffe 73, No. 10 (1983), P. 610, 620, 622, 625
10. Camani, A.: Swiss Plastics 6, 10 (1984)

2 Raw Materials

Phenolic resins are produced by the reaction of phenols with aldehydes. The simplest representatives of these types of compounds, phenol and formaldehyde, are by far the most important.

2.1 Phenols

Phenols are a family of aromatic compounds with the hydroxyl group bonded directly to the aromatic nucleus. They differ from alcohols in that they behave like weak acids and dissolve readily in aqueous sodium hydroxide, but are insoluble in aqueous sodium carbonate. Phenols are colorless solids with the exception of some liquid alkylphenols. The most important phenols are listed in Table 2.1 below. Data regarding the molecular structure of phenols and cresols are listed in Sect. 3.1.

Table 2.1. Physical properties of some phenols [1-8]

Name		MW	MP °C	BP °C	pK$_a$ 25 °C
Phenol	hydroxybenzene	94.1	40.9	181.8	10.00
o-Cresol	1-methyl-2-hydroxybenzene	108.1	30.9	191.0	10.33
m-Cresol	1-methyl-3-hydroxybenzene	108.1	12.2	202.2	10.10
p-Cresol	1-methyl-4-hydroxybenzene	108.1	34.7	201.9	10.28
p-tert. Butylphenol	1-tert-butyl-4-hydroxybenzene	150.2	98.4	239.7	10.25
p-tert. Octylphenol	1-tert-octyl-4-hydroxybenzene	206.3	85	290	–
p-Nonylphenol	1-nonyl-4-hydroxybenzene	220.2	–	295	–
2,3-Xylenol	1,2-dimethyl-3-hydroxybenzene	122.2	75.0	218.0	10.51
2,4-Xylenol	1,3-dimethyl-4-hydroxybenzene	122.2	27.0	211.5	10.60
2,5-Xylenol	1,4-dimethyl-2-hydroxybenzene	122.2	74.5	211.5	10.40
2,6-Xylenol	1,3-dimethyl-2-hydroxybenzene	122.2	49.0	212.0	10.62
3,4-Xylenol	1,2-dimethyl-4-hydroxybenzene	122.2	62.5	226.0	10.36
3,5-Xylenol	1,3-dimethyl-5-hydroxybenzene	122.2	63.2	219.5	10.20
Resorcinol	1,3-dihydroxybenzene	110.1	110.8	281.0	–
Bisphenol-A	2,2-bis(4-hydroxyphenyl)propane	228.3	157.3	–	–

2.1.1 Physical Properties of Phenol

The melting point of pure phenol (40.9 °C) is lowered considerably by traces of water, approximately 0.4 °C per 0.1% of water. A water content of $\leq 6\%$ renders it liquid even at room temperature. To produce phenolic resins, a mixture of 90% of phenol/ 10% of water is preferably used. Above 65.3 °C phenol can be mixed with water at any ratio. During the cooling period of those solutions, which may contain 28–92% of water, two phases develop, phenol/water and water/phenol.

Phenol [3–5)] is quite soluble in polar organic solvents, but not very soluble in aliphatic hydrocarbons. Phenol crystallizes in the form of colorless prisms. When exposed to air, phenol rapidly develops a reddish color, especially if it contains traces of copper and iron. This happens if phenol is utilized in copper clad or iron reactors, or if phenolic resins are stored in iron drums. Additional safety and technical data of phenol are listed in Table 2.2 below:

Table 2.2. Technical and safety data of phenol[3–5)]

Flash point	°C	79
Ignition point	°C	605
Explosion limits	vol.-%	2–10
MAK-value	ppm	5
MAK-value	mg/m^3	19
Vapor pressure, 20 °C	mbar	0.2
Threshold odor concentration	ppm	0.5

The toxicological and physiological characteristics of phenol are described in Chap. 6.

2.1.2 Supply and Use of Phenol

The largest use of phenol is the production of phenol-formaldehyde resins[9, 10)] as is shown in the following Table 2.3:

Table 2.3. Breakdown of phenol consumption 1982

%	USA	West-Europe	Japan
Phenolic resins	37	35	55
ε-Caprolactame	17	26	–
Bisphenol-A	21	16	24
Adipic acid	3	7	–
Others	22	16	21

In 1983, about 3% of the world production of phenol was derived from coal. Among the synthetic processes, the cumene process is the most prevalent. The basic raw materials for the cumene process and for the production of phenol are benzene and propylene. To better understand the dependence, availability, and price of phenolic resins in relation to crude oil, raw material flow diagram is shown in Fig. 2.1. In 1983 the phenol capacities amounted to 1.5 million tons in West-Europe and 1.7 mil-

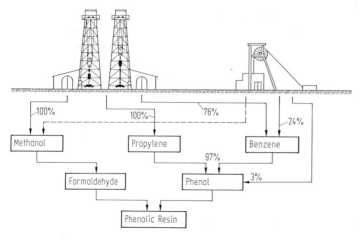

Fig. 2.1. Raw materials origin for the production of phenolic resins. West-Germany 1983

lion tons in the USA. The actual US production for that year amounted to 1.16 million t[11], world production was about 3.5 million t.

2.1.3 Phenol Production Processes

The cumene process is by far the most important synthetic process[3–5] for the production of phenol, and today, probably accounts for more than 90% of the synthetic phenol capacity in the Western world.

A two step oxidation process based on toluene was developed by Dow Chemical. Kalama Chemical in USA and DSM in The Netherlands are manufacturing phenol according to this process (2.1).

$$
\underset{\text{Toluene}}{\text{CH}_3\text{-C}_6\text{H}_5} + 1,5\ O_2 \xrightarrow[\text{(Co)}]{140°C/\ 3\,bar} \underset{\text{Benzoic acid}}{\text{COOH-C}_6\text{H}_5} \ (+H_2O) \longrightarrow \text{OH-C}_6\text{H}_5 + CO_2 \qquad (2.1)
$$

The hydrolysis of halogenated aromatic compounds developed in 1930 by Raschig, was later improved to the Raschig-Hooker process. The first step, the oxychlorination of benzene with hydrochloric acid in the presence of a copper-on-alumina catalyst at 275 °C is followed by hydrolysis of the chlorobenzene with water (steam) on a copper-promoted calcium phosphate catalyst at 400–450 °C.

The sulfonation process, the earliest one, is no longer utilized because of waste disposal and energy considerations.

Cumene Process (Hock Process)

The phenol synthesis based on cumene was discovered by H. Hock in Germany and published jointly with Sho Lang[12].

Soon after World War II the first pilot plant was constructed jointly by Rütgerswerke and Bergwerksgesellschaft Hibernia with Hock's assistance. Commercial production was first developed by the Distillers Co. (GB) and Hercules Powder (USA).

The cumene (isopropylbenzene) required for the Hock process is produced[13] by alkylation of benzene with propylene over a solid phosphoric acid catalyst [UOP-Process, Eq. (2.2)].

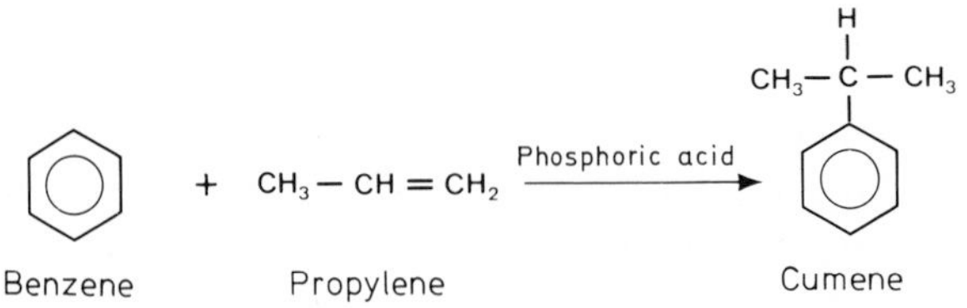

$$CH_3-CH=CH_2 \xrightarrow{\text{Phosphoric acid}} \tag{2.2}$$

Benzene   Propylene   Cumene

Cumene is oxidized with oxygen in air in the liquid phase to cumene hydroperoxide (CHP) according to reaction (2.3), yielding small amounts of dimethylbenzyl alcohol and acetophenone as by-products. The mechanism shown for the acid-catalyzed peroxide decomposition [Eq. (2.4)] was postulated by Seubold and Vaugham[14].

CHP, a liquid with relatively low vapor pressure, is stable at normal temperature and conditions, but decomposes very rapidly under acidic conditions

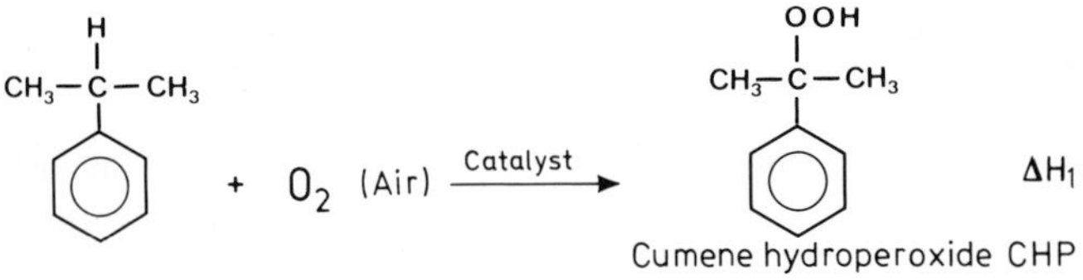

$$+ \ O_2 \ \text{(Air)} \xrightarrow{\text{Catalyst}} \qquad \Delta H_1 = -116 \ kJ/mol \tag{2.3}$$

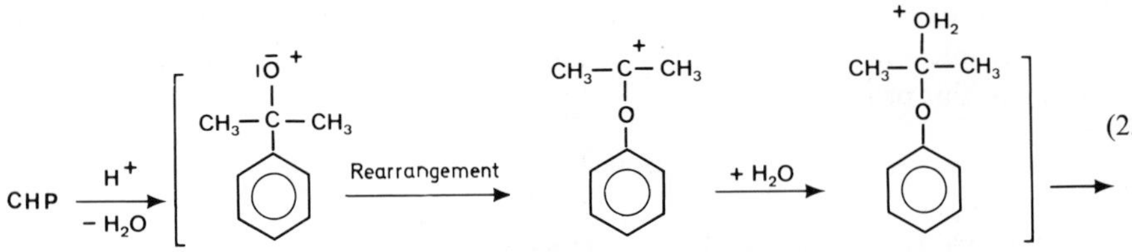

Cumene hydroperoxide CHP

$$CHP \xrightarrow[-H_2O]{H^+} \left[\xrightarrow{\text{Rearrangement}} \xrightarrow{+H_2O} \right] \longrightarrow \tag{2.4}$$

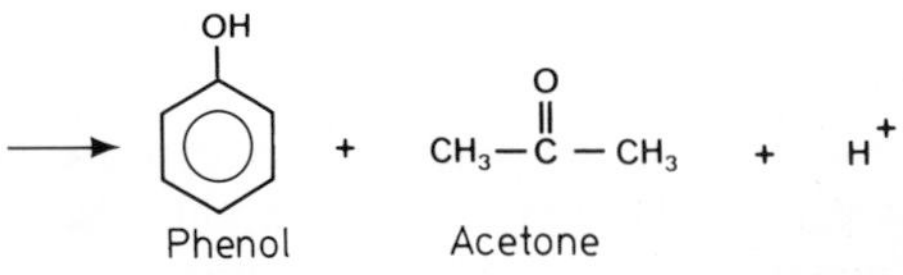

$$\longrightarrow \quad \text{Phenol} \quad + \quad CH_3-\overset{O}{\overset{\|}{C}}-CH_3 \quad + \quad H^+ \qquad \Delta H_2 = -253 \ kJ/mol \tag{2.5}$$

Phenol   Acetone

and elevated temperatures. During the second stage of the process, the concentration of CHP and separation of unreacted cumene occur. The concentrated reaction product is then converted by use of sulfuric acid as catalyst to a crude mixture of phenol and acetone, which also contains α-methylstyrene as by-product. Then, various purification and distillation steps follow. α-Methylstyrene can also be hydrogenated and returned to the process.

CHP is a potentially hazardous material. Therefore, many safety precautions must be observed and safety equipment must be installed in the plants[15].

The economics of phenol manufacture are sensitive to the market requirements of both phenol and acetone. The diminished use of acetone as a solvent in coatings and adhesives and greater dependence as starting material for methyl methacrylate (MMA) suggests that more of price burden in the cumene process will be borne by phenol. A new, more economical process for MMA is based on isobutylene or *t*-butylalcohol, air and methanol.

2.1.4 Cresols and Xylenols – Synthesis Methods

Cresols[6, 7, 16)], hydroxy derivatives of toluene, commonly designated as methyl phenols, exist as three isomers depending on the relative position of the methyl in relation to the hydroxyl group. The molecular configuration is described in Sect. 3.1. The main source of cresols was originally coal tar. Today, however, synthetic processes predominate, mainly based on toluene and phenol. The importance of the petroleum industry as

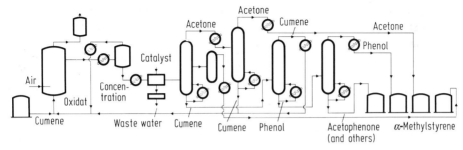

Fig. 2.2. Flow diagram of the cumene process, (Drawing: Phenolchemie, D-4390 Gladbeck)

Fig. 2.3. Phenol production via the cumene process, oxidation plant (Photo: Phenolchemie D-4390 Gladbeck)

a source of cresols and xylenols as by-products is relatively insignificant. Starting with toluene, the cresols are obtained either by sulfonation, by alkylation with propylene, or by chlorination. In the sulfonation process, the main product is the *para* derivative together with some *ortho* derivative. In the chlorination process the *meta* isomer prevails (about 50%) with an approximately equal *o/p* ratio. This route is advantageous for resin-grade cresols. The chemistry of the toluene alkylation is very similar to the cumene process with differences in the oxidation step. Toluene is reacted first with propylene in the presence of $AlCl_3$ or other catalysts to obtain a mixture of cymenes. In this process the *m/p* ratio of approximately 2:1 but less than 5% *o*-cymene is reported[17]. Mitsui and Sumitomo operate plants in Iwakuni and Ohita respectively, each with a capacity of 10,000 tons per year.

$$\text{Toluene} + \text{Propylene} \xrightarrow{\text{Catalyst}} \text{Cymene} \xrightarrow{\text{Oxidation}} \tag{2.6}$$

$$\text{Cymene hydroperoxide} \xrightarrow{H^+} \text{m-Cresol} + \text{Acetone} \tag{2.7}$$

More important among the synthetic processes is the production of cresols and xylenols based on alkylation of phenol with methanol[6, 7, 16]. In the gas phase process (Koppers and Pitt-Consol, USA; Croda, Great Britain), methanol and phenol vapors pass over aluminium oxide catalyst at approximately 350 °C under moderate pressure. Mainly *o*-cresol and 2,6-xylenol are obtained. If 2,6-xylenol is desired as main product which is used for PPO production, magnesium oxide is employed as catalyst.

Other processes operated by Chemische Werke Lowi and UK-Wesseling, are conducted in the liquid phase. The Lowi process is carried out at 300–350 °C at a pressure of 40–70 bar, Al-methylate is used as catalyst. Thus, mainly *o*-cresol is obtained. Higher methanol ratio favors the formation of 2,4 and 2,6-xylenol. By transalkylation of xylenols in the presence of phenol the yield of cresols can be increased.

UK-Wesseling produces *o*- and *p*-cresol of 99% purity, 2,6-xylenol of 98% and 2,4-xylenol of 92% purity by use of zinc bromide as catalyst. The synthesis of *p*-cresol, used mainly for BHT (butylated hydroxy toluene) or similar antioxidants, is also obtained by the sulfonation of toluene (Sherwin Williams, USA).

2.1.5 Alkylphenols

Phenolic compounds with a saturated carbon side chain containing a minimum of three carbon atoms should be termed alkylphenols. These alkylphenols are produced from phenols or cresols by Friedel-Craft alkylation with olefins, mainly isobutene, di-isobutene or propylene[6, 8]. At low temperatures, e.g. below 50 °C, *o*-substitution predominates. o-Alkylphenols can be rearranged to p-isomers by heating up to 150 °C with acid catalysts. A high yield of o-derivatives is achieved by use of Ca-, Mg-, Zn- or Al-phenoxides as catalysts at a temperature of about 150 °C.

A process of mainly meta alkyl substituted phenol has been reported (US-PS 4,405,818) and relies on a special zeolite to stabilize the meta isomer while the para alkyl phenol reverts to olefin and phenol in a temperature range of 150–600 °C.

In the resin area, alkylphenols are used for the production of coating resins because of their good compatibility with natural oils and increased flexibility or as crosslinking agents in the rubber industry. Other uses include adhesives, carbonless paper, antioxidants, surfactants and phosphoric acid esters.

2.1.6 Phenols from Coal and Petroleum

An average of approximately 1.5% crude phenols, mainly phenol ($\sim 0.5\%$) as well as o-, m-, and p-cresols, 2,3-, 2,4-, 2,5-, 2,6-, and 3,5-dimethylphenol, is found in coal tar[18]. Phenols are further obtained from condensates of coke oven gases and waste waters of coal gasification plants. The extraction is performed either according to the older Pott-Hilgenstock process using benzene/sodium hydroxide[19] or according to Lurgi's Phenosolvan process with diisopropyl ether as solvent[20]. The extraction of phenols from coal tar is performed with dilute sodium hydroxide (8–12%), followed by precipitation of the crude phenols with carbon dioxide. The flow diagram of an extraction plant is shown in Fig. 2.4. The crude phenoxide solution contains approximately 0.5% non-phenolic components – neutral oils (hydrocarbons) and pyridine bases – which are removed prior to precipitation, mostly by steam distillation[18].

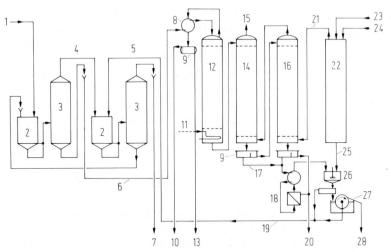

Fig. 2.4. The recovery of phenols from coal tar by alkali-extraction according to Franck/Collin[18]. *1* Carbolic oil; *2* Mixing equipment; *3* Separating column; *4* 1st Extraction; *5* 2nd Extraction; *6* Raw phenoxide oil; *7* Phenol-free carbolic oil; *8* Condenser; *9* Separator; *10* Steam distilled oil; *11* Steam; *12* Steam distillation column; *13* Waste water; *14* Precipitation column; *15* Exhaust gas; *16* 2nd Precipitation column; *17* Calcium hydroxide; *18* Cooling equipment; *19* Sodium hydroxide; *20* Raw phenols; *21* Carbon dioxide; *22* Lime kiln; *23* Limestone; *24* Coke; *25* Calcium oxide; *26* Causticizing; *27* Lime sludge filter; *28* Calcium carbonate

Other processes recommend selective solvents to extract phenol, e.g. aqueous methanol (Metasolvan process). However, only Lurgi's Phenoraffin process which uses an aqueous sodium phenoxide solution as selective solvent has achieved technical and commercial prominence. The selectivity of NaOH is superior to all other recommended solvents.

By-products and effluents from coal gasification processes, especially the Lurgi process, offer greater potential for phenols. The dephenolization can be performed by processes described above, yielding a complex fraction containing a large variety of mono- and di-substituted phenols[21]. The composition of the raw phenol fraction from Sasol II/III (South Africa) is shown in Table 2.4.

However, it is not presently an economically viable process for the production of phenols; the fraction is transformed into BTX-aromatics.

Table 2.4. Composition of the raw phenol fraction from Sasol II/III[22]

	%
Phenol	45
o-Cresol	12
m- and p-Cresol	22
Xylenols	10
Higher phenols	8
Hydrocarbons	0.5
N-compounds	2.5

Low rank coal can be easily depolymerized to phenols using acid catalysts, e.g. a phenol-BF_3-complex or p-toluenesulfonic acid[23].

Hydrocarbon Research Inc. developed[24] a two-step process to convert lignin (i.e. derived from kraft pulping) to phenol and its derivatives. Lignin is first depolymerized by hydrocracking in a fluidized bed reactor and the resulting monoaromatic compounds dealkylated to phenols and benzene.

A further source of cresols is the petroleum industry, particularly in the USA. During the catalytic cracking process various phenolic compounds are formed. Like the recovery of phenolic compounds from coal tar, the extraction is carried out with dilute sodium hydroxide solution. After the separation of the phenols by precipitation with carbon dioxide, further treatment follows by distillation. Thus, technically pure phenol (BP 181.8 °C) and o-cresol (BP 191.0 °C) can be recovered in a straightforward manner.

The separation of m- and p-cresol is only possible with special chemical or physico-chemical methods due to the similar boiling points. This is also applicable to xylenols. In the urea process, the mixture of m- and p-cresol is heated with urea. Upon cooling a crystalline addition compound of m-cresol and urea is formed. Further, p-cresol when gently heated to 90 °C forms a crystalline addition compound with anhydrous oxalic acid. m-Cresol with sodium acetate results in adducts with low solubility. The chemical processes use differences in sulfonation rate (sulfuric acid process) followed by crystallization and hydrolysis or the different behaviour in the alkylation with isobutylene (isobutylene process) and following dealkylation with sulfuric acid.

2.1.7 Other Phenolic Compounds

Cashew Nut Shell Liquid (CNSL)

An important phenolic compound from natural sources is cashew nut shell liquid (CNSL). This liquid from the shells of cashew nuts, which grow mainly in Southern India, has become a useful raw material in the manufacture of special phenolic resins used in coatings, laminates, and brake lining resin formulations[25]. These particular resins possess outstanding resistance to the softening action of mineral oils and high resistance to acids and alkali.

The CNSL, which is obtained by a special heat treatment and decarboxylation, contains a mixture of mono- and diphenols (2.8) with an unsaturated C_{15} side chain in the meta position, thereby exhibiting high reactivity towards formaldehyde.

$$
\text{(2.8)}
$$

Cardanol (90%)   Cardol (10%)

$R_1 \quad = -(CH_2)_7-CH=CH-(CH_2)_5-CH_3$
$R_1, R_2 = -(CH_2)_7-CH=CH-CH_2-CH=CH-(CH_2)_2-CH_3$
$R_1, R_2 = -(CH_2)_7-CH=CH-CH_2-CH=CH-CH_2-CH=CH_2$

The reacted CNSL-resin, via a hardening reaction that includes polymerization and polyaddition, yields a solid infusible product which in powdered form („friction dust") retains high binding power at raised temperatures and is used in brake lining formulations[26].

Resorcinol

Resorcinol, as a dihydric phenol (1,3-dihydroxybenzene), is a very interesting intermediate material for the production of thermosetting resins. However, it is only used for special applications due to its relatively high price. The reaction rate with formaldehyde is considerably higher compared with that of phenol[27, 28]. This is of great technical importance for the preparation of cold setting adhesives[29, 30]. Resorcinol or resorcinol-formaldehyde prepolymers can be used as accelerating compounds for curing phenolic resins. The addition of 3–10% of such compounds permits shorter cure cycles in particle board and grinding wheel production. Furthermore, the adhesion of textile materials, e.g. tire cord to rubber, are greatly improved by pretreating them with resorcinol-formaldehyde resin.

Resorcinol can also be used as intermediate material for azo- and triphenylmethane and other dyes, pharmaceuticals, cosmetics, tanning agents, textile treating agents and antioxidants[31].

Until now the most important process[32, 33] for the production of resorcinol has been the alkali fusion of m-benzenedisulfonic acid according to Eq. (2.9).

$$
\text{(2.9)}
$$

Benzene   Benzene disulfonic acid   Resorcinol

At an intermediate stage, the disodium salt of the acid is formed. It is then fused with sodium hydroxide in a nickel alloy tank. The melt is dissolved in water and the resulting slurry acidified with sulfuric acid. Resorcinol is then recovered by counter-current extraction and purified by distillation. Plants with a capacity of 10,000 tons per year each are operated by Koppers, USA, and Hoechst, West-Germany. A more recent method, used by Mitsui and Sumitomo, Japan, is the familiar Hock process starting with benzene followed by alkylation with propene (2.10).

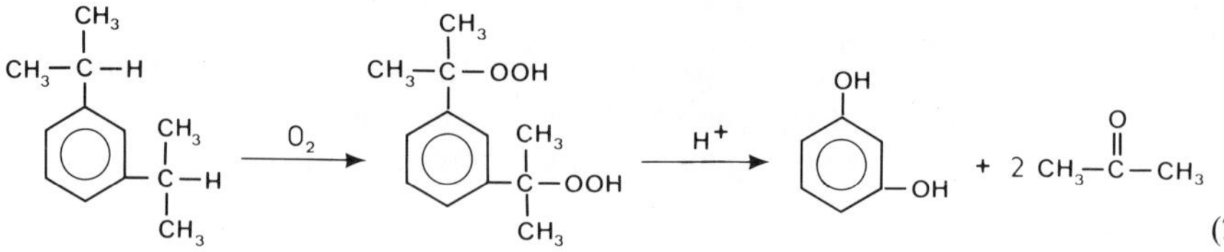

$$(2.10)$$

The air oxidation is carried out at about 90 °C, the decomposition with diluted sulfuric acid in acetone or a similar solvent.

Bisphenol-A

Bisphenol-A is the common name for 2,2-*bis*(4-hydroxyphenyl) propane[34].

The importance of the colorless BPA/F-resins for the coating industry will be discussed in Chap. 15. Today, the main use for BPA is the production of epoxide resins (45%) and engineering plastics like polycarbonate (40%), and polysulfone. Sulfuric acid is used in the newer process because of problems associated with the volatility and corrosiveness of hydrochloric acid. Sulfur compounds like thioglycolic acid or mercaptans promote the reaction rate of the acid-catalyzed addition of carbonyl compounds to phenols[35].

While the purity of BPA made by the sulfuric acid process is satisfactory for the use in formaldehyde resins, high purity BPA is needed for the production of epoxy resins and especially for polycarbonate. This is normally accomplished by recrystallization from toluene or by crystallization as phenol-BPA adduct.

2.2 Aldehydes

Formaldehyde[36] is virtually the only carbonyl component for the synthesis of technically relevant phenolic resins. Special resins can also be produced with other aldehydes, for example acetaldehyde, furfural or glyoxal, but have not achieved much commercial importance. Ketones are rarely used. The physical properties of aldehydes[1, 37, 38] are compiled in Table 2.5.

2.2.1 Formaldehyde, Properties and Processing

Formaldehyde is produced by dehydrogenation of methanol, over either an iron oxide/molybdenum oxide catalyst or over a silver catalyst. Because of hazards in hand-

Table 2.5. Physical properties of some aldehydes[1, 37, 38]

Type	Formula	MP °C	BP °C
Formaldehyde	$CH_2=O$	−92	−21
Acetaldehyde	$CH_3CH=O$	−123	20.8
Propionaldehyde	$CH_3-CH_2-CH=O$	−81	48.8
n-Butyraldehyde	$CH_3(CH_2)_2-CH=O$	−97	74.7
Isobutyraldehyde	$(CH_3)_2CH-CH=O$	−66	61
Glyoxal	$O=CH-CH=O$	15	50.4
Furfural	$\begin{array}{c} CH \!\!-\!\! CH \\ \| \quad \| \; H \\ CH \quad C\text{-}C=O \\ \diagdown\!\!O\!\!\diagup \end{array}$	−31	162

ling mixtures of pure oxygen and methanol, air is used as oxidizing agent[39]. Oxygen is used to burn the hydrogen by-product.

$$CH_3-OH + 1/2\,O_2 \xrightarrow{\text{Catalyst}} HC\!\!\begin{array}{c}\diagup O \\ \diagdown H\end{array} + H_2O \qquad \Delta H = -159 \text{ kJ/mol} \tag{2.11}$$

Methanol   Formaldehyde

When the silver catalyst is used, the reaction mixture of methanol and air is prepared so as to be below the flammability limit. With the oxide catalyst, a higher gas flow can be accommodated, but methanol concentration is lean and below flammability limit[39]. Reactor effluent passes to the absorption train where formaldehyde and other condensables are recovered by condensation and absorption in recirculating formalin streams. The raw formaldehyde solution is then purified by removal of the unreacted methanol. Formaldehyde composition may be adjusted by monitoring the amount of water added to the absorber column or by subsequent dilution in storage tanks. Inhibitors are added to retard the formation of paraformaldehyde during storage.

The Formox process occurs with a mixture of iron oxide and molybdenum oxide as catalyst. The reaction proceeds at relatively low temperatures, between 250–400 °C, almost to completion (95–98%). As a side reaction, formaldehyde is oxidized to carbon monoxide and water (2.12).

$$CH_2=O + 1/2\,O_2 \longrightarrow CO + H_2O \qquad \Delta H = -215 \text{ kJ/mol} \tag{2.12}$$

The Perstorp/Reichhold, Montecatini, Nissui-Topsoe, CdF, Lummus and Hiag/Lurgi processes operate in accordance with this method[40–42].

In the BASF and Monsanto processes a silver catalyst is used[38, 43]. Methanol is partially oxidized and dehydrogenated at 330–450 °C on silver crystals or silver nets. The BASF process uses a vapor/methanol/air mixture. The conversion is considerably high or approximately 90%. Silver catalyst processes with an incomplete methanol conversion performed at 330–380 °C have been developed by Degussa and ICI.

A further process is based on the direct oxidation of methane with air at approximately 450 °C and 10–20 bar on an aluminum phosphate contact. This process, however, has yet to achieve any commercial importance.

Table 2.6. Properties of formaldehyde[38]

Molecular weight	30.03	
Boiling point	−19.2	°C
Melting point	−118	°C
Enthalpy of solution in H_2O	62	kJ/mol
Enthalpy of formation (25 °C)	−116 ± 6	kJ/mol
Free energy	−110	kJ/mol
Self-ignition temperature	300	°C
Explosion limits in air	7−72	vol % CH_2O
Dissociation constant in H_2O 0 °C	$1.4 \cdot 10^{-14}$	
Dissociation constant in H_2O 50 °C	$3.3 \cdot 10^{-13}$	

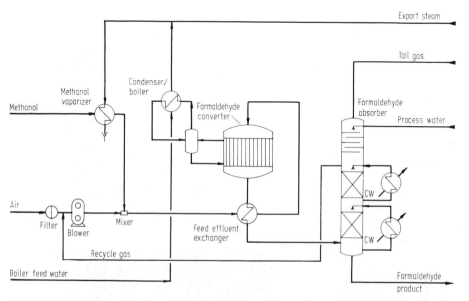

Fig. 2.5. Flow diagram of the iron oxide-molybdenum oxide catalyzed formaldehyde production process[40]. (Drawing: Haldor Topsoe A/S DK-2800 Lingby)

Formaldehyde, a colorless, pungent, irritating gas, exists in aqueous solution almost exclusively as a mixture of oligomers of polymethylene glycol (2.14). Their distribution has been determined for 6–50 wt% CH_2O solutions with low CH_3OH using NMR and GPC techniques.

$$HO-CH_2-OH \qquad\qquad HO-[-CH_2-O-]_n-CH_2-OH$$

$$(2.13) \qquad\qquad\qquad\qquad (2.14)$$

The portion of formaldehyde $CH_2=O$ in aqueous solution[36] is very low (<0.01%). The MWD is indicated in Sect. 3.2. Further equilibria exist in the presence of methanol[37], which aids stabilization by forming a hemiformal terminus (2.15).

$$CH_3OH + HO-CH_2-OH \rightleftharpoons CH_3-O-CH_2-OH + H_2O \qquad\qquad (2.15)$$

Under the influence of acids, hemiacetals react to form acetals by eliminating water. A very important reaction is the formation of HMTA from ammonia and formaldehyde. The overall reaction is indicated in Eq. (2.21); the reaction mechanism is discussed in Sect. 3.3.2. By the catalytic action of strong bases, e. g. sodium hydroxide, formaldehyde undergoes a disproportionation reaction, known as the Cannizzaro reaction, yielding methanol and formic acid according to Eq. (2.16).

$$2\ CH_2{=}O + NaOH \rightleftharpoons CH_3OH + CH_3{-}C{\overset{\displaystyle O}{\underset{\displaystyle ONa}{<}}} \qquad (2.16)$$

Formaldehyde solutions always contain trace amounts ($\sim$0.05%) of formic acid due to the Cannizzaro reaction. The formic acid content can easily be determined by titration with sodium hydroxide.

The Kriewitz-Prins reaction [44] may have some importance in modification reactions of phenolic resins with unsaturated compounds (Chap. 19). Olefins can react with carbonyl compounds under non-free radical conditions to yield predominantly unsaturated alcohols (2.17), m-dioxanes (2.18) and 1.3 glycols (2.19).

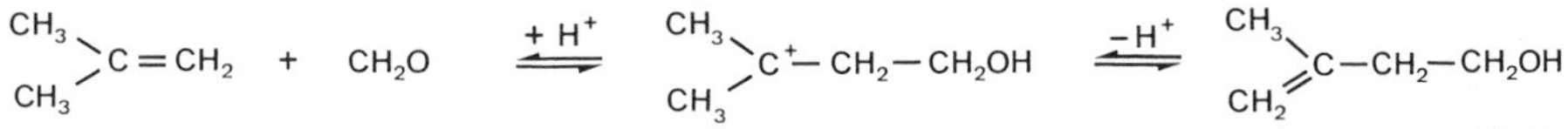

$$(2.17)$$

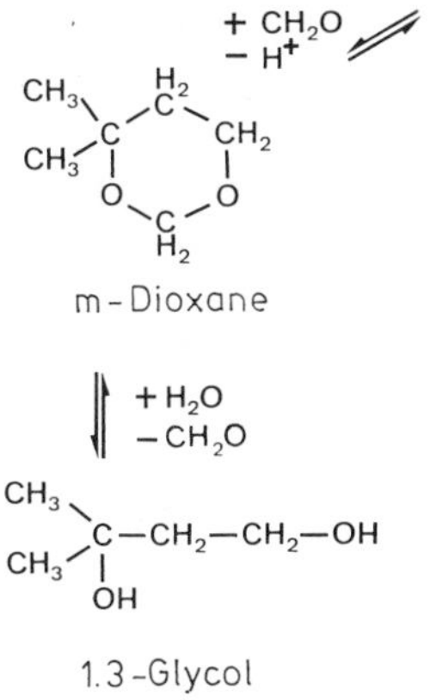

m – Dioxane

$$(2.18)$$

1.3 -Glycol

$$(2.19)$$

As in the hydroxymethylation of phenols, the hydroxymethylene carbonium ion $^{\oplus}CH_2{-}OH$ is the alkylating agent (Chap. 3).

When storing formaldehyde solutions, care should be given to the fact that at lower temperatures paraformaldehyde will increase in concentration and precipitate. The stabilization of aqueous formaldehyde solutions can be achieved with alcohols, preferably methanol. Urea, melamine, methylcellulose and guanidine derivatives are also recommended for stabilization among other compounds. Proper storage containers should be of stainless steel; iron containers are not suitable. Containers with plastic lining or RP containers may also be used.

The major use of formaldehyde[45] is in the production of thermosetting resins based on phenol, urea and melamine. The actual US formaldehyde production in 1983 was 2.47 million t[11].

Table 2.7. Formaldehyde consumption in the USA, Europe and Japan
(1983)

%	USA	West-Europe	Japan
Urea resins	27	52	36
Phenolic resins	21	10	8
Melamine resins	6	7	7
Polyacetal resins	9	5	13
Pentaerythrol	7	7	9
Hexamethylenetetramine	7	3	5
Others	23	6	22

As shown in Table 2.7, approximately 55% of the total formaldehyde consumption
is directed to the production of thermosetting resins. Since 37–50% aqueous solutions
are being used in resin production, it follows that many resin manufacturers install
their own small formaldehyde plants which are supplied with methanol by a large central
methanol plant. In spite of the high transportation costs of formaldehyde
solutions the relatively small capital investment is soon recovered.

Formaldehyde is the more reasonably priced raw material component of phenolic
resins. As an average, about 1.6 mol formaldehyde including HMTA per mol phenol
are used for all application areas.

Instead of formaldehyde solutions of 37–50%, higher concentrated aqueous or alcoholic
solutions may be used. These solutions are produced by dissolving paraformaldehyde
in water at 80–100 °C by addition of a small quantity (1%) of NaOH or tertiary
amines as depolymerization catalyst. It is also possible to concentrate formaldehyde
solutions by adding paraformaldehyde. Resin grade formaldehyde solution
(37%) specifications are shown in Table 2.8.

Table 2.8. Formaldehyde solution specification, resin
grade

Formaldehyde		37.0 ± 0.1%
Methanol	max.	0.5%
Formic acid	max.	0.02 %
Chloride	max.	0.5 ppm
Fe	max.	0.12 ppm
Al	max.	0.25 ppm

2.2.2 Paraformaldehyde

Paraformaldehyde[38, 46] is a white, solid, low molecular polycondensation product of
methylene glycol with the characteristic odor of formaldehyde. The degree of polymerization
ranges between 10 and 100. Commercially available paraformaldehyde products
contain approximately 1–6.5% of water. The preparation of paraformaldehyde
is performed by distillation of 30–37% aqueous formaldehyde solutions. Depending
on conditions (temperature, time, pressure) different types of paraformaldehyde are

Table 2.9. Properties of paraformaldehyde[46]

Content of formaldehyde	90–97%
Content of free water	0.2–4%
Specific weight	1.2–1.3 g/cm^3
Flash point	71 °C
Melting range	120–170 °C

obtained. The values in Table 2.9 show that paraformaldehyde is not a uniform compound.

Paraformaldehyde is seldom used for resin production because of its high price compared with aqueous formaldehyde solutions and because of problems associated with the exothermal heat evolution (Chap. 5). Paraformaldehyde is used in special circumstances by resin manufacturers seeking low water content or high solids content resins. This may change due to availability of paraformaldehyde in „prill" form which is more easily handled than powder in solids handling equipment. Further incentives for paraformaldehyde use are less energy required for distillation (water removal) and cost of distillate disposal. Paraformaldehyde with or without an acid catalyst may be used to cure novolak resins[47]. However, the odor and uneconomical use of formaldehyde make it unattractive. Products obtained are inferior to those obtained when HMTA is used.

Yet, paraformaldehyde is used almost exclusively to crosslink resorcinol prepolymers, e.g. in cold setting structural wood adhesives. Lower curing temperatures are adequate because of the higher reactivity of resorcinol, thus formaldehyde evolution is greatly reduced. The reactivity of paraformaldehyde depends on the degree of polymerization. A fairly accurate reactivity test method is the resorcinol test. This test measures the period of time in minutes in which an alkaline resorcinol/paraformaldehyde mixture heats up to 60 °C due to the „condensation" reaction.

2.2.3 Trioxane and Cyclic Formals

Trioxane, a cyclic trimer of formaldehyde or methylene glycol, is a colorless solid (MP 62–64 °C, BP 115 °C) and can be prepared by heating

$$
\underset{\text{Trioxane}}{
\begin{array}{c}
\text{O} \quad \text{O} \\
| \qquad | \\
H_2C \qquad CH_2 \\
\text{O}
\end{array}}
\xrightarrow[H_2O]{BF_3}
\underset{\text{Poly-oxymethylene}}{HO-CH_2\!\!\left[O-CH_2\right]_n\!\!OH}
\longleftrightarrow
\underset{\text{Formaldehyde}}{CH_2O \;+\; H_2O}
\qquad (2.20)
$$

paraformaldehyde or a formaldehyde solution (60–65%) in the presence of 2% of sulfuric acid[38]. Trioxane can be used as a source of formaldehyde or a curing agent for phenolic resins. It is used as starting material for one of the commercial processes to manufacture polyacetal resins.

Cyclic formals: 1,3-dioxolane, 4-phenyl-1,3-dioxolane and 4-methyl-1,3-dioxolane have been recommended as novolak curing agents and because of their solvent action, for low-pressure laminating resins with reduced viscosity[48, 49].

2.2.4 Hexamethylenetetramine, HMTA

HMTA, used to crosslink novolak resins, is prepared from formaldehyde and ammonia according to Eq. (2.21).

$$6 \, CH_2O + 4 \, NH_3 \rightleftharpoons (CH_2)_6N_4 + 6 \, H_2O \tag{2.21}$$

The reaction is reversible. HMTA decomposes at elevated temperatures[49], generally above 250 °C. In aqueous solutions, HMTA is easily hydrolyzed; the corresponding reaction mechanism is indicated in Sect. 3.3.2. HMTA is often used as catalyst in the resol formation reaction rather than ammonia with equivalent results.

Table 2.10. Physical properties of hexamethylenetetramine[50]

Molecular weight	140.2
Specific weight g/cm^3	1.39
Behavior when heated	sublimation at 270–280 °C
Solubility in 100 g H_2O/20 °C	87.4 g
Solubility in 100 g H_2O/60 °C	84.4 g

HMTA is quite soluble in water, relatively easy to dissolve in chloroform, and less soluble in methanol or ethanol. The aqueous solution shows a weak alkaline action with a pH range between 7–10. Finely ground HMTA causes dust explosions. As the particle size of a material is decreased, the finely divided material or dust is quite susceptible to a sudden, unexpected dust explosion. HMTA dust is rated as a severe explosion hazard (see further discussion on dust explosions, Chap. 5.5).

A further application for HMTA is in the explosives industry; cyclotrimethylenetrinitramine is obtained by nitration (Hexogen, Cyclonite, RDX). In the pharmaceutical field, it is used as disinfectant (Urotropin). In the rubber industry HMTA is used as an activator in the vulcanization reaction[31]. A large use of HMTA is in the manufacture of nitrilotriacetic acid/salts as chelating agents and builders in synthetic detergents.

2.2.5 Furfural

Furfural, sometimes called furfurol, is a colorless liquid with chemical properties similar to benzaldehyde. Commercial production starts with residues of annual plants like maize cobs, bagasse or rice hulls. These naturally occurring pentosans are hydrolyzed by dilute sulfuric acid to furfural which is then isolated by steam distillation. With alkaline catalysts the first step in the reaction with phenol is similar to that of formaldehyde, yielding α-(o- or p-hydroxyphenyl)-furfuryl alcohol[51] [Eq. (2.22)].

$$\tag{2.22}$$

The further reaction mechanism is not fully understood. Furan ring scission and reaction with another phenol nucleus may occur leading to a relatively wide range of products.

Also the labile furan ring system may be involved in the polymerization reaction and is very likely cleaved under acidic conditions. Phenol-furfural resins show enhanced flexibility, low melt viscosity and a low viscosity index. Furfural is used in combination with formaldehyde for the preparation of resins for grinding and friction materials. Phenol-furfural resins may be prepared by continuous addition (30 min) of furfural to a phenol melt at 135 °C and refluxing for 3.5 h[52]. Similarly mixed formaldehyde-furfural resins can be prepared[53].

An important derivative of furfural is furfuryl alcohol. It is obtained from furfural by hydrogenation (2.23). Furfuryl alcohol/PF resin blends and acidic catalysts are used in the foundry industry for the no-bake and hot-box core-making process and for the preparation of acid resistant cements. Furfuryl alcohol is also used for the production of furan resins.

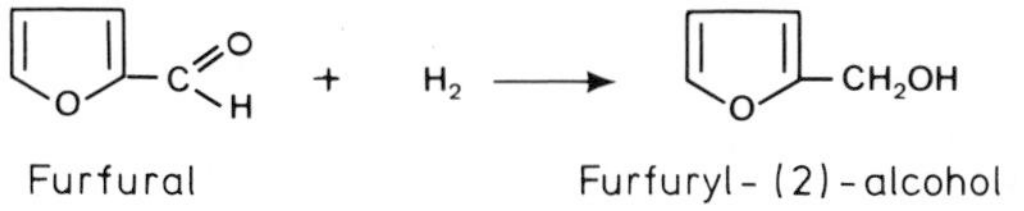

$$\text{Furfural} \quad + \quad H_2 \quad \longrightarrow \quad \text{Furfuryl-(2)-alcohol} \tag{2.23}$$

Furfural   Furfuryl-(2)-alcohol

Starches, starch hydrolysates, sugar and byproducts of sugar production can be used in phenolic resin preparation as low cost additives/modifiers[54]. These carbohydrate compounds are decomposed to 5-hydroxymethyl furfural at novolak production via acid catalyzed hydrolysis with strong inorganic acids in situ (2.24) and (2.25). For resol modification, a two step process is necessary.

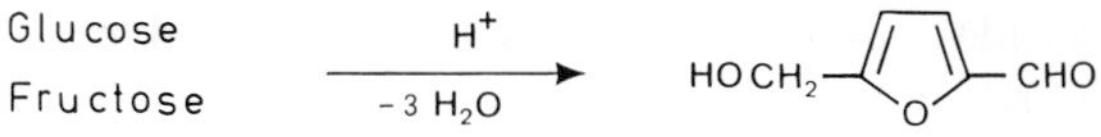

$$\begin{array}{c} \text{Glucose} \\ \text{Fructose} \end{array} \xrightarrow[-3\,H_2O]{H^+} \quad HOCH_2 - \text{furan} - CHO \tag{2.24}$$

$$\text{phenol(OH)} + CH_2O + HOCH_2 - \text{furan} - CHO \xrightarrow{H^+}$$

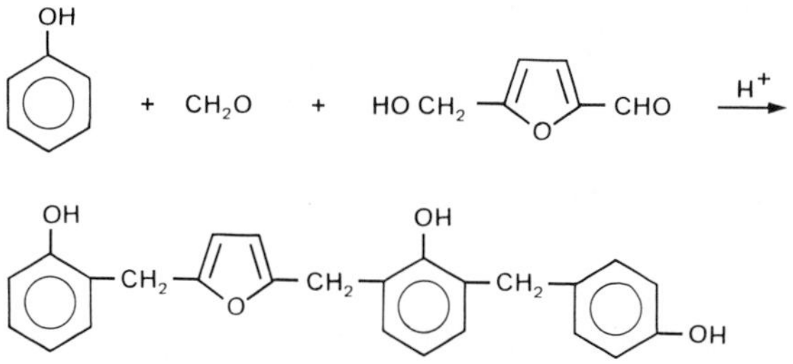

$$\tag{2.25}$$

Hydroxymethylfurfural reacts via the hydroxymethyl and aldehyde group similarly to formaldehyde (2.25) and, because of the higher molecular weight (126), allows savings up to 40% phenol and 65% formaldehyde[54].

2.2.6 Other Aldehydes

Higher aldehydes react with phenol at considerably lower rates. Acidic catalysts are preferred[55], e.g. for the preparation of certain dinuclear antioxidants (Chap. 20). A base-catalyzed reaction is not practical with acetaldehyde or higher aldehydes since they undergo rapid aldol condensation and self-resinification reactions. Recently, a facile method to prepare phenol acetaldehyde novolak resins was described[56, 57]. The method consists of a non-transition metal phenolate (Mg, Zn, Al) catalyzed reaction with acetaldehyde in an aprotic solvent medium for 12 h at 110 °C. Yields in excess of 90% with exclusive *ortho* substitution are obtained (2.26).

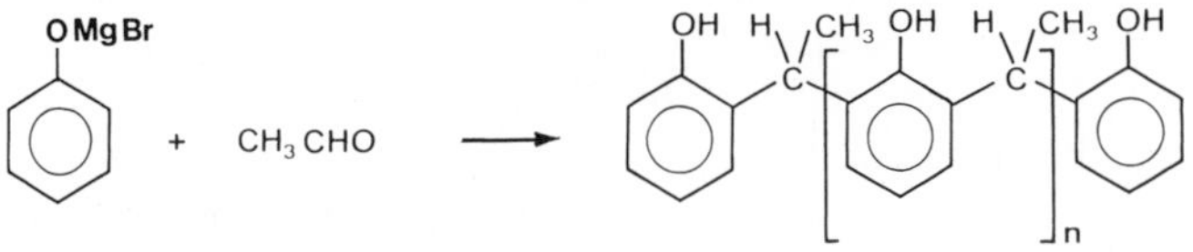

$$(2.26)$$

Novolak resins with higher aldehydes in general are prepared under strong acidic conditions, preferably in a water-free system, by continuous aldehyde addition to the phenol melt. The preferred mol ratio phenol/aldehyde is between $1:0.8$ to $1:1.3$[58]. Only acetaldehyde and butyraldehydes are, however, of limited commercial importance, i.e. for rubber modification, antioxidants and wood binders. The chemical structure of acetaldehyde novolak resins corresponds to those obtained by the reaction of acetylene and phenol with cyclohexylamine as catalyst[59].

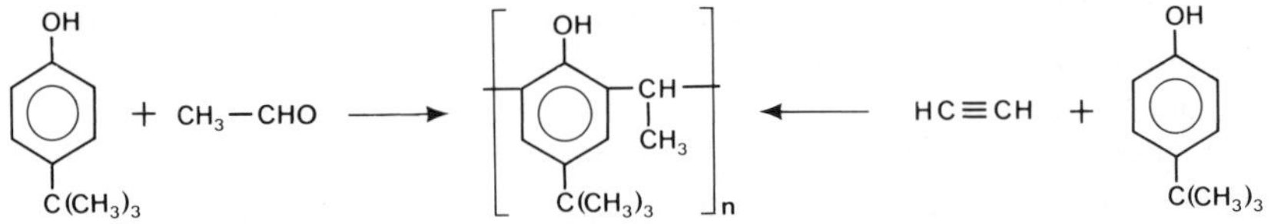

$$(2.27)$$

The reaction of unsaturated aldehydes, acrolein and crotonaldehyde, in acid medium was studied[60] by means of ^{13}C-NMR and GPC. The alkylation reaction via the double bond seems to be the dominant pathway.

2.3 References

1. Weast, R. C. (ed.): Handbook of Chemistry and Physics, 63rd ed. Cleveland: The Chemical Rubber Co. 1982
2. Conrad, F. M.: Phenols. In: Encyclopedia of Polymer Science and Technology, Vol. 10, P. 73. New York: Interscience Publishers 1969
3. Thurman, C.: Phenol. In: Kirk-Othmer, Encyclopedia of Chemical Technology, Vol. 17, P. 373, J. Wiley & Sons, New York 1982
4. Jordan, W., Cornils, B.: Phenole. In: Methodicum Chimicum, Vol. 5, P. 105, Stuttgart: Thieme 1975
5. Jordan, W. et al: Phenol. In: Ullmanns Encyclopädie der technischen Chemie, Vol. 18, P. 177, 4. Ed. Weinheim: Verlag Chemie 1979
6. Fiege, H. et al: Phenol-Derivate. In: Ullmanns Encyclopädie der technischen Chemie, Vol. 18, P. 191, 4. Ed. Weinheim: Verlag Chemie 1979
7. Clouts, K. E., McKetta, R. A.: Cresols and Cresylic Acids. In: McKetta J.J. Ed: Encyclopedia of Chemical Processing and Design, Vol. 13, P. 212, New York: Marcel Dekker Inc., 1981
8. Reed, H. W. B.: Alkylphenols. In: Kirk-Othmer, Encyclopedia of Chemical Technology, Vol. 2, P. 72, New York: J. Wiley & Sons, 1978
9. N. N.: Chem. Ind. 33, 565 (1981)
10. Suehiro: Jpn. Chem. Week No. 1082, 15.1.1981, P. 27
11. Schmidt, K. H.: Chem. Ind. 36, 78 (1984)
12. Hock, H., Lang, S.: Chem. Ber. 77, 257 (1944)

13. Pujado, P. R., Salazar, J. R., Berger, C. V.: Hydrocarbon Proc., March 1976, P. 91
14. Seubold, F. H. Vaugham, W. E.: J. Amer. Chem. Soc. 75, 3790 (1953)
15. Flemming, J. B., Lambrix, J. R., Nixon, J. R.: Hydrocarbon Proc., Jan. 1976 (185)
16. McNeil, D.: Cresols. In: Kirk-Othmer: Encyclopedia of Chemical Technology, 2. Ed., Vol. 6, London: Interscience 1966
17. Ito, K.: Hydrocarbon Proc., August 1973, P. 89
18. Franck, H.-G., Collin, G.: Steinkohlenteer. Berlin, Heidelberg, New York: Springer 1968
19. Wurm, H.-J.: Chem. Ing.-Techn. 48, 840 (1976)
20. Lurgi GmbH: Dephenolization of Effluents by the Phenosolvan Process, Technical Bulletin
21. Macak, J., Burgan, P., Nabirach, V. M.: Coke and Chem. USSR, No. 3, 37 (1979)
22. Rütgerswerke AG, unpublished
23. Heredy, L. M.: The Chemistry of Acid-Catalyzed Coal Depolymerization, Coal Structure, ACS No. 192, P. 179 (1981)
24. Hydrocarbon Research Inc.: Chem. Eng. News, P. 35, Nov. 3, 1980
25. Nylen, P., Sunderland, E.: Modern Surface Coatings: New York: Interscience Publishers 1965
26. BP Chemicals Int. Ltd.: Cellobond Products for Friction Materials, Technical Bulletin
27. Van Gils, G. E.: I & EC Product Research and Development 7/2, June 1968, P. 151
28. Šebenik, A. et al: Polymer 22, 804 (1981)
29. Rhodes, P. H.: Mod. Plastics, August 1947, 145
30. N. N.: Jpn. Chem. Week, No. 1148, 15. 4. 1982, P. 8
31. Koppers Co. Inc.: Resorcinol, Technical Bulletin
32. N. N.: Chemical Age, 19. 6. 1981, P. 10
33. Topp, A.: Resorcin. In: Ullmanns Encyclopädie der technischen Chemie, Vol. 20, P. 189, 4. Ed. Weinheim: Verlag Chemie 1981
34. Conrad, F. M.: Bisphenol-A. In: E. G. Hancock (ed.): Benzene and its Industrial Derivatives. London: Ernst Benn Ltd. 1975
35. Barclay, R., Sulzberg, Th.: Bisphenols and their Bis (Chloroformates). In: J. K. Stille, T. W. Campbell (ed.): Condensation Monomers. New York: Willey 1973
36. Walker, J. F.: Formaldehyde. ACS Monograph No. 159, 3. Ed. (1964)
37. Zabicka, J. (ed.): The Chemistry of the Carbonyl Group, Interscience, 1970
38. Diehm, H., Hilt, A.: Formaldehyde. In: Ullmanns Encyclopädie der techn. Chem., Vol. 11, 4. Ed., Weinheim: Verlag Chemie 1976. Diehm, H.: Chemical Enging. 27, 83 (1978)
39. Maux, R.: Air Best for Formaldehyde and Maleic, Hydrocarbon Proc., March 1976, P. 90
40. Haldor Topsoe A/S: Nissui Topsoe Formaldehyde Process, Technical Bulletin
41. Chauvel, A. R.: Hydrocarbon Proc., Sept. 1973, 1979
42. N. N.: Hydrocarbon Proc., Nov. 1965, 216; Nov. 1969, 183; Nov. 1975, 150
43. N. N.: Hydrocarbon Proc., Nov. 1973, 135
44. Krauch, A., Kuntz, W.: Reaktionen der Organischen Chemie, Heidelberg: Hüthig 1976
45. N. N.: Chem. Eng. News, Jan. 30, 1984, P. 14; Feb. 4, 1985, P. 15
46. Degussa: Paraformaldehyde. Technical Bulletin
47. Adabbo, H. E., Williams, R. J. J.: J. Appl. Polymer Sci. 27, 893 (1982)
48. Heslinga, A., Schors, A.: J. Appl. Polymer Sci. 8, 1921 (1964)
49. Keutgen, W. A.: Phenolic Resins. In: Kirk-Othmer: Encyclopedia of Chemical Technology, Vol. 15, 2. Ed., New York: Interscience 1968
50. Degussa: Hexamethylentetramin, Technical Bulletin
51. Porai, A. E., Koshitz et al: Kunststoffe 23, 27 (1933)
52. Brown, L. H.: Ind. Engng. Chem. 44, 2673 (1952)
53. Mikes, J. A.: J. Polymer Sci 53, 1 (1961)
54. Koch, H., Krause, F., Steffan, R., Woelk, H. U.: Starch/Stärke 35, 304 (1983)
55. Martin, J. C.: J. Org. Chem. 35, 2904 (1970)
56. Casiraghi, G.: Polymer Preprints 24 (2), 183 (1983)
57. Casiraghi, G., et al: Macromolecules 17, 19 (1984)
58. BASF AG: DE-OS 2605 482 (1976)
59. I. G. Farb.: DE-PS 645 112 (1932)
60. Šebenik, A., Osredkar, U.: ACS Polymer Preprints 24/2, 185 (1983)

3 Reaction Mechanisms

Phenolic resins are obtained by step-growth polymerization of difunctional monomers (aldehydes) with monomers of functionality greater than 2 (phenol). Phenols of lower functionality are used to incorporate special properties in the resin (coatings, adhesives). Phenols as monomers possess a functionality of 1 to 3 depending on substitution (Table 3.1); aldehydes are difunctional.

In practice, a decreasing "functionality" is found with increasing MW. An average phenol novolak functionality value of 2.31 has been determined experimentally by Drumm and Le Blanc[1]. The lower functionality value is attributable to a "molecular shielding" phenomenum.

Temperature and pH conditions under which reactions of phenols with formaldehyde are carried out have a profound effect on the characteristics of the resulting products.

Three reaction sequences must be considered: formaldehyde addition to phenol, chain growth or prepolymer formation and finally the crosslinking or curing reaction. The rate of the phenol-formaldehyde reaction at pH 1 to 4 is proportional to the hydrogen ion concentration, above pH 5 it is proportional to the hydroxyl ion concentration, indicating a change in reaction mechanism. Two prepolymer types are obtained depending on pH.

Novolaks are obtained by the reaction of phenol and formaldehyde in a strongly acidic pH region. The reaction is carried out at a molar ratio of 1 mol phenol to 0.75–0.85 mol of formaldehyde. Novolaks are linear or slightly branched condensation products linked with methylene bridges of a relatively low MW up to approximately 2,000. These resins are soluble and permanently fusible, i.e. thermoplastic, and are cured only by addition of a hardener, almost exclusively formaldehyde supplied as HMTA, to insoluble and infusible products.

Table 3.1. Functionality of different phenols in the reaction with aldehydes

Functionality 1	2	3
1,2,6-Xylenol	o-Cresol	Phenol
1,2,4-Xylenol	p-Cresol	m-Cresol
	1,3,4-Xylenol	1,3,5-Xylenol
	1,2,5-Xylenol	Resorcinol
	p-tert-Butylphenol	
	p-Nonylphenol	

Resols are obtained by alkaline reaction of phenols and aldehydes, whereby the aldehyde is used in excess. P/F molar ratios between 1:1.0 to 1:3.0 are customary. These are mono- or polynuclear hydroxymethylphenols (HMP) which are stable at room temperature, but are transformed into three dimensional, crosslinked, insoluble, and infusible polymers by the application of heat and rarely with acids. Resols with two or more rings are moderately branched molecules due to competing reactions between the formaldehyde addition reaction and methylol condensation reaction.

The progressive or finite polymerization which is commonly referred to as curing is distinguished by crosslinking of mainly linear chains with the occurrence of gelation at some intermediate stage in the polymerization reaction. At the gel point the system loses fluidity, since the gel is insoluble in all solvents at elevated temperatures. The gel corresponds to the formation of an infinite network in which crosslinked polymer molecules are transformed into macroscopic molecules. The non-gel portion of the polymer remains soluble (sol). As the polymerization proceeds beyond the gel point, the amount of gel increases at the expense of the sol. According to the Flory Stockmayer statistical model which describes the condensation of a trifunctional monomer (phenol) with a difunctional monomer (formaldehyde), gelation should occur at a P/F ratio of 0.75. The model assumes equal reactivity of sites, absence of substitution effects and absence of self condensation. Williams[2] has reported that resol gelation occurs at a much lower P/F ratio or 0.5. The lower value is attributable to unequal reactivity at functional sites, substitution effects and methylol self condensation. A higher P/F ratio of 0.85–0.90 for novolak gelation is determined[1, 2]. An even higher P/F ratio of 1 is calculated[2] for high o,o' linear novolak with the proposed phenol monomer reactivity being difunctional.

Further discussion concerning molecular size distribution and mathematical treatment is mentioned in Chap. 3.10.

The methylene bridge is thermodynamically the most stable crosslink site. It is prevalent in cured phenolic resins. Theoretically, 1.5 mol of formaldehyde is required for the complete three dimensional crosslinking of 1 mol of phenol (3.1). However, not all reactive sites are accessible to formaldehyde as the oligomer increases in size due to steric reasons or molecular shielding. Generally an excess of formaldehyde is supplied in resin manufacture to meet individual product requirements or specifications.

(3.1)

3.1 Molecular Structure and Reactivity of Phenols

The molecular configuration in solution and crystal structure[3] of phenol is determined by a strong propensity to form hydrogen bonds[4]. In the solid state phenol forms H-bonded chains in the form of a threefold spiral[4]. In solution e.g. in benzene

Table 3.2. Acidity of phenols[9−12)] at 25 °C

Compound	pK$_a$(25 °C)
Phenol	10.00
o-Cresol	10.33
m-Cresol	10.10
p-Cresol	10.28
2-Hydroxymethylphenol	9.84
4-Hydroxymethylphenol	9.73
2,4-Dihydroxymethylphenol	9.69
2,4,6-Trihydroxymethylphenol	9.45

containing small amounts of water, trimolecular species, Ph$_3$, Ph$_2 \cdot$H$_2$O, and Ph$\cdot$2H$_2$O were identified[5)].

For Ph$_3$ a cyclic structure is proposed. Hydroxymethyl phenol forms a strong intramolecular hydrogen bond[6)]. Recent X-ray crystallographic and NMR data of related methylol phenols support this view (Chaps. 4 and 7). The tendency and extent of H bonding of phenols can be easily detected by NMR or by IR-spectroscopy. Often a linear relationship exists between the thermodynamic characteristics for H-bond formation and pK$_2$ values.

In Table 3.2 a comparison of various substituted phenols shows that alkyl substituted phenols are slightly less acidic than phenol. The differences between phenol and cresols are very small. More pronounced is the effect of bulky substituents in *ortho*-position because of steric factors. Hydroxymethylphenols are stronger acids than phenol. Phenols in their electronically excited states are more acidic than the ground state molecules as deduced from spectroscopic data[7, 8)].

The hydroxyl group is in the benzene ring plane even for 2,6-di-tert-butylphenol[13)]. Crystallographic data (Chap. 4) of calixarenes (cyclic oligomers) and high-*ortho* novolaks indicate a high degree of hydroxyl group coplanarity with benzene ring and H-bond formation.

The hydroxyl group is an inductive electron withdrawing ($-$I) and conjugatively electron releasing ($+$R) group. Both effects favour *para*-substitution. Steric reasons also decrease the accessability of the *ortho* positions. In comparison with other activating groups the order with decreasing activating power is:

NMe$_2$ > NHMe > NH$_2$ > OH > OMe > Me

The standard CNDO/2 (Complete Neglect of Differential Overlap)-Method has been applied to phenol, cresols, hydroxymethylphenols and corresponding anions by Knop[14)] (Tables 3.3–3.5). CNDO/2 and ab initio molecular calculations (using the Gaussian 70 program with a STO-3b basis set) have been used by Dietrich et al.[15)] to study intramolecular hydrogen bonding and substituent interactions of *ortho*-substituted phenols and thiophenols. Theoretical results were compared with IR-spectra.

The oxide group of the phenoxide ion is a very strong activating substituent, stronger than NR$_2$, and more *ortho* directing than the hydroxyl group. This differentiation is not so marked in the neutral molecules (Table 3.4). The direct experimental

Table 3.3. CNDO/2 calculations. Ground state energy and dipole moment of phenols[14]

Compound	Ground state energy [eV]	Dipole moment [Debye]
Phenol	−1512.32185	1.52
Phenoxide ion	−1486.42812	7.89
ortho-Cresol *(cis)*	−1720.89644	1.75
ortho-Cresol *(trans)*	−1720.68707	1.46
meta-Cresol *(cis)*	−1720.64869	1.92
meta-Cresol *(trans)*	−1720.65463	1.60
para-Cresol	−1720.64234	1.79
ortho-Hydroxymethylphenol	−2111.80143	1.55
para-Hydroxymethylphenol	−2123.93774	1.92

Table 3.4. CNDO/2 calculations. Electron densities of the ground state of phenols[14]

Position		Phenol	*o*-Cresol	*m*-Cresol	*p*-Cresol	*o*-Hydroxy-methyl-phenol	*p*-Hydroxy-methyl-phenol
	1	3.836	3.851	3.843	3.847	3.842	3.861
	2	4.005	3.964	4.015	4.007	3.962	4.003
	3	3.995	4.001	3.944	4.005	4.001	4.020
	4	4.011	4.016	4.018	3.959	4.010	3.894
	5	3.997	4.005	3.999	4.005	4.001	4.026
	6	4.015	4.011	4.023	4.017	4.004	4.011
	7	6.378	6.374	6.367	6.368	6.380	6.381

Table 3.5. CNDO/2 calculations. Electron densities of the ground state of phenoxide ions[14]

Position		Phenoxide ion	*o*-Hydroxymethyl-phenoxide ion	*p*-Hydroxymethyl-phenoxide ion
	1	3.901	3.848	3.910
	2	4.048	3.970	4.047
	3	4.022	4.030	4.031
	4	4.063	4.054	3.944
	5	4.022	4.029	4.036
	6	4.048	4.050	4.042
	7	6.765	6.594	6.761

comparison of phenol substitution rates with benzene that would indicate the relative activating power of the hydroxyl group on the benzene nucleus is impracticable because of the extremely large difference[13] in the reaction rates in the order of 10^{10}.

The formaldehyde-phenol reaction corresponds to an electrophilic aromatic substitution in acidic media for novolak preparation whereas the highly nucleophilic phenoxide reacts with methylene glycol for resol preparation. It is generally assumed that this reaction type involves the formation of a π-complex in the rate determining step followed by rapid loss of a proton. The actual reaction pathway, however, is much more complicated with phenols because of solvent interactions and inter- and intra-

Table 3.6. Cresol anions; ab initio total atomic charges calculation[14]

Position		o-Cresol	m-Cresol	p-Cresol
	1	5.925	5.925	5.926
	2	6.075	6.145	6.149
	3	6.072	5.999	6.072
	4	6.155	6.157	6.081
	5	6.075	6.074	6.073
	6	6.148	6.150	6.147

molecular hydrogen bond formation. The abnormally divergent variation in *ortho/para* ratio supports this.

Reaction at the *para* position is favoured by polar solvents and acidic conditions, while reaction at the *ortho* position is favoured by nonpolar solvents, alkaline conditions and group II metal oxide-, hydroxide-, or acetate catalysts. The results of kinetic studies are described in Sect. 3.3.3.

The phenolic-OH torsional frequencies (IR) for a large number of substituted phenols have been reported[16]. The shift of the torsional frequency from its position in phenol itself was found to be directly related to the electron-donating or withdrawing power of the substituent group.

Ab initio molecular orbital theory of the STO-3b level was applied to phenol by Hehre et al.[17] and phenoxide ion by Taft et al.[18] and included a variety of substituents to study conformations, stabilities, acidities, and charge distributions. Calculations also confirm that π-donors (e.g. CH_3, NH_2, OH) interact favourably with the OH group in phenol at the *meta* position, but unfavourably at the *para* position. On the other hand, π as well as σ acceptors as aromatic substituents generate positive charges in the π system at the *ortho* and *para* positions.

Calculated total energies and dipole moments of phenols including different conformations (*cis, trans*) are shown in Table 3.3. In *ortho* and *meta* methyl and alkyl phenols the hydroxyl group can be oriented (3.2) *cis* or *trans* to the methyl (alkyl) group[14, 19].

cis   trans

(3.2)

CNDO/2 and ab initio calculations[14, 15] provide reasonable CH_3 rotational barriers for the *cis* and *trans* conformers of o-cresol and o-cresol anion. CNDO/2 results indicate that the (most stable) *cis* conformer is 3.5 kJ/mol more stable than the (most stable) *trans* conformer (each with the CH_3 group staggered with respect to the OH group). This is in contrast to an expected repulsive interaction between the methyl and hydroxyl group in the *cis* conformer. On the other hand, experimental evidence shows that the *trans* conformer is slightly more stable than the *cis* conformer. Theoretical results[15] of o-tert-butylphenol are more reasonable regarding repulsive interac-

Table 3.7. Ab initio conformational dependence of energies of cresol anions [14]

		Δ Energy kJ/mol	Angles minimum	(α) maximum
o-Cresol		2.93	90	30
m-Cresol	α	0.55	30	90
p-Cresol		0.21	30, 90	0, 120
	Ring plane			

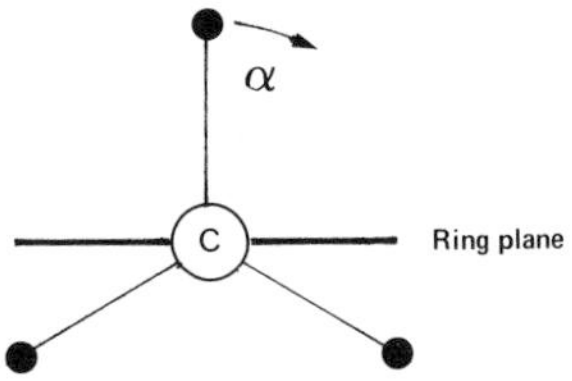

tion showing the *trans* conformer to be 18.8 kJ/mol stable as compared to the *cis* conformer.

Rotational barriers (Δ energy) and angles of minimum/maximum energy conformations depending on the rotation of the methyl group in cresol anions calculated [14] by ab initio method using the Gaussian 80 quantum chemistry program are included in Table 3.7.

The electron density distribution [14] of phenol and related compounds and corresponding anions is presented in Tables 3.4–3.6. An increased differentiation of the electron density distribution is found for the corresponding ions (Table 3.5). The electron density of the *para*-position in the phenoxide ion is higher than that of the *ortho*-position and provides a plausible explanation for the *ortho/para* ratio obtained experimentally.

The different effect of the hydroxymethyl group on the electron density distribution (CNDO/2) compared to the phenoxide ion is shown in Table 3.5. The electron density of the remaining *ortho'*-position (6) is increased by the introduction of the hydroxymethyl group into the *ortho*-position (2), whereas that of the *para*-position (4) is decreased. The electron density of both *ortho*-positions is decreased by the introduction of the methylol group into the *para*-position.

Total atomic charges of cresol anions calculated [14] by the ab initio method are compiled in Table 3.6. Only relatively small differences are evident in the *ortho*- and *para*-position as they relate to the position of the methyl group.

3.2 Formaldehyde-Water and Formaldehyde-Alcohol Equilibria

Formaldehyde is by far the most reactive carbonyl compound. In aqueous medium a very fast acid and base catalyzed hydration reaction of formaldehyde to methylene glycol occurs [20, 21]. The equilibrium indicated in Eq. (3.3) lies far on the side of methylene glycol and can be estimated by UV-spectroscopy (n–π^* transition of the carbonyl group), by NMR or by polarographic methods [22, 23].

$$CH_2 = O + H_2O \underset{k_2}{\overset{k_1}{\rightleftharpoons}} HO-CH_2-OH \tag{3.3}$$

$$K_d = \frac{[CH_2O]}{[HOCH_2OH]} = 1.4 \cdot 10^{-14}$$

Methylene glycol is found in aqueous solutions as the key monomeric species. It is also obtained by dissolving paraformaldehyde. The concentration of monomeric non-hydrated formaldehyde is very low, generally less than 0.01%[20].

Methylene glycol is present as monomer only in dilute aqueous formaldehyde solutions. The depolymerization of aqueous polyoxymethylene

$$HO + CH_2 - O]_{\overline{n}} H + H_2O \rightleftharpoons HO + CH_2O]_{n-1} H + HO - CH_2 - OH \tag{3.4}$$

glycol (3.4) in the presence of acidic and basic catalysts is of importance for the overall reaction rates for resol and novolak formation.

Alcohols are often present in the PF reaction. Methanol is present at least in small amounts ($\sim 1\%$) because the formaldehyde manufacturing process commences with methanol. In addition it can be formed from formaldehyde during storage by disproportionation (Cannizzaro reaction). Furthermore, methanol may be added because it is very efficient in stabilization of concentrated aqueous formaldehyde solutions. Chain termination prevents the formation of low soluble polymers so that precipitation or turbidity can be avoided. Alcohols can react with aqueous formaldehyde in a neutral pH to form hemiformals (3.5). Formals are not formed under these conditions.

$$R - OH + HO - CH_2 - OH \rightleftharpoons R - O - CH_2 - OH + H_2O \tag{3.5}$$

The reaction between hydroxymethylphenols and methylene glycol must also be considered. The extent of this reaction with the hydroxymethyl group (3.6) as well as with the phenolic hydroxyl group (3.7) has been studied by high-resolution NMR[24].

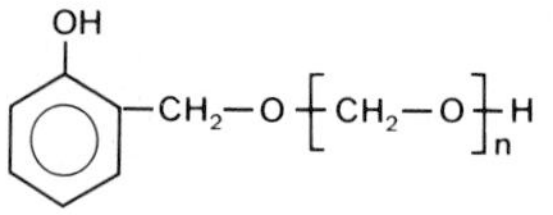

$$\tag{3.6}$$

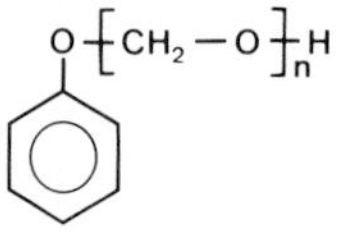

$$\tag{3.7}$$

Peaks for $n = 0, 1, 2, 3$ have been identified. In a mixture of 70 parts of 40% formalin and 100 parts of phenol approximately 10% of the phenol has reacted with methylene glycol to form phenol hemiformal.

The overall competitive characteristics of the hydroxyl group of methanol : benzyl alcohol : phenol is 70 : 20 : 1 for hemiformal formation. Benzyl alcohol is proposed as the model compound for the methylol functionality of HMP[24]. Hence formaldehyde is also consumed in the PF reaction to form hemiformals which constitute a potential source of formaldehyde and can be detected by the usual titrimetric methods. It is, however, uncertain whether the hemiformal is transformed into methylol. The consumption of formaldehyde to hemiformal is viewed as a plausible explanation for the apparent reduction in reaction rate during conversion as stated by Zsavitsas[25].

3.3 Phenol-Formaldehyde Reaction under Alkaline Conditions

The reaction between formaldehyde and phenol in the alkaline pH-range was first observed in 1894 by L. Lederer[26] and O. Manasse[27] and is usually referred to as the Lederer–Manasse reaction. At a pH above 5, *bis*- and *tris* alcohols are formed as well as *mono* alcohols and other compounds. The simplest product of this reaction, 2-hydroxybenzylalcohol (saligenin), was previously isolated from the glucoside salicyn by dilute acid hydrolysis in 1845[28].

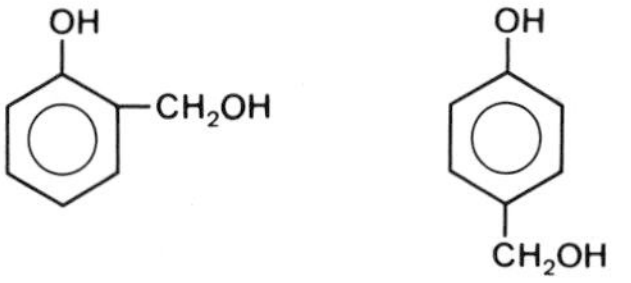

$$(3.8)$$

o-Hydroxymethylphenol   *p*- Hydroxymethylphenol
MP86°C   MP124...126°C

$$(3.9)$$

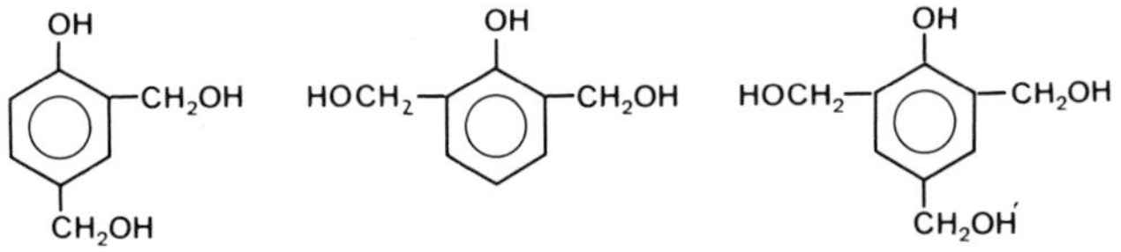

o,p-Dihydroxymethylphenol   *o,o'*-Dihydroxymethylphenol   Trihydroxymethylphenol
MP 93 °C   MP 101 °C   MP 79...82 °C

3.3.1 Inorganic Catalysts and Tertiary Amines

Sodium hydroxide, ammonia and HMTA, sodium carbonate, calcium-, magnesium-, and barium hydroxide and tertiary amines are used as catalysts in the alkaline hydroxymethylation reaction. Formaldehyde is present as methylene glycol in aqueous solution in commercial processes. Phenol reacts rapidly with alkali to form the phenoxide ion which is resonance stabilized according to Eq. (3.10).

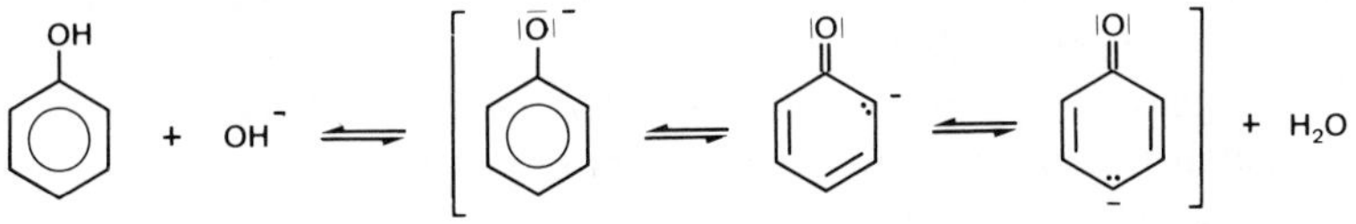

$$(3.10)$$

In the ensuing reactions, catalyzed by alkali, C-alkylation in *ortho* and *para* positions occurs almost exclusively. *Meta* substitution is not detected.

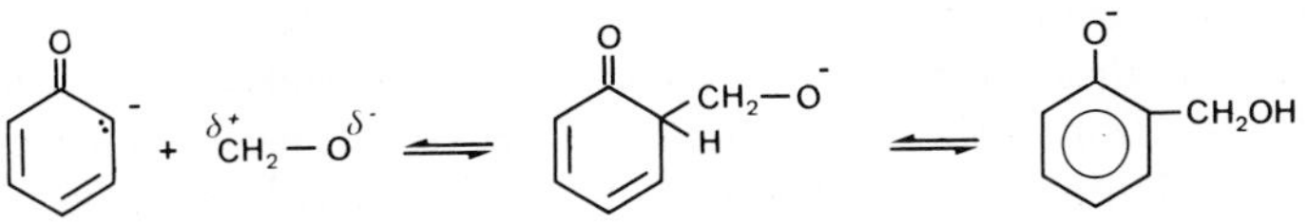

$$(3.11)$$

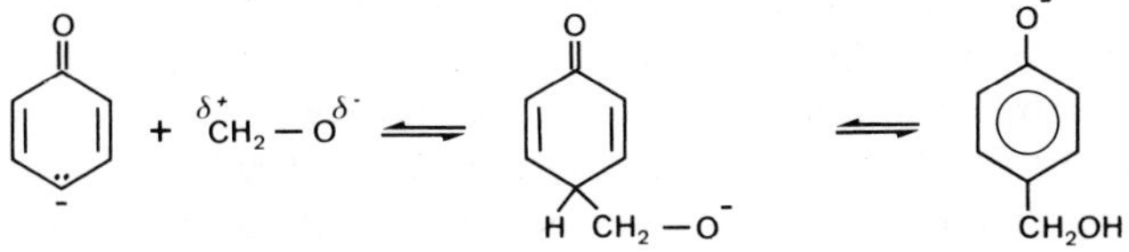

$$(3.12)$$

The quinoid transition state is stabilized by proton shift as indicated in the Eqs. (3.11) and (3.12). These reaction mechanisms were proposed for dilute solutions[30]. The monomethylol derivative continues to react with formaldehyde, forming two dimethylol- and one trimethylol compounds.

The kinetics of the base-catalyzed phenol-formaldehyde reaction have been thoroughly examined[25, 29-33] and are relatively well understood. In general, a second order reaction was observed with the exception of the ammonia catalyzed reaction, which surprisingly corresponds to a first order reaction. The general expression for the overall reaction rate is:

reaction rate = $k[Ph^-][\text{methylene glycol}]$

It must be pointed out, however, that the actual structure or reaction sequence of the hydroxyalkylating agent in the alkaline catalyzed reaction is not fully understood. It is not clear how methylene glycol reacts with the phenoxide ion[25]. The concentration of non-hydrated formaldehyde is too low to explain the reaction rates.

$$HO-CH_2-OH \rightleftharpoons {}^+CH_2-OH + OH^- \qquad (3.13)$$

$$HO-CH_2-OH \rightleftharpoons CH_2^{\delta+}-O^{\delta-} + H_2O \qquad (3.14)$$

A divergent reaction mechanism was proposed quite early by Claisen, later by Walker[20] and others. The presence of hemiformals (3.7) in aqueous phenol-formaldehyde solutions has been supported by NMR[24].

$$\text{(phenol with } O-CH_2OH) \rightleftharpoons \text{(phenoxide } O^- \text{ with ring } -CH_2OH) \qquad (3.15)$$

Claisen and others claimed that this hemiformal rearranges to hydroxymethyl phenoxide (3.15). The reaction will not occur if the phenolic group is etherified. This indirect evidence suggested a plausible explanation for the intermediacy of the hemiformal. Unfortunately this line of reasoning is not entirely correct since the highly nucleophilic phenoxide ion is the vital reactive intermediate in the alkaline hydroxymethylation.

A series of experimental results indicates that the composition of the transition state is considerably more complex than indicated in the Eq. (3.11). The crucial evidence is the dependence of the *ortho/para* substitution ratio on the type of catalyst[32, 34]. The *ortho/para* ratio decreases from 1.1 at pH 8.7 to 0.38 at pH 13.0. It has been recognized that the *ortho* substitution is considerably enhanced if metal hydroxides of the first and second main group among the series K < Na < Li < Ba < Sr < Ca < Mg are used as catalysts (Fig. 3.1) as well as an intermediate pH. Even more subtle is the effect of transition metal hydroxides. *Ortho* substitution is highly favored, the higher the chelating strength of the cation. The directing effect of the Fe, Cu, Cr, Ni, Co, Mn, and Zn ions is explained by Peer as due to formation of chelates as transient intermediates according to formula (3.16). Boric acid also has a strong *ortho*-directing effect (3.17).

$$(3.16)$$

$$(3.17)$$

High-*ortho* novolaks[35, 36] are prepared with MgO or ZnO catalysts. Thus conditions favoring high *ortho* substitution consist of an intermediate pH, catalyst with che-

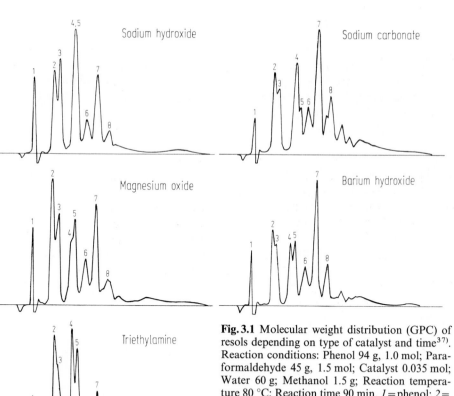

Fig. 3.1 Molecular weight distribution (GPC) of resols depending on type of catalyst and time[37]. Reaction conditions: Phenol 94 g, 1.0 mol; Paraformaldehyde 45 g, 1.5 mol; Catalyst 0.035 mol; Water 60 g; Methanol 1.5 g; Reaction temperature 80 °C; Reaction time 90 min. *1* = phenol; *2* = *o*-hydroxymethylphenol; *3* = *p*-hydroxymethylphenol; *4* = 2,6-*bis*(hydroxymethyl)phenol; *5* = 2,4-*bis*(hydroxymethyl)phenol; *6* = diphenylmethane derivative; *7* = *tris*-(hydroxymethyl)phenol; *8* = diphenylmethane derivative

lating capability and in some instances a less polar solvent environment (Sect. 3.4.3); all favoring a cyclic transition state.

The different activity of some of the more commonly used catalysts and their effect on MWD as examined by GPC is shown in Fig. 3.1. Identical reaction conditions have been used for the production of the resols. The catalyst was neutralized with hydrochloric acid and the resin analyzed without distillation. A high degree of *ortho* orientation is observed[35, 36] when zinc acetate is used (Fig. 3.6) followed by magnesium oxide and triethylamine as catalysts (Fig. 3.1). Even thought a lower F/P ratio is used in the zinc acetate experiment (Fig. 3.6) nevertheless high *ortho* substitution is obtained.

Discussion of the orientation and directing effects of any particular catalyst must be analyzed and interpreted with care. The concentration of the resulting methylol phenol in the reaction mixture depends upon rate of formation as well as its disappearance due to further reaction. Thus it is dependent upon molar ratio of phenol to formaldehyde and reaction time. In GPC analysis, the order of appearance of individual methylol phenols is related to the number of OH groups despite molecular size and is attributable to solvent interactions (Chap. 7).

3.3.2 Ammonia, HMTA and Amine-Catalyzed Reactions

Ammonia catalyzed resols differ significantly from all other resols by their characteristic yellow color, which is attributable to the azomethine group –CH=N–, and also by their higher average MW. This color is also typical of novolaks cured with HMTA. It has been shown[24, 38, 39] that these prepolymers contain secondary amine (3.18), tertiary amine (3.19) and benzoxazine (3.20) structures.

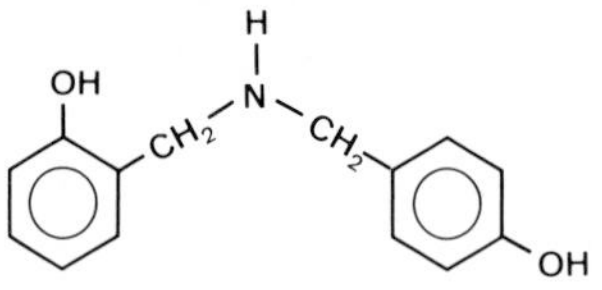

(3.18)

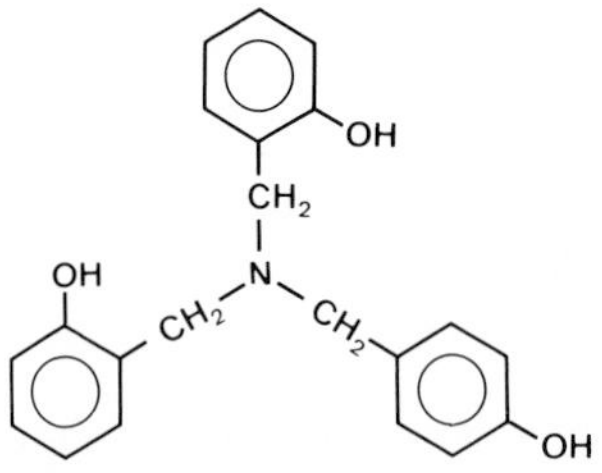

(3.19)

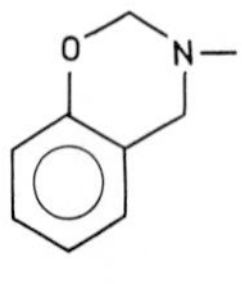

(3.20)

By monitoring the reaction of HMTA with phenol by ^{13}C-NMR and FTIR, Sojka has shown[40] that there is an initial rapid decay of HMTA with concurrent benzylamine formation. The amine is predominantly secondary with some tertiary amine and trace amounts of primary amine. The substitution pattern of the amine is exclusively *ortho* and suggests the apparent intermediacy of benzoxazine which rearranges to the amine(s).

The reaction between phenols, aldehydes, and amines can also be considered as an aminoalkylation according to the reaction sequence shown in equation (3.21), commonly known as the Mannich-reaction[41].

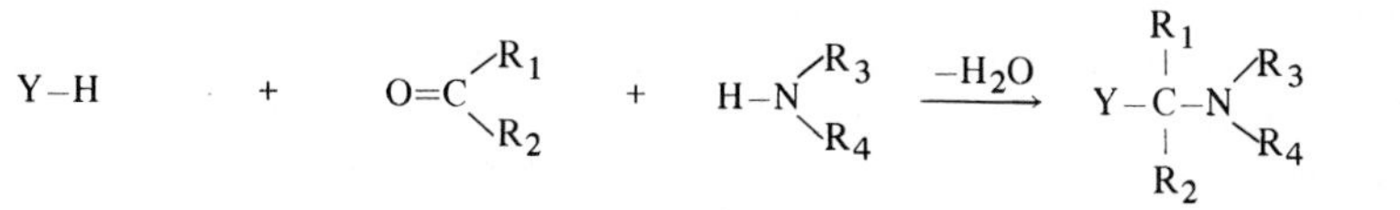

$$Y-H \quad + \quad O=C\begin{smallmatrix}R_1\\R_2\end{smallmatrix} \quad + \quad H-N\begin{smallmatrix}R_3\\R_4\end{smallmatrix} \xrightarrow{-H_2O} \quad Y-\underset{R_2}{\overset{R_1}{C}}-N\begin{smallmatrix}R_3\\R_4\end{smallmatrix} \qquad (3.21)$$

| H-acidic compound (Phenol) | Carbonyl compound (Formaldehyde) | Amine (Ammonia) | α-Aminoalkylated compound |

Phenols are O–H as well as C–H acidic compounds. However, the substitution always takes place on the aromatic nucleus.

If amines are present concurrently, the carbonyl component is confronted with 2 nucleophiles. For this reason it must be determined whether formaldehyde initially reacts with phenol or with amine.

The ammonia or hexamethylenetetramine catalyzed reaction of PF is much more complex in view of the various model compound studies and various equilibria between ammonia, formaldehyde, and HMTA as well as α-aminoalcohol intermediates. Kinetic studies indicate the ammonia catalyzed reaction of PF corresponds to a first order reaction.

The reaction of NH_3 or aliphatic amines with formaldehyde leads to HMTA and hexahydrotriazine; the latter product emerges from formaldehyde and amine reaction. The formation of HMTA from F and NH_3 is a very fast reaction in which the formation of *bis*-hydroxymethyl-amine [Reaction (3.22)] is the slowest step and rate determining. The proposed[42] reaction mechanism is indicated below (3.22/23/24).

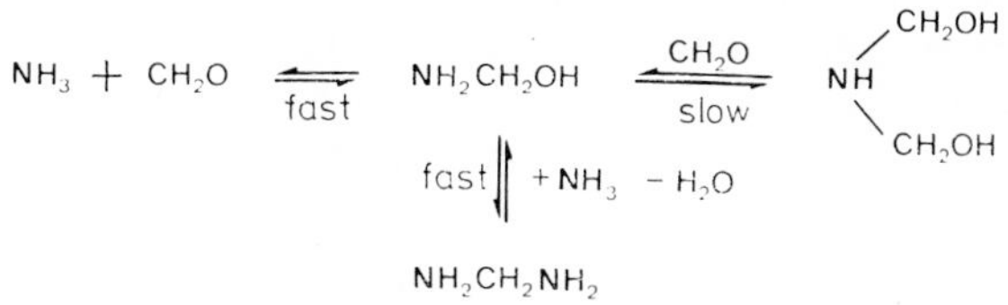

$$NH_3 + CH_2O \underset{fast}{\rightleftharpoons} NH_2CH_2OH \underset{slow}{\overset{CH_2O}{\rightleftharpoons}} NH\begin{smallmatrix}CH_2OH\\CH_2OH\end{smallmatrix} \qquad (3.22)$$

$$fast \updownarrow +NH_3 \quad -H_2O$$

$$NH_2CH_2NH_2$$

$$\begin{matrix}NH_2\\|\\CH_2\\|\\NH_2\end{matrix} + \begin{matrix}HOCH_2\\|\\NH\\|\\HOCH_2\end{matrix} \overset{-2H_2O}{\rightleftharpoons} \begin{matrix}NH-CH_2\\|\quad\quad|\\CH_2\quad NH\\|\quad\quad|\\NH-CH_2\end{matrix} \qquad (3.23)$$

$$+NH(CH_2OH)_2 \updownarrow$$

$$\begin{matrix}CH_2-N-CH_2\\|\quad\quad|\quad\quad|\\NH\quad CH_2\quad NH\\|\quad\quad|\quad\quad|\\CH_2-N-CH_2\end{matrix} \overset{CH_2O}{\rightleftharpoons} \text{(HMTA cage structure)} \qquad (3.24)$$

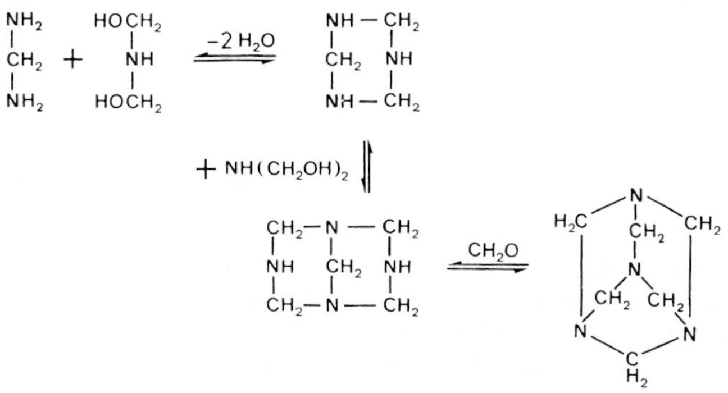

In dilute aqueous systems the reaction stops at the α-aminoalcohol stage. Aldimines are only formed in trace amounts. All reaction steps are reversible. The concentration of formaldehyde shifts the equilibrium to aminoalcohols and ultimately to HMTA. Thus the PF reaction with NH_3 can be envisioned as one conducted with HMTA.

Besides encountering formaldehyde with two nucleophiles, HMTA is equally confronted with two species: phenol and water with either undergoing a different reaction with HMTA.

The hydrolysis of HMTA also leads to aminomethylated products [Reaction (3.23)] so that HMTA or an equivalent amount of ammonia may be used interchangeably as catalysts giving equivalent results. HMTA[43] is used as hardener for curing novolaks (Sect. 3.4.4).

The reaction of phenols with formaldehyde and formation of methylol compounds occurs in parallel, now catalyzed by hydroxybenzylamines. The reaction of m-cresol with HMTA was found to be first order. The reaction rate constant was evaluated to be 1.9×10^{-4} l mol^{-1} s^{-1}, HMTA hydrolysis and formation of the aminomethylene ions was the slowest and rate determining step[44]. The activation energy was found to be 64.9 kJ/mol.

The MWD of resols catalyzed by high amounts of ammonia deviates considerably from other resols. Tertiary amines lead to prepolymers which are similar to sodium hydroxide-catalyzed resols in their structure (Fig. 3.1), except higher *ortho* orientation.

Ammonia-catalyzed resol solutions are of considerable importance for the production of impregnating resins for electrical laminates and coatings. In these cases, the removal of the catalyst (by precipitation or washing) is not required to obtain favorable electrical properties with high corrosion resistance.

The structure of ammonia catalyzed resols allows the commercial production of solid resols with a melting range between 40 and 60 °C. Solid, pulverized resols are required for many applications, such as the production of molding compounds, brake lining mixtures, abrasive materials and coating resins. Resins with even higher melting points would be desirable as storage stable and free-flowing powders. However, it is difficult to realize a high batch volume since the curing reaction which commences between 90–100 °C precludes a higher residence time. Cooling and discharging the molten resol requires considerable time. The newly developed dispersion technique allows a facile method to prepare higher melting solid resols (Chap. 4).

3.3.3 Reaction Kinetics of the Base-Catalyzed Hydroxymethylation

It is the aim of any kinetic study to determine how a reaction proceeds. However, the determination of reliable kinetic data for phenolic resin formation is complicated because the reaction conditions, including temperature, type and amount of catalyst, and mol ratios and polarity of the solvents exert a profound influence on the results obtained. The purity of the raw materials, and even the nature of the reaction vessel, will play significant roles. The identification of the reaction products is also relatively difficult[45]. Therefore, it is not surprising that the reported kinetic data differ considerably. For the interpretation of kinetic data, it must be remembered that there are two *ortho*-positions in phenol and p-hydroxymethylphenol. It is more reasonable, for the above mentioned reasons, to compare the relative reaction rates, which consider

Table 3.8. Relative positional reaction rates in the phenol-formaldehyde reaction, *ortho*-substitution set as unity

		Relative reaction rates		
		Freeman and Lewis[31] 1954	Zsavitsas and Beaulieu[25] 1967	Eapen and Yeddana-palli[33] 1968
Phenol	→ *o*-methylolphenol	1.00	1.00	1.00
Phenol	→ *p*-methylolphenol	1.18	1.09	1.46
o-Methylolphenol	→ *o-o'*-dimethylolphenol	1.66	1.98	1.75
o-Methylolphenol	→ *o-p*-dimethylolphenol	1.39	1.80	3.05
p-Methylolphenol	→ *o-p*-dimethylolphenol	0.71	0.79	0.85
o-p-Dimethylolphenol	→ trimethylolphenol	1.73	1.67	2.04
o-o'-Dimethylolphenol	→ trimethylolphenol	7.94	3.33	4.36

the availability of nuclear positions ("concentration"). The NaOH catalyzed reaction has been the most thoroughly examined reaction. The relative reaction rates are compiled in Table 3.8. The rate constant of the formation of *o*-HMP divided by two because two *ortho*-positions are available was arbitrarily set as unity for ease of data comparison.

One of the most complete studies was published by Freeman and Lewis[31]. The reaction was performed at 30 °C with a molar ratio of phenol/formaldehyde/NaOH of 1:3:1. To avoid the involvement of different reacting species arising from the higher acid strength of the hydroxymethylphenols generated, one mol of sodium hydroxide per mol of phenol was used. Each individual reaction has been assumed to be of the second order (Chap. 3.3.1), i.e., first order with respect to each reaction component, formaldehyde, and phenol anion. A plot of the disappearance of phenol and formaldehyde in conjunction with the appearance of various methylol phenols is shown in Fig. 3.2.

The *para*-position in phenol shows a slightly higher relative reactivity towards formaldehyde than the *ortho*-position (electron density value). However, *o*-HMP is produced at a higher rate due to the fact that two *ortho*-positions are available. The *ortho/para* ratio was found to be 1.7. Since the methylol group is regarded as electron attracting, i.e. ring deactivating, a decrease in the reactivity of the remaining positions towards further formaldehyde addition is expected. This has also been observed in the case of *p*-HMP (Table 3.8).

However, a considerable increase in reactivity has been observed at the incorporation of the methylol group into the *ortho*-position. *o*-HMP is found to be twice as reactive as *p*-HMP with respect to either position. Increased reactivity is also found with higher methylolated phenols and is especially remarkable with 2,6-BHMP.

CNDO calculations of the electron densities of the HMP anions (Table 3.5) show a reduced electron density compared to the phenate ion of the *ortho*-position of *p*-HMP which is in accordance with reported lower positional reaction rates. On the other hand, in *o*-HMP anion the electron density in *para* position is increased, in the *ortho* position slightly decreased. In this case, the reported substitution rates only par-

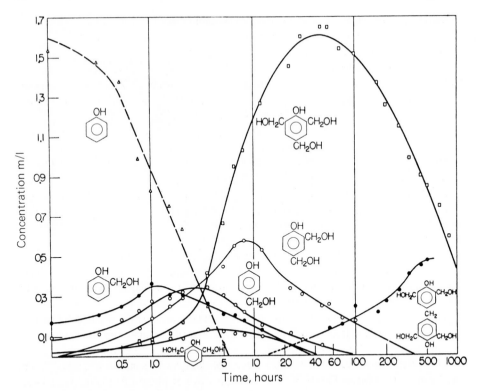

Fig. 3.2. Change in concentration of phenol and methylol phenols with time[31)] (formaldehyde and phenol at 30 °C with NaOH catalyst)

$k_1 = 10.5 \cdot 10^{-6}\,\mathrm{l\,mol^{-1}\,s^{-1}}$
$k_2 = 6.2 \cdot 10^{-6}\,\mathrm{l\,mol^{-1}\,s^{-1}}$
$k_3 = 7.5 \cdot 10^{-6}\,\mathrm{l\,mol^{-1}\,s^{-1}}$
$k_4 = 7.3 \cdot 10^{-6}\,\mathrm{l\,mol^{-1}\,s^{-1}}$
$k_5 = 8.7 \cdot 10^{-6}\,\mathrm{l\,mol^{-1}\,s^{-1}}$
$k_6 = 9.1 \cdot 10^{-6}\,\mathrm{l\,mol^{-1}\,s^{-1}}$
$k_7 = 41.7 \cdot 10^{-6}\,\mathrm{l\,mol^{-1}\,s^{-1}}$

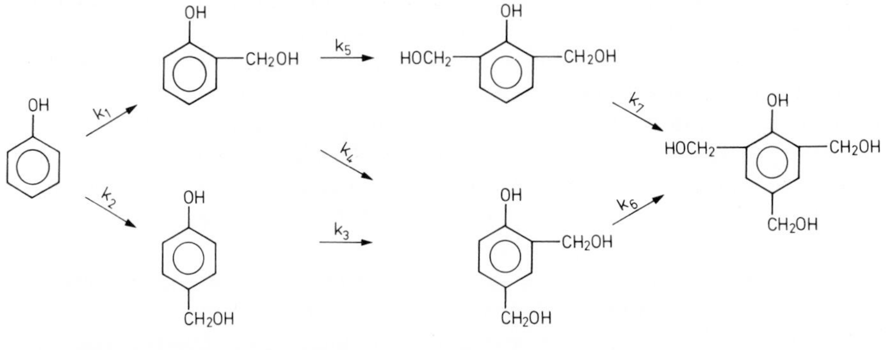

Fig. 3.3. Reaction steps and individual rate constants of the phenol-formaldehyde reaction according to Freeman and Lewis[31)]

tially relate to the calculated electron density values. The positional reactivities determined by Freeman and Lewis were confirmed later by studies of Zsavitsas[25, 30] who improved the identification procedure by GC separation of phenol alcohols[46]. In addition, the reaction was performed at lower pH using only catalytic amounts of NaOH to ascertain a second order reaction. Reaction rates of about 50% higher have been found compared to Freeman and Lewis. A theoretical calculation based upon pk_a-values, resulted in excellent agreement with experimental kinetic results. In a later study[25], formaldehyde equilibria (Chap. 3.2) were incorporated in the kinetic model.

The increasing positional reactivity of the PF system as the methylol substitution proceeds also has considerable practical importance. Due to the marked tendency towards the formation of polyalcohols, commercial resols contain considerable amounts of unreacted phenol, even if a relatively high formaldehyde ratio is used. This is a disadvantage not only for reasons of resin efficiency in the commercial processes but because it may also cause environmental problems.

Rates of reaction, entropy of activation and Arrhenius parameters of the reaction between *m*-cresol[48], *o*-cresol[49], and 2,5-dimethylphenol[50] and formaldehyde have been investigated by Malhotra et al. An increased rate of reaction was observed with an increase in the alkali concentration.

The influence of methyl substitution on the relative reactivity of phenols towards formaldehyde was studied by Sprung[47]. This study encompassed the reaction of phenol, cresols, and xylenols with paraformaldehyde without addition of water using triethanolamine as catalyst. The results must be considered as relative rates of formaldehyde consumption (Table 3.9). It was found that *m*-cresol adds formaldehyde at nearly three times the rate of phenol. The methyl group introduced in *ortho* or *para* position renders the phenolic nucleus less reactive.

Later studies[33, 51], however, indicate significantly higher relative positional reactivities of *para* and *ortho* cresol and xylenols compared to phenol. The rate of formaldehyde conversion with phenols was investigated[51] in water/methanol because of poor solubility of alkylphenols in water. Mole ratios applied were 0.5 mol formaldehyde for each reactive position and 0.1 mol sodium hydroxyde per 1 mol phenol leading to a pH between 10.1 and 11.5. Results which partially disagree with industrial experience are shown in Table 3.10.

Table 3.9. Relative reaction rates of various phenols with formaldehyde[47]

Compound	Relative reactivity
2,6-Xylenol	0.16
o-Cresol	0.26
p-Cresol	0.35
2,5-Xylenol	0.71
3,4-Xylenol	0.83
Phenol	1.00
2,3,5-Trimethylphenol	1.49
m-Cresol	2.88
3,5-Xylenol	7.75

Table 3.10. Formaldehyde conversion rate[51] in water/methanol; phenol = 1

3,5-Dimethylphenol	4.1
m-Cresol	3.1
2,6-Dimethylphenol	2.7
3,4-Dimethylphenol	2.6
2,4-Dimethylphenol	1.9
o-Cresol	1.8
p-Cresol	1.2
p-tert-Butylphenol	1.1
Phenol	1.0
p-Cumylphenol	0.9
Bisphenol A	0.8
p-Octylphenol	0.8
p-Phenylphenol	0.2

Thus further research work taking into account common reaction conditions is desirable. Halogens, though *ortho*- and *para*-directing, deactivate all nuclear positions in the phenol molecule. All *meta*-directing groups, such as nitro and sulfonic group, reduce the reactivity to a great extent.

The rate of methylolphenol formation is a function of pH above pH 5. Tertiary amines and quaternary ammonium hydroxides are comparable to sodium hydroxide in their activity, while mono- and diethylamine or ammonia (Fig. 3.1) are essentially weaker catalysts due to lower basicity.

The kinetics of the alkali-catalyzed addition of formaldehyde to dihydroxydiphenyl methanes (DPM) has also been examined[52]. The order of reactivity of the three DPM's is found to be *o,p'* > *p,p'* > *o,o'*-DPM. The low reactivity of *o,o'*-DPM is explained on the basis of the formation of a chelate ring in its singly ionized form, but considerably different from HMTA gel data (Table 3.13). The rate constants for the addition of formaldehyde to DPM's are about the same order of magnitude as those of the condensation of methylolphenols at low alkali concentration. But at higher alkali concentration, the rate of formaldehyde addition is increased while that of the condensation reactions is reduced (Fig. 3.5).

3.3.4 Prepolymer Formation

Apart from formaldehyde addition, condensation reactions between methylolphenols or condensation with phenol frequently occur during normal reaction conditions such as a temperature range of 60 to 100 °C and lead to prepolymer formation. Below 60 °C and at high pH, the condensation reaction is negligible. The condensation reactions of methylolphenols have been investigated in particular by three research groups led by Zinke, von Euler, and Hultzsch. The findings of these groups only partially agree. The analytical procedures were comparable. Model compounds (mono- and di-methylolphenols) were synthetized, and, as they separate upon heating, the formaldehyde and water were estimated quantitatively.

A comprehensive description of these studies can be found in publications by Megson[45] and Martin[53]. Two prime reactions (3.25) and (3.26) have been proposed:

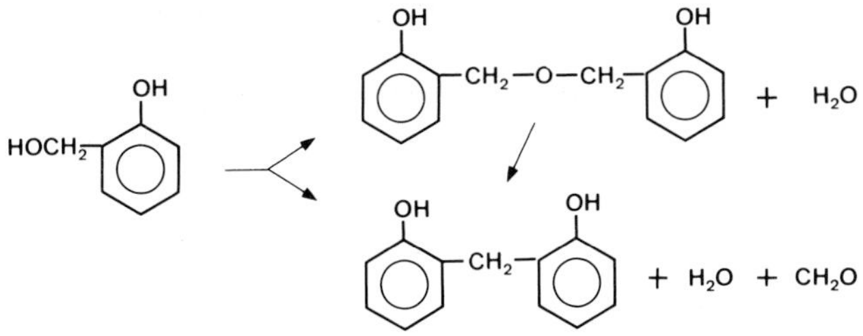

$$\text{(3.25)}$$

$$\text{(3.26)}$$

Attempts to distinguish either reaction pathway by model compound studies were not very illuminating.

It was later determined[54] that the most important reaction under strong alkaline conditions is the formation of diphenylmethanes according to Eq. (3.26).

The formation of dihydroxydibenzyl ether [Eq. (3.25)] is very unlikely under strong alkaline conditions. This reaction, however, is considered the prevalent one under neutral or weak acidic conditions and temperatures up to 130 °C as they normally exist at the curing of resols[55]. Above 130–150 °C methylene bridge formation becomes predominant[45] and, at still higher temperatures, a number of ill defined reactions become important[56].

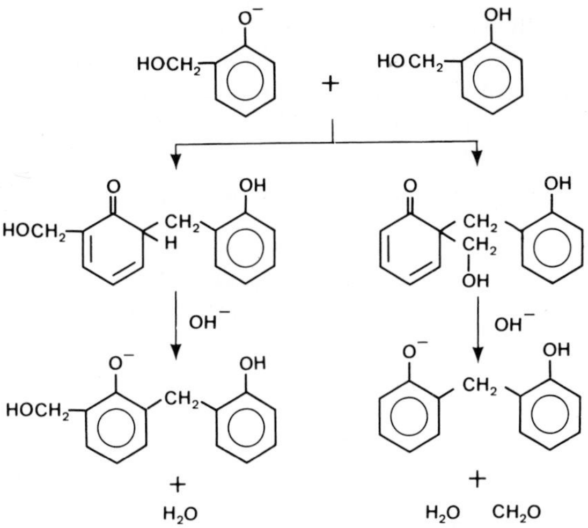

$$\text{(3.27)} \qquad\qquad \text{(3.28)}$$

It was shown that the condensation reaction between two HMP's proceeds substantially faster than the reaction between HMP and phenol[54]. The methylol group has a strong activating effect on the condensation reaction. The reaction between HMP's may occur by displacement of a proton (3.27) or formaldehyde (3.28). Both reactions are believed to be S_N2 mechanisms, involving attack of a methylol substituent with displacement of a hydroxyl group[57] which is, in fact, a poor leaving group. The ratio of products, dimethylene ether and DPM, depends on the structure of the methylol phenols involved and the reaction conditions.

Recently Jones[58] has proposed a quinone methide intermediate (Chap. 3.3.5) during high pH conditions in the transformation of methylol phenol to DPM products. Whether a S_N2 displacement or quinone methide addition mechanism occurs requires further study.

Under neutral or weakly acidic conditions and at lower temperatures linear dimethylene ether containing polymers have been prepared from 2,6-*bis*(hydroxymethyl)-4-methylphenol[59] (3.29).

$$HOCH_2 \overset{OH}{\underset{CH_3}{\bigcirc}} CH_2OH \quad \xrightarrow{130\,°C} \quad \left[CH_2 \overset{OH}{\underset{CH_3}{\bigcirc}} CH_2-O \right]_n + \; n\,H_2O \qquad (3.29)$$

Recently Harwood et al.[60] have examined the self-condensation of dimethylol *p*-cresol and related compounds by NMR. A transition state consisting of hydrogen bonding between a phenolic group and a neighboring methylol group is proposed and causes the benzylic alcohol moiety present in *o*-methylol phenols to be susceptible to nucleophilic displacement under very mild conditions.

3.3.5 Resoll Crosslinking Reactions. Quinone Methides

Heat curing, by far the most important crosslinking process for resols, is conducted at temperatures between 130 and 200 °C. Since this is a polycondensation reaction, the molecular weight increases with conversion and shows a different MW advancement in comparison to a polymerization reaction (Fig. 3.4)[61].

Since the curing reaction occurs under different conditions than prepolymer formation in aqueous solutions, divergent reaction mechanisms are possible. This is con-

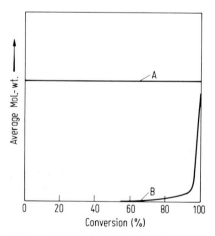

Fig. 3.4. Dependence of the average molecular weight on conversion, A = radical polymerization, B = polycondensation reaction

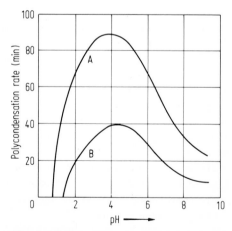

Fig. 3.5 Effect of pH and PF-ratio on polycondensation rate of resol resins[44] at 120 °C. Resin A: 1 mol F/1 mol P Resin B: 2 mol F/1 mol P

firmed by the composition of the bridging units and functional groups present in cured resols. The methylene bridge prevails as the most thermodynamically stable linkage.

Solid-state ^{13}C-NMR study of resol resins by Maciel[62] has resulted in the generation of useful information related to the thermal curing of resols. The authors propose the direct involvement of the hydroxyl group of phenols (3.30), condensation of methylene bridge with hydroxymethyl group (3.31) and crosslinking of methylene bridges with formaldehyde liberated during cure (3.32) as proposed by Zinke[63]. More detailed studies are in progress and necessary to support these suggestions.

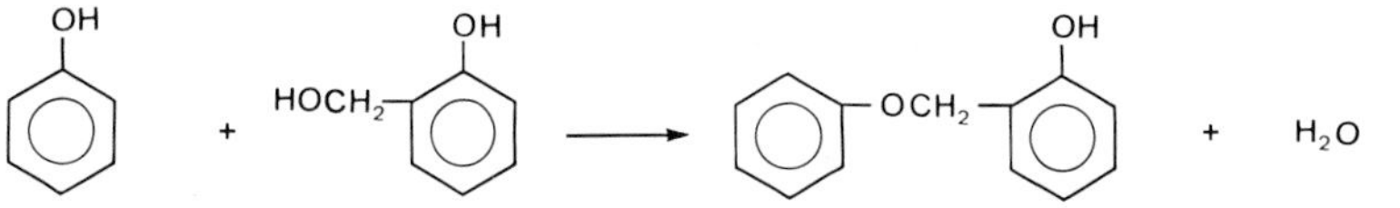

$$(3.30)$$

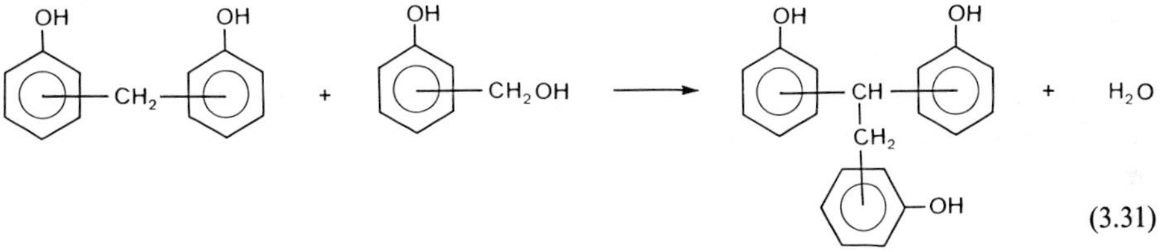

$$(3.31)$$

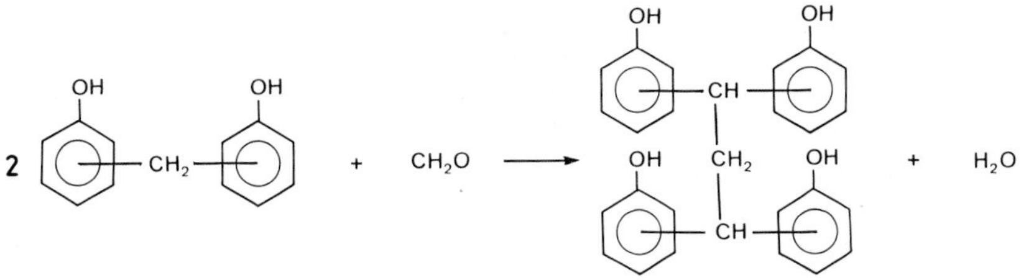

$$(3.32)$$

Acid Curing

The acid curing of phenol resols is of some commercial importance especially in phenolic foam, RIM, coatings, and laminates. Resols can be cured by the addition of a variety of strong organic and inorganic acids at ambient temperature. Organic sulfonic acids such as toluene sulfonic acid and phenol sulfonic acid are utilized with the latter preferred because of excellent compatibility and anticipated incorporation into the polymer microstructure. Acid incorporation into the resulting crosslinked PF would minimize acid migration and subsequent corrosion problems. The products resulting from acid curing are well established[64]. Hultzsch[65] found that numerous hydroxymethylphenols under acidic conditions result in dihydroxydiarylmethane derivatives. The reaction mechanism corresponds to the second step in the hydroxylmethylation in acidic medium, e. g. novolak formation, and is mentioned in detail in the following chapter. Applications for acid cured phenolics are mentioned in Chaps. 10.2, 10.3, 13.2, 15.2, and 19.3.

Heat Curing

Hultzsch[66, 67] and von Euler[68, 69] proposed quinone methides (QM) as intermediates in the heat curing of resols (and prepolymer formation). Regarding chemical

structure[70], quinone methides are intermediate in structure to quinones and quino-
dimethanes, but differ in their π electron charge distribution[71]. Thus, they show a high
reactivity to electrophiles as well as nucleophiles.

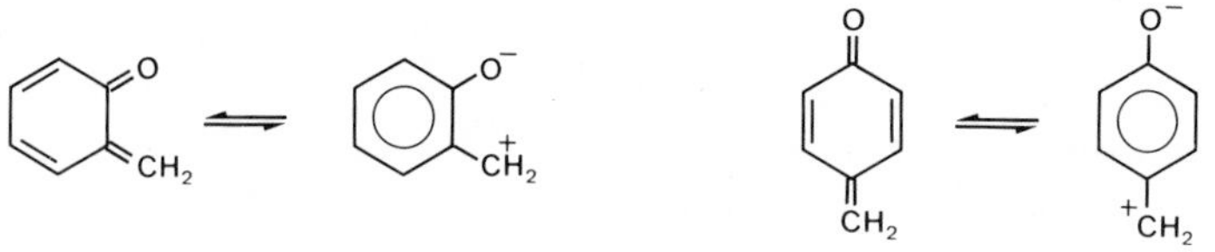

(3.33)

o-Benzoquinone methide   ***p***-Benzoquinone methide
(6-methylene-2,4-cyclohexadiene-1-one)   (4-methylene-2,5-cyclohexadiene-1-one)

The structure of QM (*ortho* or *para*) can be written as resonance hybrid between
the quinoid and benzenoid structure (3.33).

Closer theoretical examination of *ortho* and *para* QM's as well as other substituted
QM's by Koutek[72] by HMO method suggests that the contribution of aromatic stabilization provided by the ionic resonance species is insufficient for overall stability of
the highly reactive QM. Recent ^{13}C-NMR spectra of a variety of substituted QM's[73]
have established various chemical shift and long range carbon-proton coupling data
and the tetraene structure. The reactivity of QM's is therefore generally so high that
it cannot be isolated under normal conditions. In the absence of reactive compounds,
they undergo internal reaction forming colorless dimers (3.34), trimers (3.35) or
oligomers.

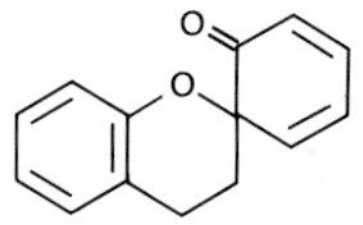

(3.34)

o- Quinone methide dimer

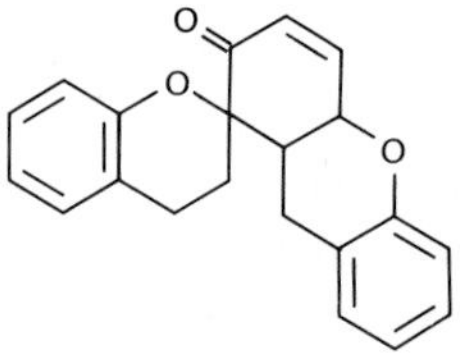

(3.35)

o- Quinone methide trimer

The preferred reaction is the Michael type addition with any hydroxyl containing
compound (3.36)

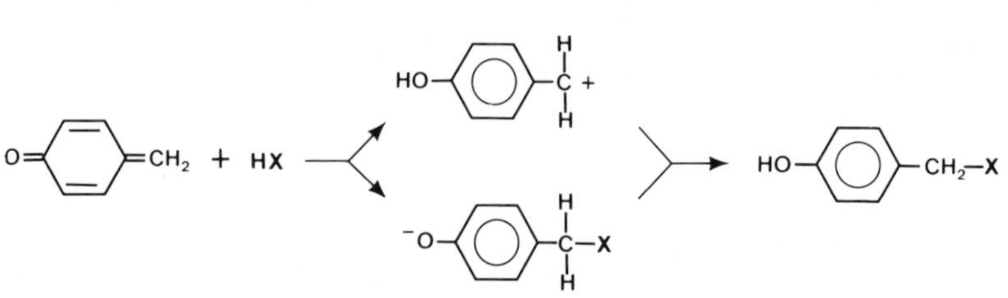

(3.36)

With hydroxyl compounds the reaction rate, which is greatly accelerated by traces of acids, corresponds to the following sequence[71]:

$$PhOH > H_2O > CH_3OH > C_2H_5OH \qquad (3.37)$$

In the presence of olefins, an addition reaction occurs yielding chroman derivatives, e.g. flavan (3.38) in the case of styrene.

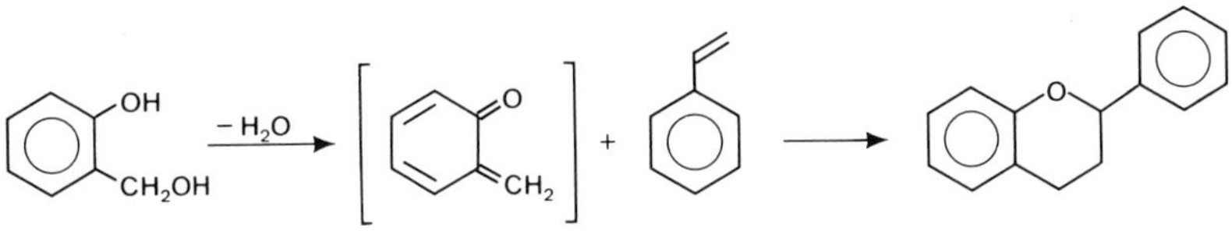

$$(3.38)$$

In the past this reaction was regarded as very important for the vulcanization of rubber and modification of phenolic resins with olefinic compounds, for instance rosin, tung oil and others. The reaction mechanism proposed by Hultzsch and separately by von Euler is mentioned in Sect. 19.1.

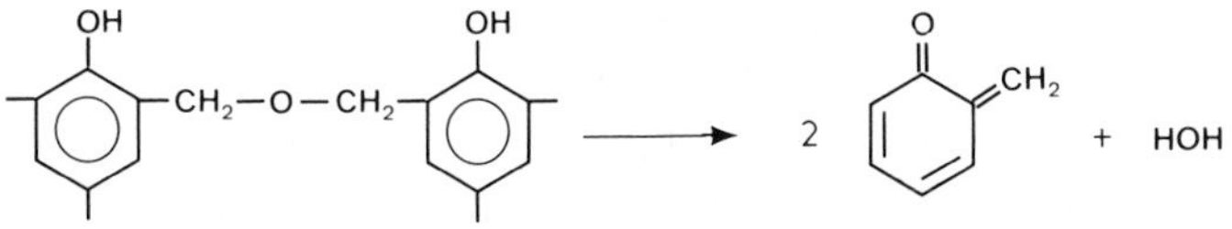

$$(3.39)$$

Hultzsch assumed that QM's are derived from dibenzyl ethers (3.39). Von Euler suggested the direct generation of QM from methylol phenol (3.38). However it is difficult to predict an appropriate temperature range for generating these elusive and highly reactive species. Many factors must be considered when attempting to identify satisfactory conditions for QM generation. This is especially true in the case of pre-polymer formation in aqueous medium and competing reactions (3.36).

Other conditions (solvent- and water-free etc.) must also be considered during the formation of quinone methides, and this could explain the presence of some different linkages identified in cured resins according to the following formulae:

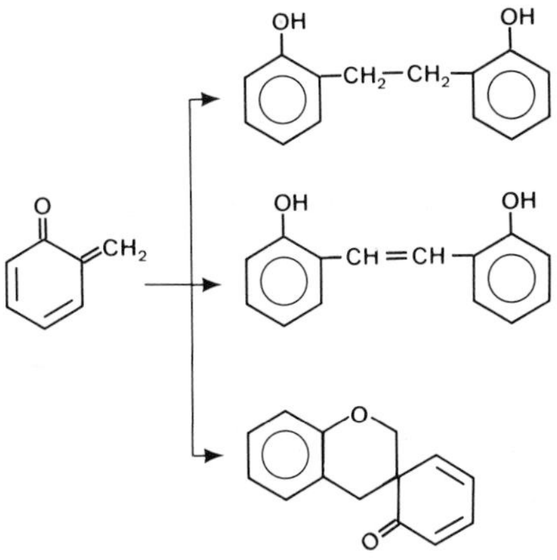

$$(3.40)$$

$$(3.41)$$

$$(3.42)$$

It was suggested that aldehydes also found in cured resins, were formed as a result of a redox reaction (3.43).

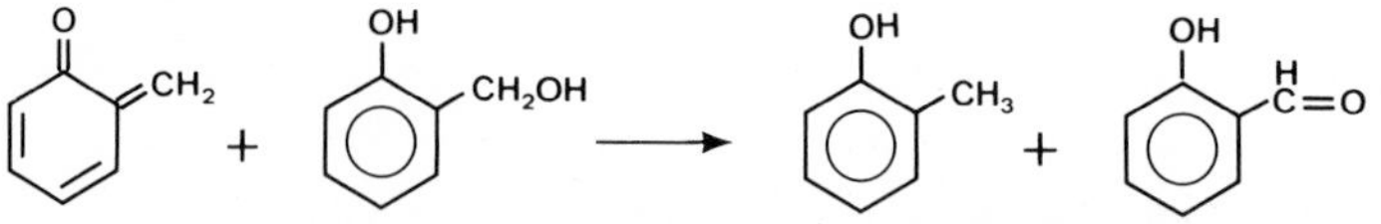

$$(3.43)$$

Aldehydes on the other hand, can develop, according to Zinke, by splitting of di-methylene ether linkages in a disproportionation reaction (3.44).

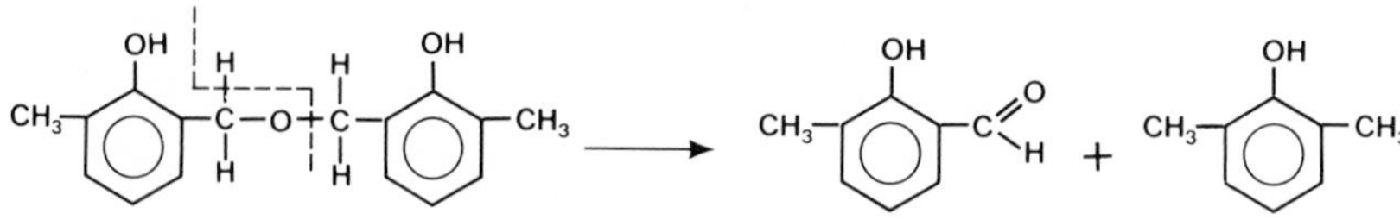

$$(3.44)$$

The content of aromatic aldehydes found in cured resols amounts up to 2%. How-ever, it can be considerably higher for some model compounds.

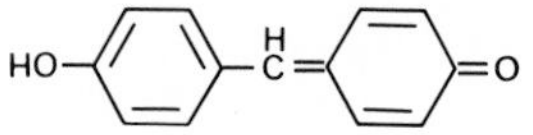

$$(3.45)$$

The yellowish-red color of cured resols can be attributed to a relatively stable quinone methide (3.45) which may form by oxidation of dihydroxydiphenyl-methane.

However, it is not necessary to postulate the formation of quinone methides as reac-tion intermediates in the heat curing of resols. The most important crosslinking reac-tions at temperatures between 130–180 °C can occur via carbonium ions as indicated in the Eqs. (3.27), (3.28), (3.29). The generation of carbonium ions followed by stabi-lization under chroman ring formation also takes place in the reaction of resols with olefinic compounds.

Milder conditions capable of generating QM have been proposed by Casiraghi[74] in the preparation of high-*ortho* novolaks. The reaction of phenol magnesium complex with paraformaldehyde in an aprotic solvent (benzene) leading to the high-*ortho* novo-lak invokes a QM intermediate (Sect. 3.4.3).

3.4 Phenol-Formaldehyde Reactions under Acidic Conditions

3.4.1 Strong Acids

The reaction between phenol and formaldehyde in the strongly acidic pH range occurs as an electrophilic substitution. The catalysts most frequently used are oxalic acid, sul-furic acid or *p*-toluene sulfonic acid. In the first step according to Eq. (3.46) the for-mation of a hydroxymethylene carbonium ion from methylene glycol occurs. This is the hydroxyalkylating agent.

$$HO-CH_2-OH \overset{H^+}{\rightleftharpoons} {}^+CH_2-OH + H_2O \tag{3.46}$$

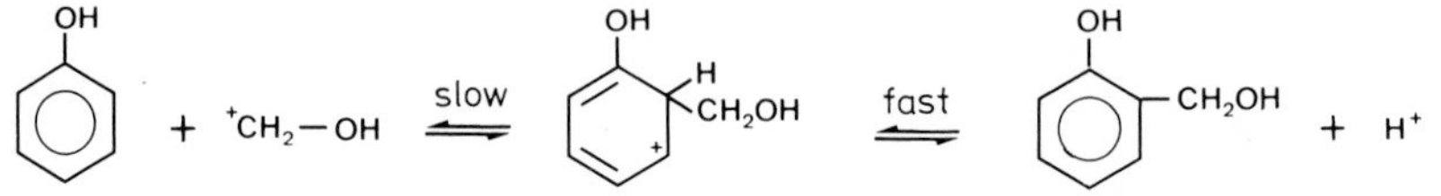

$$(3.47)$$

The subsequent addition of the hydroxymethylene carbonium ion to phenol (3.47) is relatively slow and rate determining. *Meta*-substitution is also not observed in the acidic medium.

However, the methylol group is a transient intermediate under acidic conditions. Benzylic carbonium ions result (3.48) under these conditions, which then react very fast with phenol, yielding dihydroxydiphenylmethane[45, 53, 55] according to Eq. (3.49).

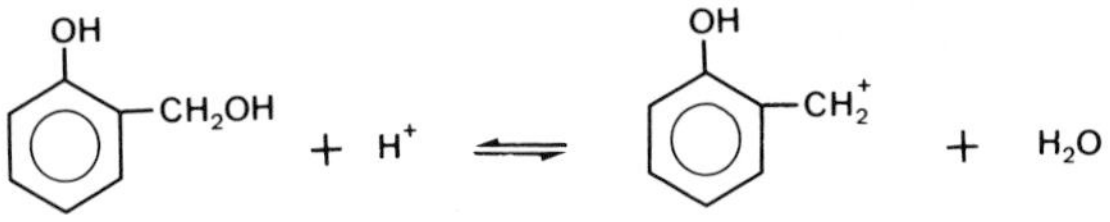

$$(3.48)$$

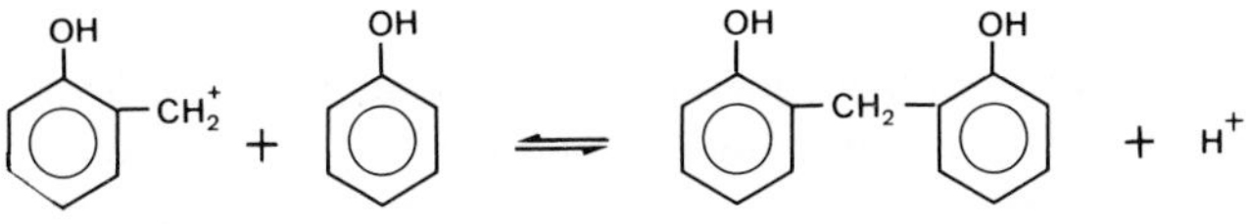

$$(3.49)$$

Thus, phenol alcohols cannot be isolated as intermediates contrary to the alkaline hydroxymethylation or under weak acid conditions(high-*ortho* resins). However, their existence can be detected for instance by NMR-spectroscopy[24]. Under acidic conditions, methylol substitution and methylene bridge formation both occur preferably at the *para* position. *Ortho*-substitution extent can be further reduced by high acidity, by the use of higher aldehydes, for instance acetaldehyde, or reaction conducted in aqueous/alcoholic solution. In the presence of alcohols a sterically hindered carbonium ion is formed from the hemiacetal which will be the reacting species (3.50).

$$R-O-CH_2OH \xrightarrow{H^+} R-O-CH_2^+ + H_2O \qquad (3.50)$$

The high selectivity of thioalcohols or thioglycolic acid in the formation of 2,2-*bis*-(4-hydroxyphenyl)propane (bisphenol-A) from phenol and acetone[75] occurs by a similar sequence involving hemithioketal intermediate (3.51).

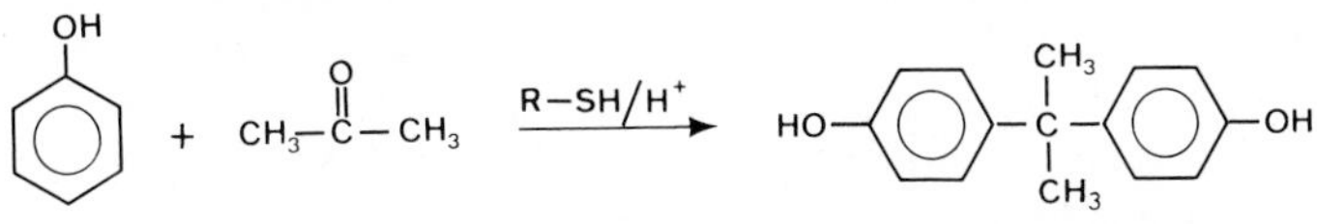

$$(3.51)$$

95% p–p′ Isomer

Hemithioketals formed from ketones and thioalcohols are considered to be more reactive than the hydrated ketones. Therefore, the reaction can be conducted at lower temperature.

In the hydrogen chloride-catalyzed reaction, the intermediate formation of chloromethanol (3.52) has been frequently suggested and chloromethylation (3.53) has been proposed as the main reaction.

$$CH_2O + HCl \rightleftharpoons HO-CH_2-Cl \qquad (3.52)$$

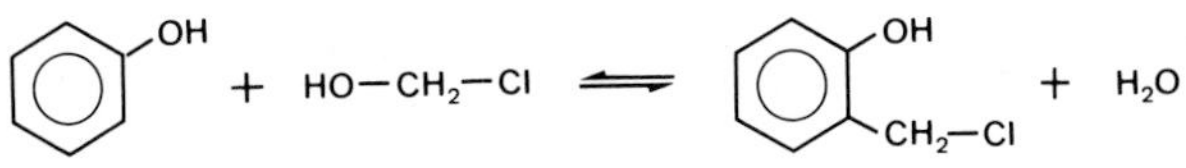

$$(3.53)$$

But there is no evidence for the formation of chloromethanol under these conditions. The C–Cl bond would be immediately hydrolyzed in aqueous medium and consequently, it is highly improbable that even chloromethanol is formed as an intermediate[76].

3.4.2 Reaction Kinetics in Acidic Medium

To obtain reliable kinetic data, the condensation reaction rate of individual phenol alcohols conducted in separate reactions under acidic conditions must be studied as well as the overall substitution/condensation reaction. In most early studies[29, 77–79] the overall reaction rate was in most cases found to follow second-order reaction kinetics.

The rate of formation of dihydroxydiphenylmethane (condensation reaction) was found to be more than five times[53] and 10–13 times, as reported in a second study[80], as fast as the formation of HMP (substitution reaction). The reaction rate is proportional to the H-ion concentration. The activation energy and activation entropy for the overall reaction[80] increase with increasing pH, indicating a possible change in mechanism at higher pH. Hydrochloric acid was used as catalyst in this study (Table 3.11).

In general, a second formaldehyde addition does not occur at usual PF molar ratios of 1 : (0.70–0.85). Mainly linear chain molecules are formed (MW generally between 500–1 000) containing 5 to 10 phenol units connected by methylene bridges. Branching[81, 82] is proposed as occurring above 10 units and is observed by ^{13}C-NMR. Computer simulation studies (Sect. 3.10) reveal that 2 branches are predicted for 10 units and 3 branches for 15 units[83].

The MWD in relation to PF molar ratio and catalyst type is shown in Fig. 3.6.

The reaction conditions are noted in Table 3.12. The acidic catalyst was neutralized with NaOH and the resin dissolved in DMF (ca. 3%) without prior distillation. No significant deviations in MWD are found if oxalic or HCl acid is used as catalyst. The individual DPM-isomers are not separated in the GPC spectrum. NMR spectroscopy, particularly ^{13}C-NMR, identifies DPM isomers and provides guidance regarding o-p, p-p', and o-o' substitution patterns during novolak preparation. Computer simulation of novolak formation generates novolak substitution similar to structures obtained at normal reaction conditions and also specifies the generation of cyclic oligomers, calixarenes, when substituted phenols are used.

Table 3.11. Arrhenius energy of activation and entropy of activation of the overall phenol-formaldehyde reaction in acidic medium according to Malhotra and Avinash[80]

pH	k at 80 °C (1 mol^{-1}sec^{-1})	E_A(kJ)	S_A(J/K · mol)
1.14	$0.52 . 10^{-2}$	250	−581
1.32	$1.53 . 10^{-3}$	331	−390
2.20	$2.59 . 10^{-4}$	681	+637
3.00	$7.05 . 10^{-6}$	794	+737

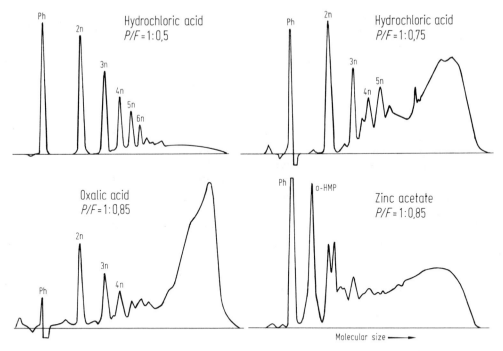

Fig. 3.6. Effect of phenol/formaldehyde ratio and catalyst type on molecular weight distribution[37] estimated by GPC (non distilled)

Table 3.12. Reaction conditions for the preparation of novolaks shown in Fig. 3.6

Designation		Catalyst	Time min.	Temp. °C	Phenol/ Formald. mol	Water g	Methanol g
1	HCl	0.020 Mol	90	100	1:0.50	35	0.8
2	HCl	0.020 Mol	90	100	1:0.75	35	0.8
3	Oxalic acid	0.020 Mol	90	100	1:0.85	35	0.8
4	Zinc acetate	0.035 Mol	210	100	1:0.85	35	0.8

3.4.3 Reaction under Weak Acidic Conditions "High-*Ortho*" Novolak Resins

High *ortho-ortho'* resins can be either solid novolaks or liquid resols depending on PF ratio. The high-*ortho* products became commercially important and of academic interest as a result of early studies of Bender and Farnham[35]. They established the unusually rapid cure rate of an *o-o'* material as compared to other isomeric DPM compounds with HMTA. The accessability of the vacant *para* position is recognized as enhancing cure rate (Table 3.13). Both high-*ortho* novolaks or resols are important in commercial applications such as molding compounds, foundry, and RIM.

Table 3.13. Reactivity of dihydroxydiphenylmethanes with hexamethylenetetramine according to Bender et al.[35, 86]

Nuclear position	MP [°C]	Gel time with 15% HMTA [sec]
o-o'	118.5–119.5	60
o-p'	119–120	240
p-p'	162–163	175

Table 3.14. Typical reaction conditions for the preparation of "high-*ortho*"-novolaks

pH	preferably between 4–6
Catalysts:	compounds of 2-valent metals: Ca, Mg, Zn, Cd, Pb, Cu, Co, Ni; preferably acetates
Molar ratio:	large phenol excess
Reaction conditions:	post-reaction at 150–160 °C necessary to convert the hydroxymethyl groups before distillation
Distillation:	120–140 °C maximum distillation temperature

Novolaks

The early preparative conditions that led to high *o-o'* novolaks consisted of an intermediate pH range of 4–7, use of divalent metals salts[34, 84] such as Ca, Mg, Zn, Cd, Pb, Cu, Co, and Ni with a large excess of phenol (Table 3.14). Characterization of *o-o'* structure was corroborated by IR and NMR[85]. ^{13}C-NMR spectrum of *o-o'* novolak is shown in Fig. 7.1.

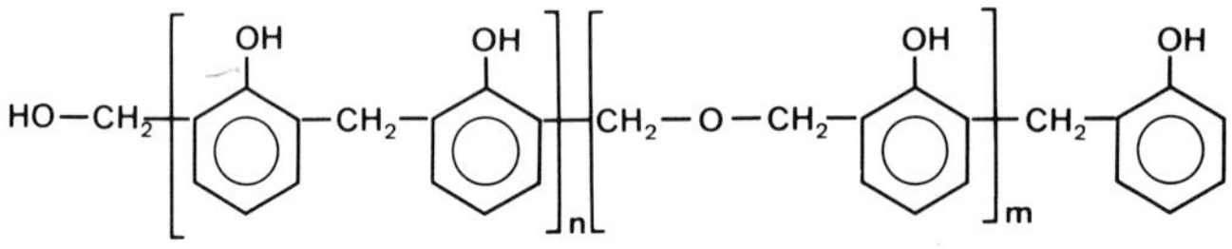

$$(3.54)$$

The mechanism of this selective *ortho*-hydroxymethylation which can be attributed to the formation of chelate-like complexes as intermediates, was already mentioned in Sect. 3.3.1. The preferred catalyst is zinc acetate. The MWD, which differs sharply from other novolaks, is shown in Fig. 3.6. The high amount of *o*-HMP (saligenin) and the presence of DPM as well is clearly indicated.

Culbertson[87] recently described a multi-sequence process for preparing high-*ortho* novolak with the initial step of methylol formation by divalent metal hydroxide, followed by benzyl ether formation and the final step being a facile decomposition of benzyl ether to aryl methylene. The latter is promoted by an organic acid co-catalyst. A key innovation is the use of an azeotropic solvent to remove water. The method reportedly increases *o-o'* content from the usual 50–75% to 75–100%. Phenol utilization is between 95–100% for this process as compared to 105 to 110% that is normally obtained in commercial novolak preparation.

A more fundamental study of the *o-o'* method of preparing novolaks has been carried out by Casiraghi and his coworkers[88-95]. By directly forming the chelated material (reaction of phenol with ethyl magnesium bromide) and then reacting it with formaldehyde in an aprotic solvent, high yields of novolaks with high-*ortho* specificity are obtained. A cyclic chelated intermediate is proposed (see Chap. 4.1.2).

A moderately non-polar solvent (benzene, xylene) is required and lends support to the proposed cyclic intermediate that undergoes intramolecular alkylation. Polar solvents inactivate the metal phenoxide intermediate and little or no product is obtained. Slightly polar solvents like dibutyl ether and dimethoxyethane result in lower yields ($\sim 80\%$) and reduced *o-o'* content, 88–90%, with 10–12% *o-p*[94]. The technique has also been extended to the preparation of all-*ortho* acetaldehyde novolak[88] [Eq. (2.26)].

Although it is difficult to determine whether the method has commercial significance for high-*ortho* novolaks, the technique has broad synthetic utility for a wide variety of organic compounds such as flavenes, chromanes, benzofurans, benzodioxins, and benzopyrylium salts as well as some optically active compounds. The reader is referred to original literature references[89].

Casiraghi has extended the high-*ortho* novolak preparation by the direct reaction of phenol and paraformaldehyde in xylene at 170–220 °C in a pressurized reactor for 12 h. Yields of 80–90% with 96% *ortho* stereoselectivity are obtained. The reaction mechanism (Chap. 4.1.2) is proposed as occurring via quinone methide intermediate[91].

A peculiarity of *ortho*-linked novolaks is their moderately high acidity[96] (hyperacidity) and the remarkable tendency to form complex compounds with di- and trivalent metals and nonmetals compared to phenol (Chap. 4). The "hyperacidity" of *o-o'*-phenols is discussed in Chap. 4.4.2.

$$(3.55)$$

Hultzsch[97] showed that for *p*-alkylphenol resins which have only *o-o*-links, anomalies also develop, even when these resins have dimethylene ether bridges ($-CH_2-O-CH_2-$) and methylol groups (3.55). It is obvious that here other nuclear distances exist and, therefore, the formation of a "hyper-acidic hydrogen" is not possible. These resins do not show hyperacidity; however, they distinguish themselves by a remarkable affinity towards divalent metals. This "base reactivity" is commercially utilized for the modification of polychloroprene adhesives with alkylphenol resins and magnesium- and/or zinc oxide to increase adhesion and bond strength at higher temperature. If alkylphenol resins, dissolved in aromatic hydrocarbons, are treated with finely distributed magnesium oxide, a certain portion of oxide will dissolve. These resins are able to bind magnesium oxide chemically up to 10%. On the other hand, it can be shown that novolaks exclusively bonded with methylene bridges cannot bind magnesium inspite of increased acidity. The reaction with magnesium leads to complexes with the coordination number 4 according to formula (3.56). Similarly, other ions can also be incorporated in phenolic resins[98] under the formation of metal containing resins, for instance the metals Ca, Ba, Cu, Ni, Co, Pb, Mn, Cr, and Fe. Some special metal incorporating reactions are mentioned in Sect. 9.3.

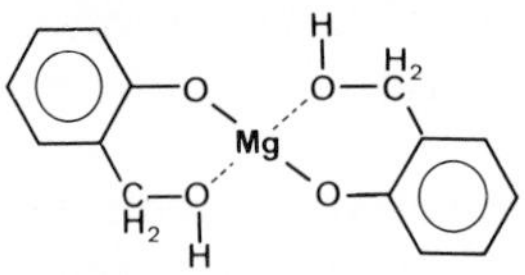

(3.56)

Resols

The preparation and use (RIM) of high-*ortho* liquid resol resins have been described in a series of patents by Brode, Chow et al.[99]. An intermediate pH of 4–7 and divalent salts are recommended conditions for the preparation of these materials. Structurally they are represented as (3.57):

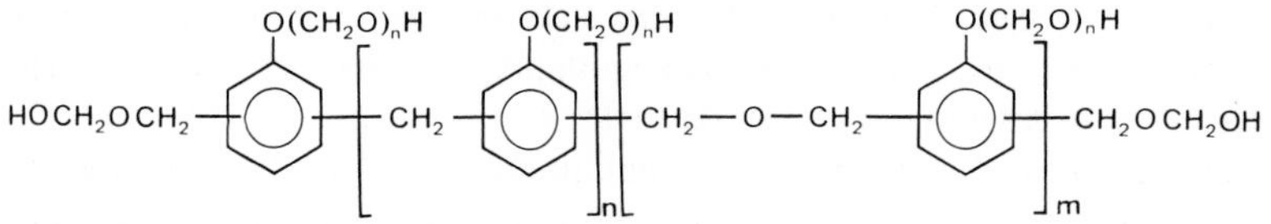

(3.57)

A temperature of 80–90 °C is maintained for 5–8 h and an azeotropic solvent is used (toluene or xylene). Phenol to formaldehyde ratio is 1:1.5–1.8. The virtual complete removal of water from added formaldehyde or by condensation reaction, apparently favors hemiformal formation for both phenolic and methylol moieties. Many resin features are reported for these hemiformal resol resins such as high solids, low viscosity, less than 5% unbound water, and better resin storage stability than conventional resol resins of similar solids.

A further procedure for preparing phenol hemiformal (uncatalyzed) entails the introduction of gaseous formaldehyde into molten phenol or substituted phenols. The hemiformal is formed without catalyst; the addition of formaldehyde to phenol is mildly exothermic with no *ortho* methylol phenol formed during hemiformal preparation. The uncatalyzed transformation of hemiformal to *ortho* methylol phenol is reported to be very slow. It is anticipated that as soon as *ortho* methylol phenol is formed, an equilibrium mixture of phenol hemiformal and *ortho* methylol phenol will occur[24].

No resin microstructure analyses are provided in describing these liquid materials. ^{13}C-NMR analyses of the phenol hemiformal at elevated temperatures would greatly aid in determining whether the intermediacy of hemiformal to *ortho* methylol phenol is on firm ground.

Thus either of these reactive liquid materials is used to dissolve a variety of thermoplastic resins (solid resol, novolak, polyester, polysulfone or unsaturated polyester) and combined with fiber glass mats for high performance composites (Chap. 10).

3.5 Novolak Crosslinking Reaction with HMTA

Curing of (thermoplastic) novolak resins requires the addition of a crosslinking compound which is mainly HMTA, and rarely paraformaldehyde or trioxane. Novolaks, generally made from phenol and formaldehyde at a PF ratio of 1:0.8 are cured by addition of 8–15% of HMTA. The most commonly employed HMTA level, yielding the best overall performance, is 9–10%. The properties of cured compositions are determined to a great extent by the ratio of the two reactants. Physical properties of HMTA are to be found in Sect. 2.2, the mechanism of HMTA formation from ammonia and formaldehyde in Sect. 3.3.2.

Nitrogen containing crosslinked resin is obtained from HMTA/novolak reaction with the generation of ammonia. Model compound studies have identified benzyl amines, benzoxazine, and azomethines as intermediates from the HMTA reaction with phenol or DPM[24, 100–102]. Trace amounts of water present in novolak are suggested as hydrolyzing HMTA which reverts to α-aminoalcohols. Due to the acidic nature of phenol, carbonium ions are generated from these α-aminoalcohols, which then react with phenol to form secondary and tertiary benzylamine containing chain molecules in a Mannich type reaction. The opinion that this is a homogeneous acid catalyzed reaction is supported by the fact that the reaction rate increases considerably at decreasing pH[24, 103]. Water and free phenol enhance the reaction rate to a great extent. The reaction of phenol with formaldehyde occurs in parallel, now catalyzed by hydroxybenzylamines with formation of methylol compounds. The crosslinking reaction is accompanied by the liberation of a considerable amount of gas, which consists of at least 95% of ammonia. The cured resin may contain up to 6% chemically bound nitrogen. At higher temperatures the benzylamine groups undergo decomposition reactions not as yet completely understood. Products containing the azomethine group –CH=N– among others, which would account for the yellow color of the oligomers, are formed.

The phenol/HMTA and novolak/HMTA reaction was studied by DTA and TGA by Orrell and Burns[104] among others. Kinetic results for the m-cresol/HMTA reaction indicate[103] a first order reaction whereby the HMTA hydrolysis and formation of aminomethylene ions are the slowest and rate determining step.

(3.58)

(3.59)

Gillham[105] has examined the effect of structure on cure for a series of novolaks with HMTA through the use of torsion braid analysis (TBA) and DSC. Novolaks consisting of a random product, a 2,2' (high-ortho) material, and an all 2,4-resin (prepared from the condensation of p-methylol phenol) with HMTA were evaluated.

Two transitions by TBA were noted during cure and were identified as gelation and vitrification. Lower gelation and exotherm values were obtained for 2,2'-resin as compared to the random or the 2,4-resin (Table 3.15).

The ring position para to the OH group forms a crosslink with HMTA more readily than an ortho position. From TBA data, the 2,2'-resin with HMTA exhibited an ac-

Table 3.15. Gelation and exotherm[105] reaction of various novolaks with HMTA (10%)

Novolak	T_{gel} TBA	$T_{exotherm}$ DSC
2,2′	113 °C	138 °C
Random	130 °C	152 °C
2,4	–	167 °C

tivation energy of 105 kJ/mol whereas the random resin/HMTA had a value of 147 kJ/mol. Thus lower activation energy, lower gelation temperature, and faster gel times favor the 2,2-novolak. Yet it is difficult to rationalize benzoxazine or benzylamine intermediates which require vacant *ortho* positions for the 2,2′-novolak. It is possible that the hyperacidity of the 2,2′-resin favors the HMTA curing reaction by a concerted hydrogen bonding-transfer mechanism. Recently an intermolecular hydrogen bonding mechanism with HMTA has been proposed[106]. Hydrogen bonding characteristics of HMTA with water, $CHCl_3$ and $CHBr_3$ have been reported[107]. Both ^{13}C- and ^{15}N-NMR chemical shift and spin relaxation data were determined for HMTA-water system as well as the closely related cyclic diamine, DABCO (3.60).

1.4. Diazabicyclooctane (DABCO)   Hexamethylenetetramine (HMTA)

(3.60)   (3.61)

A salient feature of the studies was the observed pH dependence of ^{15}N and ^{13}C chemical shifts and reflected mono basicity for HMTA and dibasicity for DABCO[108].

The hydrogen bonding mechanism occurs via complete proton transfer and cleavage of the C–N covalent bond of HMTA and is considered to be the driving force of the curing reaction.

The mechanism accommodates ammonia evolution, the expected substituted benzylamines and substituted hexahydrotriazine. The latter has been proposed by many investigators, but as yet has not been identified[24, 44].

With advent of ^{15}N-NMR and its utility in examining H bonding of HMTA as well as in UF and MF resins (Sect. 3.8) it is anticipated that ^{15}N-NMR technique will aid in a better characterization of the novolak-HMTA composition.

3.6 Reaction with Epoxide Resins

Epoxide resins[109, 110] may be used instead of HMTA to crosslink resins if the release of volatile compounds must be avoided. Epoxidized novolaks are generally utilized as solid, pulverized resins. These materials are readily obtained by the reaction of epichlorohydrine with novolak of sufficiently high MW. An alternate method which avoids epichlorohydrine is the peracetic acid oxidation of allyl novolak (Chap. 9).

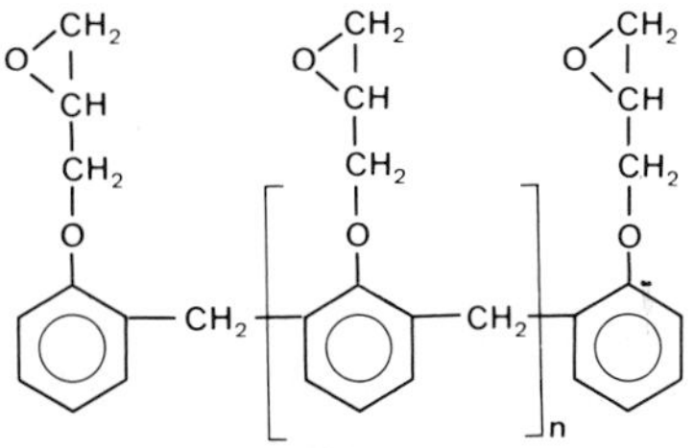

$$(3.62)$$

Both novolaks and resols may be employed. The resulting products exhibit high strength, strong adhesion, excellent dielectric properties and improved oxidation resistance. However, they also show slightly reduced thermal resistance. Additional cross-linking, thus increasing thermal stability, is facilitated by the addition of small amounts of HMTA. The ring-opening reaction of epoxides with H-active compounds, e.g. phenols, is a S_N2-type reaction which is catalzyed by inorganic and organic bases, preferably tertiary amines[111]. Similar products may be obtained if the phenolic resin is treated with epichlorohydrine and sodium hydroxide. The chlorohydrine group obtained in the first step is converted to a second epoxide group which then reacts with a phenolic hydroxyl group (3.64).

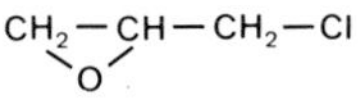

$$CH_2-CH-CH_2-Cl \qquad (3.63)$$

Epichlorohydrine

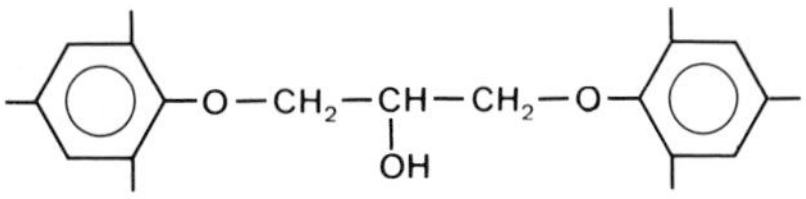

$$(3.64)$$

3.7 Reaction with Diisocyanates

The reaction of isocyanates with phenols, leading to phenylurethanes [Eq. (3.65)], is essentially a nucleophilic attack of phenol on the carbon atom of the isocyanate group. If steric factors are neglected, the reactivity of the hydroxyl compound increases as its nucleophilic character increases.

For simplicity the reaction is represented by means of monofunctional compounds.

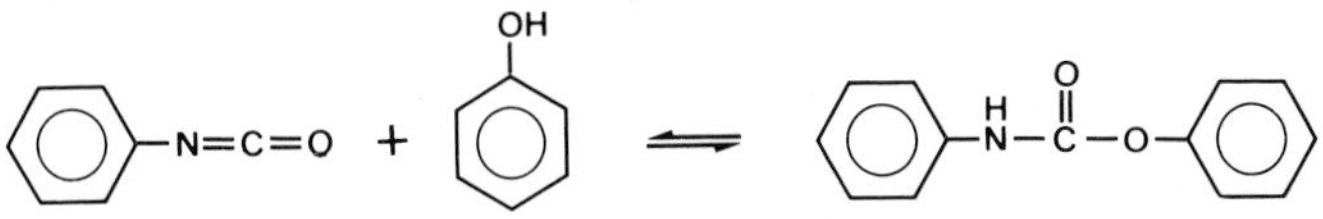

$$(3.65)$$

The most important competitive reaction in the preparation of urethane-modified phenolic resins is the reaction with the hydroxymethyl group and water. The first step in the reaction with water is the formation of the unstable carbamic acid which decomposes to form an amine and carbon dioxide [Eq. (3.66)]. The amine is more reactive with isocyanate and immediately reacts with additional isocyanate to form a substituted urea.

$$R-N=C=O + H_2O \longrightarrow \left[\begin{matrix} H & O \\ | & || \\ R-N-C-OH \end{matrix} \right] \longrightarrow R-NH_2 + CO_2 \tag{3.66}$$

The relative rates of the individual reactions are dependent on the type of catalyst used. The catalysts most commonly employed are tertiary amines (triethylenediamine, N-alkyl morpholines) and metal catalysts, especially stannous octoate[112].

The phenolic resin-diisocyanate reaction is used for fast curing "cold set" foundry resins[113]. 4,4'-Diphenylmethane diisocyanate is the preferred crosslinking compound.

However, urethanes are susceptible to thermal dissociation. The ability to dissociate depends upon chemical nature as indicated below[114, 115].

Aryl–NH–CO–O–aryl   thermostable to about 120 °C
Alkyl–NH–CO–O–aryl   thermostable to about 180 °C
Aryl–NH–CO–O–alkyl   thermostable to about 200 °C
Alkyl–NH–CO–O–alkyl   thermostable to about 250 °C

Phenol-"blocked" isocyanates are used for single component polyurethane coatings or as crosslinkers in polyester powder coatings[116].

3.8 Reaction with Urea, Melamine, and UF/MF-Resins

Copolymerized resins between phenol resins and urea- or melamine resins are utilized quite extensively in many industrial applications. The cost advantage of urea compared to phenol, the increased reaction rate and/or improvement of the flame retardant properties with amino-resin modification are the prime reasons for coreacting PF with these N containing resins. Particle boards, plywood, mineral fiber bonding, foundry resins and laminates are important application areas for amino-modified phenolic resins.

To obtain the necessary desirable properties, specific reaction parameters should be chosen to favour a copolymerization reaction instead of separate homopolymerization. The reaction between phenol, urea, and melamine and their hydroxymethyl derivatives was studied by use of ^{13}C-NMR spectroscopy[117] and chromatography[118]. The question whether co-condensation is induced by a phenol methylol group or by a urea methylol group was resolved in favour of the phenol-hydroxymethyl group[117].

The reaction rate between hydroxymethylphenols and urea (3.67) strongly relates to the pH of the reaction media. ^{13}C-NMR results indicate that a weak acidic environment favours the hydroxymethylphenol reaction with urea. It was concluded from melamine (3.68) studies that the co-condensation reaction takes place only by the reaction of phenol methylol groups and the unsubstituted amino group of methylol melamine or melamine under slightly acidic conditions (pH 5 to 6). On the other hand, in strong acidic or alkaline medium only homocondensation products are observed[118]. Recently ^{15}N-NMR spectroscopic characterization of UF or melamine formaldehyde adducts and resins was reported[119] and augments ^{1}H- and ^{13}C-NMR data for these materials.

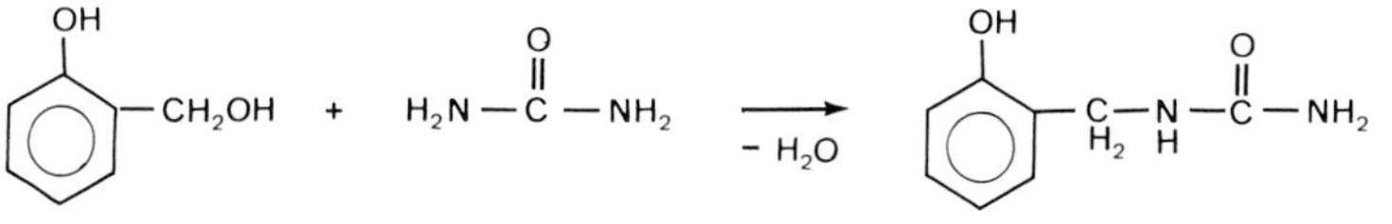

(3.67)

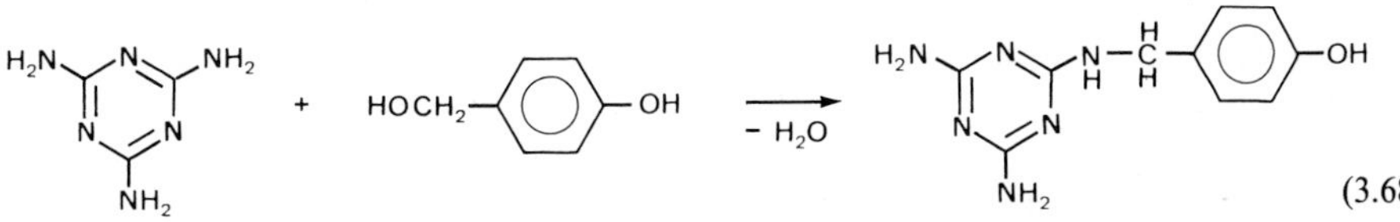

(3.68)

3.9 Reactions with Imide Precursors

The improvement in higher temperature characteristics of phenolics can be realized by coreaction of either a resol (3.69) or a novolak (3.70) with polyimide precursors[120, 121].

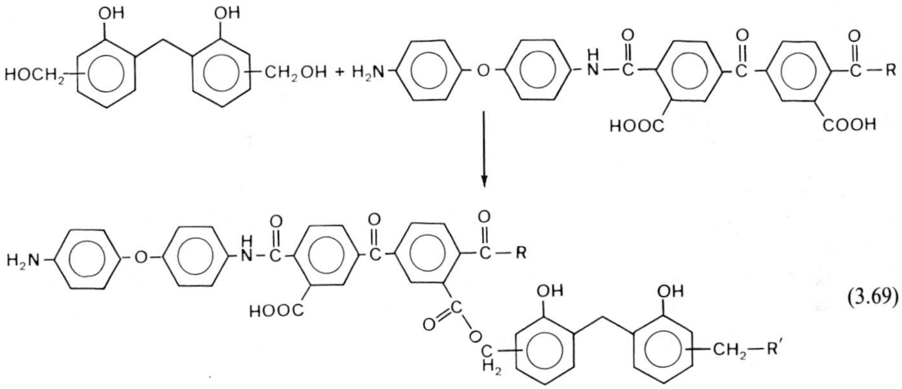

(3.69)

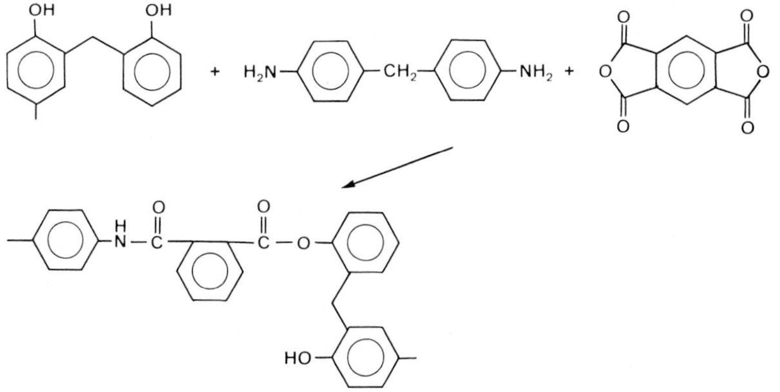

(3.70)

3.10 Statistical Approach and Computer Simulation

The polycondensation of a trifunctional monomer with a difunctional comonomer can lead to a completely gelled or crosslinked system if equimolar quantities are used. In the case of PF reaction, the mode of catalysis and the extent of reaction are further considerations as these parameters relate to soluble, branched or wholly gelled products. When the PF ratio is 1:0.95 with an acid catalyst, gelation will occur. Yet under basic conditions a PF ratio of 1:3 can lead to a soluble product as long as addition and little or no condensation occurs. Thus the PF reaction is recognized as a multifunctional polycondensation with unequal reactivity affecting conversion, polydispersity index and MWD.

Flory[122] treated the phenol-formaldehyde reaction numerically by assuming a statistical condensation of the trireactive phenol with the difunctional formaldehyde merely supplying internuclear linkages. However, Flory's simplifying assumptions: 1. equal reactivity of nuclear positions and functional groups, 2. all groups react independently of one another, and 3. no intramolecular interaction occurs as molecular size increases, do not apply to this condensation reaction. The unequal reactivity of the o- and p-position of phenol as well as of the hydroxymethyl group in methyleneglycol and hydroxymethylphenols and the general change in reactivity as reaction proceeds lead to significant differences between experimental observations and statistically derived molecular weight distribution functions and polydispersity index.

Steric effects make reactive sites less accessible as molecular size increases. Considering an effective average functionality of 2.31 for phenol because of "molecular shielding," Drumm and Le Blanc[1, 123] obtained good agreement with Stockmayer's weight distribution function[124, 125]. A recursive statistical approach extended to systems involving reactants with functional groups of unequal reactivity was developed by Macosco and Miller[126]. Distribution functions for the polycondensation of mixtures of monomers of different functionalities were derived by Stafford[127].

Theoretical or mathematical modelling studies via computer simulation[128, 129a] or kinetic data[130–132] provide predictions of the distribution of different species as well as statistical parameters in characterizing novolaks and resols. Ishida and coworkers[128, 129] have developed the computer simulation technique by examining acid and base catalyzed systems of phenol and substituted phenols with formaldehyde. Most of the kinetic statistical methods of Kumar and Gupta[130] and Williams[131, 132] have been restricted to phenol and formaldehyde.

3.11 References

1. Drumm, M. F., Le Blanc, J. R.: Kinet. Mech. Polym. 3, 157 (1972)
2. Vazquez, A., Adabbo, H. E., Williams, R. J. J.: "Statistics of Resols" In: "Phenolics Revisited, 75 Years Later," ACS Symposium, Aug. 29–31, 1983, Washington, D.C.
3. Wyckoff, R. W. G.: Crystal Structures, 2. Ed., Vol. 6. New York: Interscience 1969
4. Rochester, C. H.: Acidity and Inter- and Intramolecular H-Bonds. In: S. Patai (ed.): The Chemistry of the Hydroxyl Group, London: Interscience 1971
5. Saunders, M., Hyne, J. B.: J. Chem. Phys. 29, 1319 (1958)

6. Richards, R. E., Thompson, H. W.: J. Chem. Soc. 1260 (1947)
7. Weller, A.: Progr. Reaction Kinetics 1, 187 (1961)
8. Wehry, E. L., Rogers, L. B.: J. Am. Chem. Soc. 88, 351 (1966)
9. Bolton, P. D., Hall, F. M., Reece, J. H.: Spectrochim. Acta 22, 1149 (1966)
10. Chen, D. T. Y., Laidler, K. J.: Trans. Faraday, Soc. 58, 440 (1962)
11. Zsavitsas, A. A.: J. Chem. Engng. Data 12, 94 (1967)
12. Kortüm, G., Vogel, W., Andrusson, K.: Dissociation Constants of Organic Acids in Aqueous Solution, London: Butterworths (1961)
13. Harper, D. A., Vaughan, J.: Directing and Activating Effects. In S. Patai (ed.): The Chemistry of the Hydroxyl Group, London: Interscience 1971
14. Knop, J. V.: Unpublished
15. Dietrich, S. W., Jorgensen, E. C., Kollman, P. A., Rothenberg, S.: J. Am. Chem. Soc. 98, 8310 (1976)
16. Fateley, W. G., Carlson, G. L., Bentley, F. F.: J. Phys. Chem. 79, 199 (1975)
17. Hehre, J. W., Radom, L., Pople, J. A.: J. Am. Chem. Soc. 94, 1996 (1972)
18. Pross, A., Radom, L., Taft, R. W.: J. Org. Chem. 45, 818 (1980)
19. Schaefer, T., Chum, K.: Can. J. Chem. 56, 1788 (1978)
20. Walker, J. F.: Formaldehyde, ACS Monograph No. 159, 3 Ed. New York: Reinhold Publ. Co. 1964
21. Diehm, H., Hit, A.: Formaldehyd. In: Ullmanns Encyclopädie der techn. Chem. Vol. 11, 4. Ed. Weinheim: Verlag Chemie 1976
22. Zabicky, J. (ed.): The Chemistry of the Carbonyl Group. London: Interscience 1970
23. Moedritzer, K., Wazer, J. V.: J. Phys. Chem. 70, 2025 (1966)
24. Kopf, P. W., Wagner, E. R.: J. Polymer Sci., Polymer Chem. Ed. 11, 939 (1973)
25. Zsavitsas, A. A., Beaulieu, R. D.: Am. Chem. Soc. Div. Org. Coatings and Plastic Preprints 27, 100 (1967)
26. Lederer, L.: J. prakt. Chemie 50, 223 (1894)
27. Manasse, O.: Ber. dtsch. chem. Ges. 27, 2409 (1894)
28. Piria, R.: Liebigs Ann. Chem. 56, 37 (1845)
29. DeJong, J. I., DeJonge, J.: Rec. Trav. Chim. 72, 497 (1953)
30. Zsavitsas, A. A.: Am. Chem. Soc. Div. Org. Coatings and Plastic Preprints 26, 93 (1966)
31. Freemann, J. H., Lewis, C. W.: J. Am. Chem. Soc. 76, 2080 (1954)
32. Peer, H. G.: Rec. Trav. Chim. 78, 851 (1959)
33. Eapen, K. C., Yeddanapalli, L. M.: Makromol. Chem. 119, 4 (1968)
34. Peer, H. G.: Rec. Trav. Chim. 79, 825 (1960)
35. Bender, H. L., Farnham, A. G., Guyer, J. W., Apel, F. N., Gibb, T. B.: Ind. Engng. Chem. 44, 1619 (1952)
36. Union Carbide Corp.: US-PS 2,464,207 (1949)
37. Knop, A., Trapper, W.: Unpublished
38. Zinke, A., Hanus, F.: Monatsh. Chem. 78, 311 (1948)
39. Hultzsch, K.: Chem. Ber. 82, 16 (1949)
40. Sojka, S. A., Wolfe, R. A., Guenther, G. D.: Macromolecules 14, 1539 (1981)
41. Hellmann H., Opitz, G.: α-Amino-alkylierung. Weinheim: Verlag Chemie 1960
42. Ogata, Y., Kawasaki, A.: Equilibrium Additions to Carbonyl Compounds. In: J. Zabicky (ed.): The Chemistry of the Carbonyl Group, Vol. 2. London: Interscience 1970
43. Degussa: Hexamethylentetramin, Technical Bulletin
44. Keutgen, W. A.: Ency. Poly Sci. Vol. 17 (1969)
45. Megson, N. J. L.: Phenolic Resin Chemistry. London: Butterworths (1958)
46. Higginbottom, H. P., Culbertson, H. M., Woodbrey, J. C.: Analyt. Chem. 37, 1021 (1965)
47. Sprung, M. M.: J. Am. Chem. Soc. 63, 334 (1941)
48. Malhotra, H. C., Gupta, V. K.: J. Appl. Polym. Sci. 22, 343 (1978)
49. Malhotra, H. C., Kumar, V.: J. Macromol. Sci. – Chem. A 13, 143 (1979)
50. Malhotra, H. C., Tyagi, V. P.: J. Macromol. Sci. – Chem. A 14, 675 (1980)
51. Haller, D., Schmidt, K. H.: Diplomarbeit T. H. Essen, 1983
52. Francis, D. J., Yeddanapalli, L. M.: Makromol. Chem. 119, 17 (1968)
53. Martin, R. W.: The Chemistry of Phenolic Resins. New York: J. Wiley 1956
54. Yeddanapalli, L. M., Francis, D. J.: Makromol. Chem. 55, 74 (1962)

55. Kornblum, N., Smiley, R.A., Blackwood, R.K., Iffland, D.C.: J. Am. Chem. Soc. 77, 7269 (1955)
56. Hultzsch, K.: Chemie der Phenolharze. Berlin, Göttingen, Heidelberg: Springer 1950
57. Lenz, R. W.: Organic Chemistry of Synthetic High Polymers. London: Interscience 1967
58. Jones, R.T.: J. Poly. Sci. 21, 1801 (1983)
59. Kaemmerer, H., Kern, W., Heuser, G.: J. Polym. Sci. 28, 331 (1958)
60. Tong, S. N., Kyung, Y. P., Harwood, H. G.: ACS Polymer Preprints, 24 (2) 196 (1983)
61. Vollmert, B.: Polymer Chemistry. Berlin, Heidelberg, New York: Springer 1973
62. Maciel, G. E., Chuang, I. S., Gollob, L.: Macromolecules 17, 1081 (1984)
63. Zinke, A.: J. Appl. Chem. 1, 257 (1951)
64. Müller, H. F., Müller, J.: Kunststoffe 37, 75 (1947)
65. Hultzsch, K.: Kunststoffe 37, 205 (1947)
66. Hultzsch, K.: Ber. dtsch. chem. Ges. 74, 898 (1941)
67. Hultzsch, K.: Angew. Chem. A 60, 179 (1948)
68. von Euler, H., Adler, E., Cedwall, J. O.: Ark. Kemi. Min. Geol. 14 A, No. 14 (1941)
69. von Euler, H., Adler, E., Cedwall, J. O., Törngren, O.: Ark. Kemi. Min. Geol. 15 A, No. 11 (1942)
70. Cologne, J., Descotes, G.: α, β-Unsaturated Carbonyl Compounds. In: J. Hammer (ed.): 1,4-Cycloaddition Reactions. New York: Academic Press 1967
71. Wagner, H. U., Gompper, G.: Quinone Methides. In: S. Patai (ed.): The Chemistry of the Quinoid Compounds, Chap. 18, Vol. 2. New York: Wiley 1974
72 Musil, L., Koutek, B., Pisova, M., Soucek, M.: Coll. Czech. Chem. Comm. 46, 1148 (1981)
73. Benson, M., Jura, L.: Org. Mag. Res. 22 (2), 86 (1984)
74. Casiraghi, G., et al.: Makromol. Chem. 182, 2151 (1981)
75. Barclay, R., Sulzberg, Th.: Bisphenols and Their Bis-(Chloroformates). In J. K. Stille, T. W. Champbell: Condensation Monomers. New York: Wiley-Interscience 1973
76. Olah, G. (ed.): Friedel Crafts and Related Reactions. Interscience (1964), Vol. II, Part II, P. 762
77. von Euler, H., Dekispezy, S. V.: Z. Phys. Chem. A 189, 109 (1941)
78. Jones, T. T.: J. Soc. Chem. Ind. 69, 102 (1964)
79. Bassow, N. J. et al.: Plaste and Katschuk 6, 417 (1974)
80. Malhotra, H. C., Avinash: J. Appl. Polym. Sci. 20, 2461 (1976)
81. Kamide, K., Miyakawa, Y.: Makromol. Chem. 179, 359 (1978)
82. Sojka, S. A., Wolfe, R. A., Dietz, E. A., Jr., Dannels, B. F.: Macromolecules 12, 767 (1979)
83. Ishida, S.-I., Tsutsumi, Y., Katsumasa, K.: J. Poly. Sci., Polym. Chem. Ed., 19, 1609 (1981)
84. Bender, H. L.: Mod. Plastics 30, 136 (1953)
85. Higginbottom, H. P., Culbertson, H.M., Woodbrey, J.C.: J. Polym. Sci. Part. A, 3, 1079 (1965)
86. Brode, G. L.: "Phenolic Resins" In: Encyclopedia of Chemical Technology, Kirk-Othmer Ed., Vol. 17, John Wiley, N.Y., 1982
87. Monsanto: US-PS 4,113,700 (1978)
88. Casiraghi, G., et al.: Macromolecules 17, 19 (1984)
89. Casnati, G., Casiraghi, G., Pochini, A., Sartori, G., Ungaro, R.: Pure & Appl. Chem. 55 (11) 1677 (1983)
90. Bigi, F., Casiraghi, G., Casnati, G., Sartori, G., Zetta, L.: J. Chem. Soc. Chem. Comm., 1210 (1983)
91. Casiraghi, G., et al.: Makromol. Chem. 182, 2973 (1981)
92. Casiraghi, G., et al.: Makromol. Chem. 184, 1363 (1983)
93. Dradi, E., Casiraghi, G., Sartori, G., Casnati, G.: Macromolecules, 11, 1295 (1978)
94. Casiraghi, G., et al.: Makromol. Chem. 182, 2151 (1981)
95. Casiraghi, G., et al.: ACS Polym. Preprints 24 (2) 183 (1983)
96. Sprengling, G. R.: J. Am. Chem. Soc. 76, 1190 (1954)
97. Hultzsch, K., Hesse, W.: Kunststoffe 53, 166 (1963)
98. Nord-Aviation Société Nationale de Constructions Aeronautiques: DE-AS 1816 241 (1968)

99. Union Carbide Corp.: US-PS 4,395,521 (1983)
    Union Carbide Corp.: US-PS 4,403,066 (1983)
    Union Carbide Corp.: US-PS 4,395,520 (1983)
    Union Carbide Corp.: US-PS 4,430,473 (1984)
    Union Carbide Corp.: US-PS 4,433,119 (1984)
    Union Carbide Corp.: US-PS 4,433,129 (1984)
100. Sojka, S. A.: Macromolecules 14, 1539 (1981)
101. Zinke, A., Purcher, S.: Monatsh. Chem. 79, 26 (1948)
102. Zinke, A., Ziegler, E.: Ber. dtsch. chem. Ges. 77, 271 (1944)
103. Kamenskii, I. V., Kuznetsov, L. N., Moisenko, A. P.: Vysokomol. soyed. A 18/8, 1787 (1976)
104. Orrell, E. W., Burns, R.: Study of the Novolak-Hexamine Reaction, Plastics and Polymers 36, 125 (1968)
105. Mackey, J. H., Lester, G., Walker, L. E., Gillham, J. K.: ACS Polymer Preprints, 22 (2), 131 (1981)
106. Katovic, Z., Stefanic, M.: "Intermolecular Hydrogen Bonding in Novolacs" In: "Phenolics Revisited, 75 Years Later", ACS Symposium, August 29–31, 1983, Washington, D.C.
107. Wasylishen, R. E., Pettit, B. A.: Can. J. Chem. 55, 2564 (1977)
108. Cralk, D. J., Levy, G. C., Lombardo, A.: J. Phy. Chem. 86, 3893 (1982)
109. Lee, H., Neville, K.: Handbook of Epoxy Resins. London: McGraw-Hill 1967
110. Potter, W. G.: Epoxide Resins, London: Iliffe Books 1970
111. Ishii, Y., Sakai, S.: 1,2-Epoxides. In Frish, K. C., Reegen, S. L.: Ring Opening Polymerizations, Vol. 2. New York: Marcel Dekker Inc. 1969
112. Frisch, K. C., Rumao, L. P.: Catalysis in Isocyanate Reactions J. Macromol. Sci.-Revs. Macromol. Chem., C 5 (1) 103 (1970)
113. Ashland Oil: DE-PS 1 583 521 (1967)
114. Wicks, Z. W.: Progr. Organic Coatings 3, 73 (1975)
115. Le Thi Phai, et al.: Makromol. Chem. 185, 281 (1984)
116. Myers, R. R., Long, J. S.: Treatise on Coatings. New York: Marcel Dekker Inc. 1967)
117. Tomita, B.: ACS Polymer Preprints, Vol. 24, No. 2, 165 (1983)
118. Braun, D., Krause W., Angew. Makromol. Chem. 108, 141 (1982)
119. Ebdon, J. R., et al.: Polymer 25, 82 (1984)
120. Adduci, J., et al.: ACS Polym. Preprints 22 (2), 109 (1981)
121. Hitachi: C.A. 100 866 83 t (1984)
122. Flory, P. J.: Principles of Polymer Chemistry, Cornell University Press, Ithaca, N.Y. 1953, Chem. Rev. 39, 137 (1946)
123. Drumm, M. F., Le Blanc, J.R.: "The Reactions of Formaldehyde with Phenols" In: Step-Growth Polymerization, Solomon, D. H., Ed., Marcel Dekker, N.Y. 1972
124. Stockmayer, W. H.: J. Polym. Soc. 9, 69 (1952); 11, 424 (1953)
125. Gordon, M.: Macromolecules 17, 514 (1984)
126. Miller, D. R., Macosco, C. W.: Macromolecules 11, 656 (1978)
127. Stafford, J. W.: J. Polym. Sci.; Polym. Chem. Ed. 22, 365 (1984)
128. Ishida, S.-I., Tsutsumi, Y., Kaneko, S.: J. Polym. Sci. 19, 1609 (1981)
129. Ishida, S.-I., Nakamoto, Y.: ACS Polym. Preprints 24, 167 (1983)
129a. Steffan, K.: Ang. Makromol. Chem. 131, 25 (1985)
130. Kumar, A., Gupta, S. K., et al.: Polym. 23, 265 (1982)
131. Williams, R. J. J., et al.: ACS Polym. Preprints 24, 167 (1983)
132. Borrajo, J., Aranguren, M. I., Williams, R. J. J.: Polymer 23, 263 (1982)

4 Structurally Uniform Oligomers

by Volker Böhmer, Johannes-Gutenberg-Universität, Fachbereich Chemie, D-6500 Mainz, FRG

Model compounds as well as prototype reactions are frequently used to understand more complex chemical systems. In the case of phenolic resins a large number of model compounds have been synthesized. They are regarded as either components of a complex mixture of novolaks or resoles or as representative segments of the phenolic crosslinked network. These model compounds were formerly the basis for the correlation of physical properties to molecular structure. Presently the progress of analytical separations and identification methods is formidable; highly sophisticated chemical instrumentation allows the analysis of highly complex mixtures. Nevertheless, well defined compounds with definite structure are still required for the calibration of those analytical techniques, to test mechanistic schemes, to furnish basic data for computer simulation, etc.

The vast number of possible structures of oligonuclear phenolic compounds is best illustrated by Table 4.1. The large number of specific oligomers which have been synthesized presently are presented in Table 4.2, but this is not considered a complete survey.

If more than three phenolic units are connected by methylene bridges, branched (4.3) and cyclic compounds (4.4) result in addition to linear oligomers (4.1), (4.2).

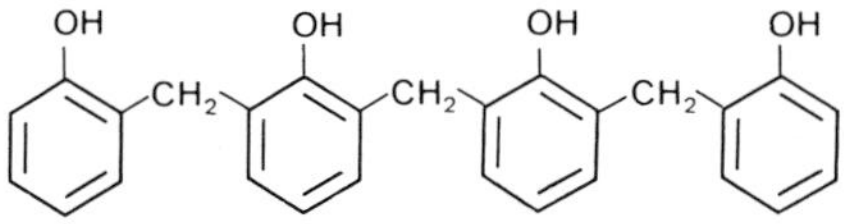

(4.1)

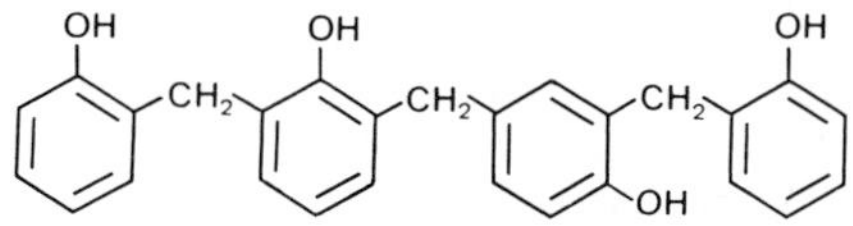

(4.2)

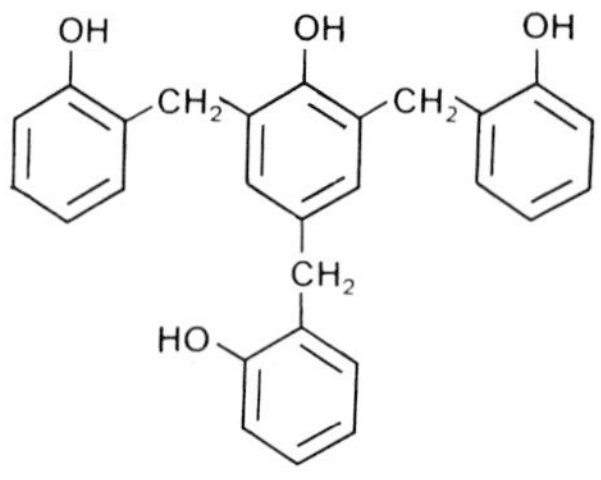

(4.3)

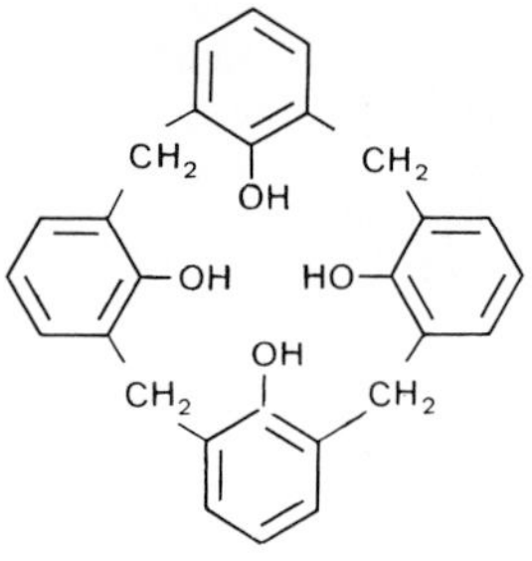

(4.4)

Table 4.1. Possible number of linear oligomers which can be obtained by linking a given number of phenol, o-cresol or p-cresol units via methylene groups in *ortho*- or *para*-position[1]

	Number of phenolic units in the molecule								
	2	3	4	5	6	7	8	9	10
Phenol	3	7	21	57	171	495	1485	4401	13203
o-Cresol	3	4	10	16	36	64	136	256	528
p-Cresol	1	1	1	1	1	1	1	1	1

Table 4.2. Examples for synthesized oligonuclear phenolic compounds, consisting of at least three phenolic units linked in *ortho*-position by the same bridge X and having the same substituent R in *para*-position. (Substituents R′ and R″ may be different)

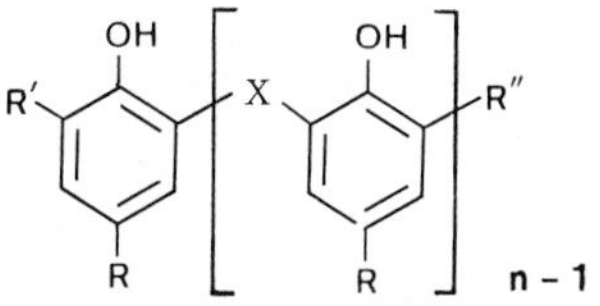

Bridge X	Substituent R	Maximal value of n	Ref.
CH_2	H	9	2, 3)
CH_2	CH_3	12	4)
CH_2	$C(CH_3)_3$	11	5, 6, 7)
CH_2	t-octyl, n-octyl	3	8)
CH_2	C_6H_5	4	9)
CH_2	F	4	10)
CH_2	Cl	7	3)
CH_2	Br	4	11)
CH_2	$COOH, COOC_2H_5$	3	12)
CH_2	$COCH_3$	3	13)
CH_2	NO_2	4	14)
CH_2	$N=N-C_6H_5$	3	15)
CH_2	OH	3	16)
CH_2-CH_2	OH	3	17)
$CH-CH_3$	H	5	18)
CO	CH_3	3	19)
CO	Cl, F	4	10)
SO_2	CH_3	4	20)

4.1 Synthesis

4.1.1 Principles, Protective Groups

The assembly of phenolic units that are connected by methylene groups is conducted primarily by two reactions:

a) The direct condensation with formaldehyde, leading to symmetrical structures:

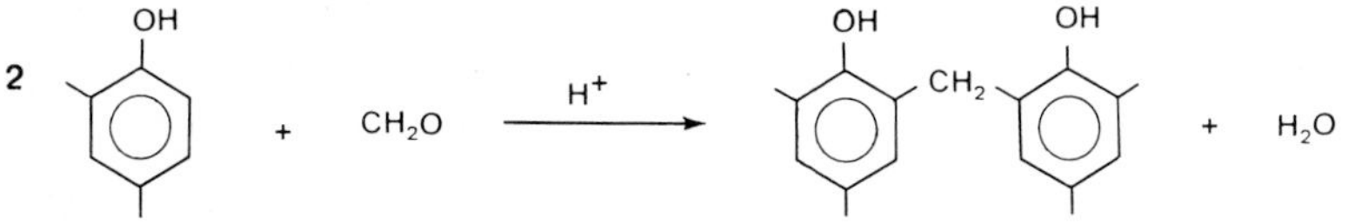

$$ (4.5) $$

b) The condensation of hydroxymethyl derivatives with other phenols can also generate non symmetrical structures:

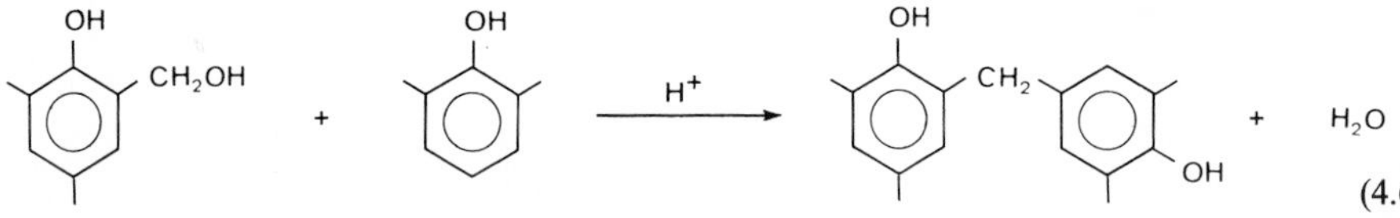

$$(4.6)$$

Instead of hydroxymethyl compounds which are available by reaction with formaldehyde under alkaline conditions[21], the use of chloromethylated (or bromomethylated) compounds is convenient or necessary in special cases[3, 14, 22].

The rational synthesis of definite oligomers (always) requires suitable protective or blocking groups. In the case of phenolics halogen (chlorine, bromine) is used to protect (or block) *ortho*- and *para*-positions from undergoing reaction. Dehalogenation is performed under very mild conditions (room temperature, normal pressure) with hydrogen in alkaline solution (4.7, 4.8 b) using Raney-nickel as catalyst[3, 23, 24].

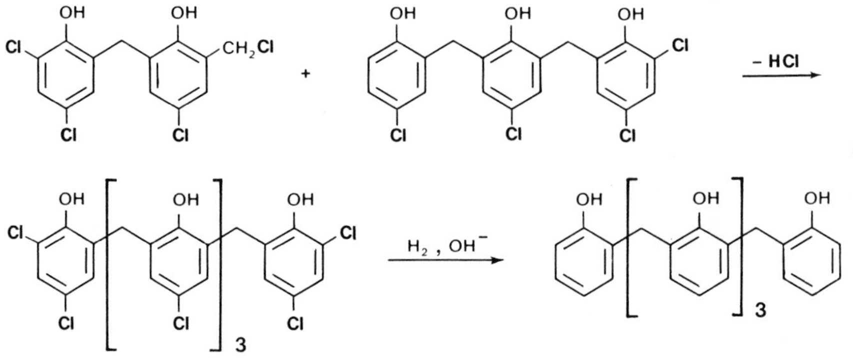

$$(4.7)$$

As an independent protecting group, the tert-butyl group[25] may be used which can be eliminated by transalkylation in toluene (4.8 a) in the presence of $AlCl_3$ at 50 °C[26].

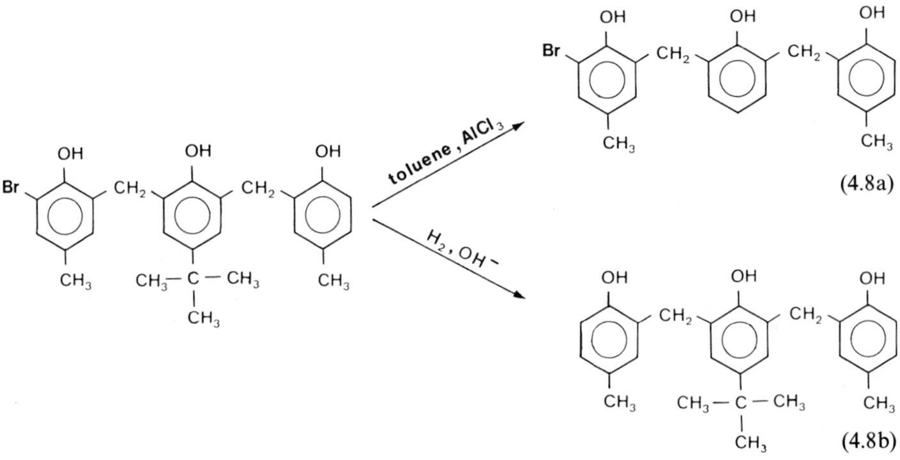

$$(4.8a)$$

$$(4.8b)$$

Bromine atoms may also be selectively eliminated with Zn/NaOH while leaving chlorine atoms intact in the same molecule[10, 27].

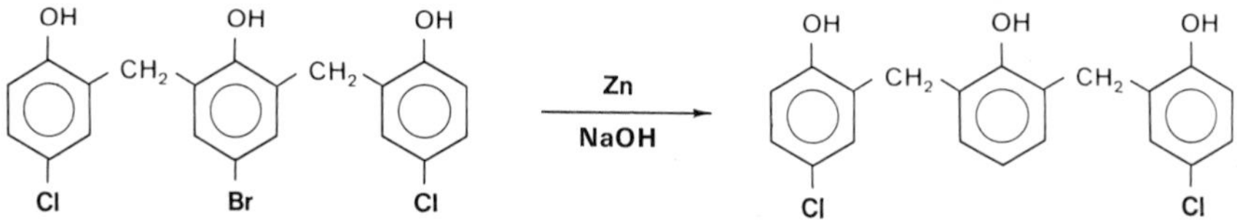

(4.9)

Among these possibilities only hydrogenation is sufficiently smooth and does not affect hydroxymethyl groups[24].

Thus, together with high *ortho*-specific reactions (see 4.1.2) different convergent strategies are known to synthesize complicated molecules, containing phenolic units linked by methylene bridges in a defined architectural manner. Furthermore, chromatographic separation techniques have been developed which allow the separation of definite compounds on a preparative scale from a mixture of similar compounds (Fig. 7.2).

4.1.2 Linear Oligomers

Selective coupling reactions[28] in *ortho*-position to the phenolic hydroxyl group are possible using bromomagnesium salts in refluxing benzene. Obviously under dry conditions magnesium acts as coordination site for formaldehyde[29] and facilitates the formation of the proposed reaction intermediate, quinone methide[30] which again is coordinated in a similar manner. The mechanism is suggested as follows:

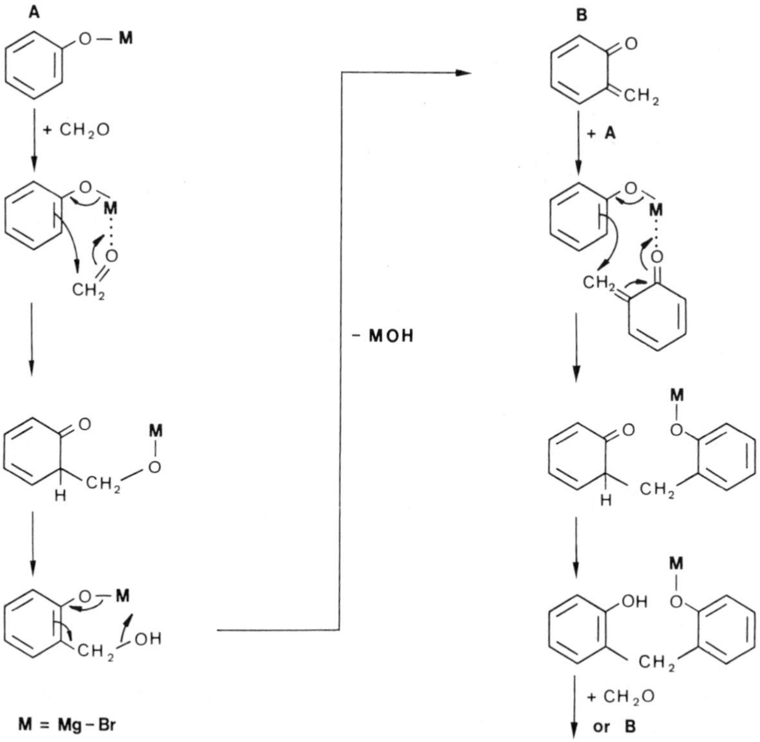

(4.10)

$M = Mg - Br$

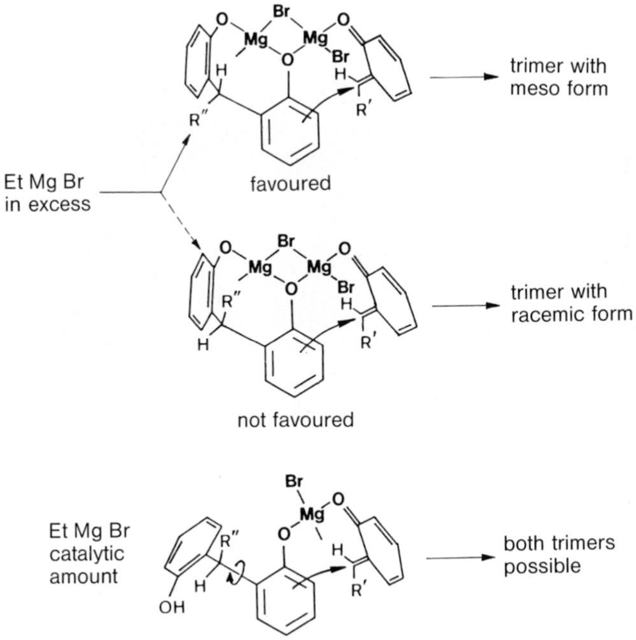

Fig. 4.1. Stereocontrolled addition of the intermediate quinone methide to a dinuclear compound. The proposed structure of the transition state explains the increasing amount of meso form which is found for increased amounts of EtMgBr[31]

The ensuing oligomerization reactions consist of a "duplication" procedure with formaldehyde or a "stepwise" extension with salicylic alcohol (generally an *ortho*-hydroxybenzyl alcohol). In this way Casiraghi et al. synthesized linear *ortho*-linked oligomers with up to nine phenol units without protecting the *para*-position[2]. The same compounds were initially obtained by Kämmerer et al. via the corresponding oligomers of *p*-chlorophenol[3] [see Eq. (4.7)].

Under similar conditions novolaks with alkylidene bridges, for instance, from acetaldehyde or isobutyraldehyde can be prepared. These compounds attract considerable theoretical interest, since stereoisomers are obtained as a consequence of the unsymmetrically substituted bridge (–CHR–). Several diastereomers with ethylidene[18] or isobutylidene[31] bridged phenolic units are obtained. As demonstrated in Fig. 4.1 the Mg not only enables regio- but also stereocontrolled synthesis. If suitable chiral alkoxy aluminium chlorides are used instead of the magnesium derivatives, enantioselective *ortho*-hydroxyalkylation[32] as well as enantiocontrolled synthesis of dinuclear compounds[33] become possible. The access to the interesting field of chiral novolaks is evident.

The synthesis of well defined oligomers containing up to three hydroquinone units was mainly studied by Manecke and coworkers[17, 34]. They may be regarded as model compounds for the corresponding hydroquinone – quinone redox polymers. Further oligomers are listed in Table 4.2; compounds consisting of different phenolic units have been prepared for special purposes (compare 4.2.2, 4.2.4).

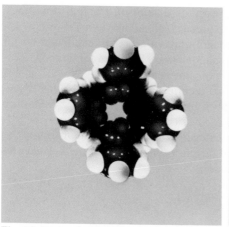

Fig. 4.2. Molecular models of calixarenes, top view: *p*-phenylcalix[4]arene (left) and *p*-phenyl-calix[8]arene (right). (Photo: D. Gutsche, Dept. of Chemistry, Washington University, St. Louis)

4.1.3 Cyclic Compounds, Calixarenes

The preparation of multi-ring compounds has attracted internationally prominent investigators in the study of a diversity of ring systems such as cyclophanes, spherands, crown ethers, catenanes, and others. Quite recently interest in cyclic condensation products of *para* substituted phenols with formaldehyde has intensified as a result of unique ring structures and chemical properties. The name for this family of cyclic compounds, calix[n]arenes, was coined by Gutsche and is derived from calix (greek, chalice) and arene (indicating a macrocyclic array of aromatic rings)[35] (see Fig. 4.2).

These oligomers demonstrate unusual properties, facilitated by their unique cavity structure (see Fig. 4.2) and their cyclic arrangement of hydroxyl groups. The latter enables them to form complexes with neutral molecules as well as with cations. Thus they may be used in conjunction with special membranes, as chelating agents, as enzyme mimics, mining chemicals, polymer stabilizers, etc.

The formation of calix [4] arenes was first recognized by Zinke et al.[36] during the alkaline condensation of *p*-alkylphenols with formaldehyde in a three stage process. Under similar conditions Cornforth later isolated two condensation products from *p*-tert-octylphenol with different melting points which he believed to be stereo isomeric cyclic tetramers[37]. Finally Gutsche showed that the direct condensation of *p*-tert-butylphenol with formaldehyde leads to a mixture of cyclic (and linear) compounds, containing not only methylene, but also dimethylene ether bridges[38] (4.11).

$$
\text{(4.11)}
$$

The composition of those mixtures is greatly dependent on reaction conditions. Synthesis conditions are now well defined to prepare calixarenes with 4, 6 or 8 p-tert-butylphenol and with 4 or 8 p-tert-octylphenol units in a preparative manner with yields up to 80%. Although even numbered oligomers are favoured for unknown reasons, small amounts of odd numbered cyclic oligomers are present[38-41].

A stepwise synthesis of calixarenes was first described by Hayes and Hunter[42] and was systematically studied and extended by Kämmerer and Happel[43-49]. Starting with o-bromo- or o-chloro-p-alkyl phenol it consists of subsequent hydroxymethylation and condensation steps. Finally the halogen is eliminated (demonstrating the value of halogen as a protective group) and the linear, monohydroxymethylated precursor is cyclized under high dilution condition.

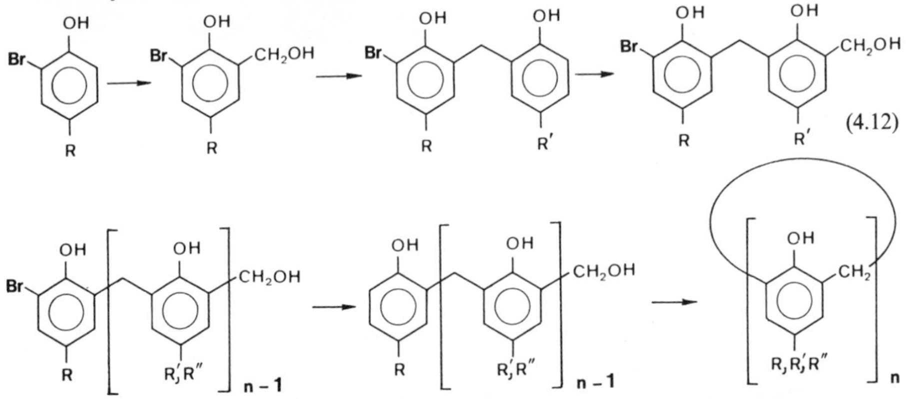

(4.12)

The ring size is here unambiguously established by the synthetic pathway which allows also some substituent variation in para-position (R, R', R"...). An obvious disadvantage is the long reaction sequence which necessarily causes rather limited overall yields.

A shorter synthesis, leading in three steps to calix[4]arenes with different substituents in para-position is outlined in the following sequence[50].

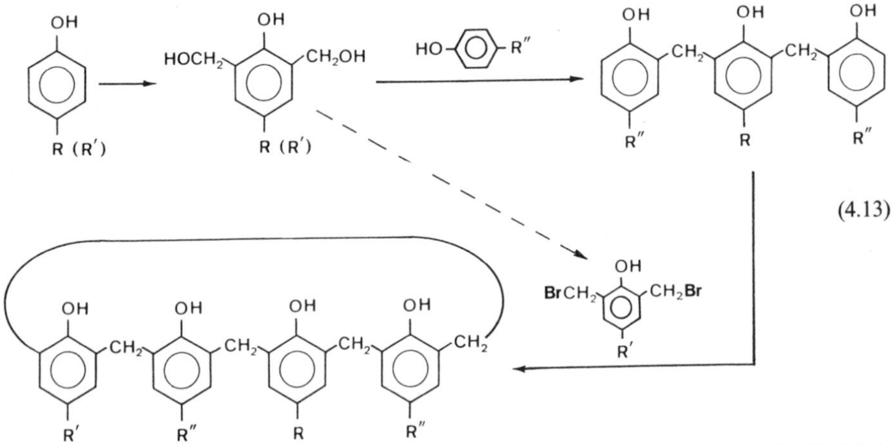

(4.13)

Although this method suffers from quite low yields in the cyclization step, despite high dilution conditions, it permits the synthesis of compounds with nitro- or halogen substituents. Using a lengthy linear oligomer in the last step, calixarenes with a higher number of phenolic units should be accessible.

Cyclic compounds in which the methylene bridges are partly or completely replaced by dimethylene ether bridges may be obtained by thermally-induced dehydration of suitable *bis*(hydroxymethyl) derivatives (simply heating in xylene) as illustrated in the following reaction[51]:

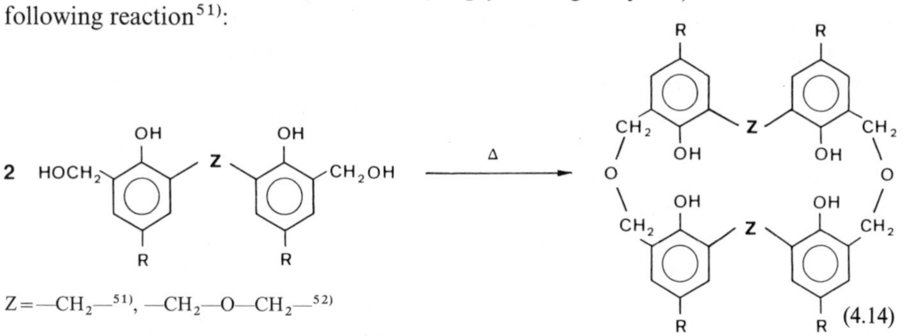

$$Z = -CH_2 - [51], \; -CH_2 - O - CH_2 - [52]$$

(4.14)

Cyclic trimers which were previously reported by Hultzsch[53] can be obtained in this way from 2,6-*bis*(hydroxymethyl) phenols[51] (4.15), while the preparation of cyclic trimers with three methylene bridges seems doubtful[54].

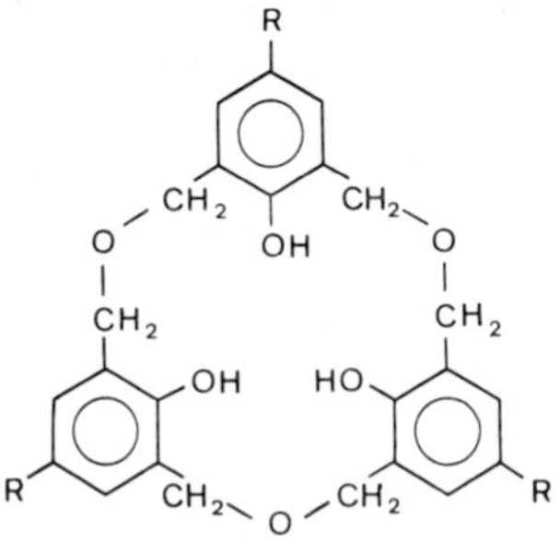

(4.15)

From resorcinol the acid catalyzed formation of cyclic tetrameric condensation products with benzaldehyde[55] or acetaldehyde[56] is reported (4.16). The more reactive formaldehyde yields an assortment of ill defined and partly crosslinked products with resorcinol. However, from veratrole (dimethylether of pyrocatechol) the cyclic trimer (4.17) (and tetramer) are obtained by acid catalyzed condensation with formaldehyde[57].

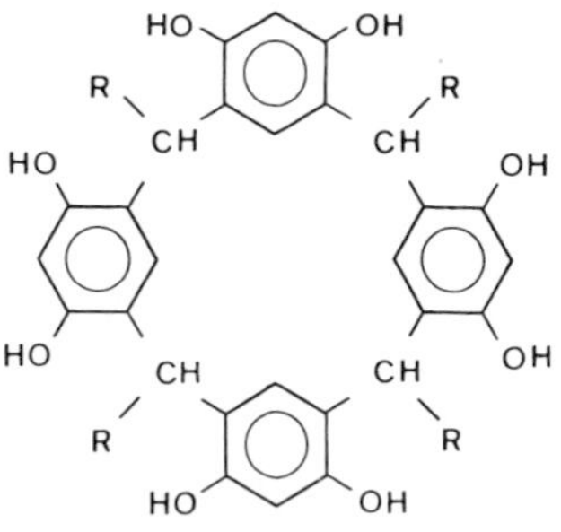

(4.16)

(4.17)

$$R = CH_3, \; C_6H_5$$

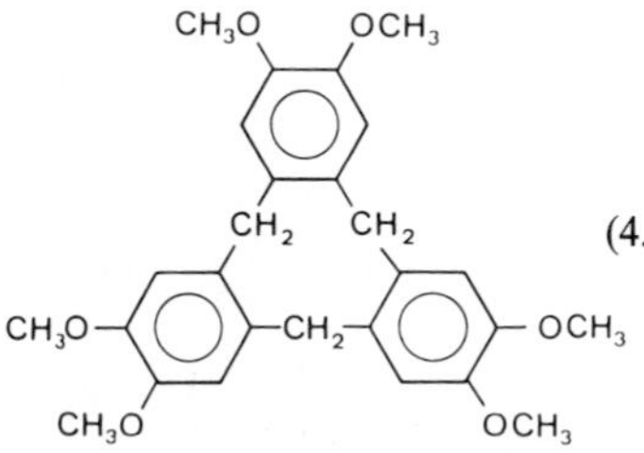

4.1.4 Chemical Modifications

Phenolic oligomers may be modified by suitable chemical reactions such as:
a) reaction of the phenolic hydroxyl groups;

b) (electrophilic) substitution of the aromatic nucleus or elimination of substituents;
c) modification of substituents or methylene bridges.

Since a rapid development was observed in the field of calixarenes during the last years, many of the reaction examples reported in the following text refer to cyclic compounds. However, analogous reactions will be possible with linear oligomers.

The phenolic hydroxyl group may be converted to ether or ester derivatives, and virtually all reactions of typical phenols can be realized. The strong interaction of the phenolic hydroxyl groups in *ortho-ortho*-linked oligomers seems to be the main reason which makes a complete conversion of all hydroxyls difficult in some cases. Thus, with diazomethane only monomethylether derivatives are obtained[58]. This relates to the observed hyperacidity (4.2.2). The complete conversion of all hydroxyl groups to methoxy groups in linear oligomers is feasible by treatment with dimethylsulfate in the presence of $Ba(OH)_2$ in DMF/DMSO[59], a procedure which yields only the trimethyl ether in the case of *p*-tert-butylcalix[4]arene[60]. Complete alkylation of all hydroxyl groups in calixarenes becomes possible by reaction with alkyl halides in THF-DMF solution in the presence of sodium hydride. Methyl, ethyl, allyl, and benzyl ethers have been prepared by this procedure in excellent yields[60] showing that even in cyclic compounds steric hindrance of hydroxyl groups is only a minor factor.

Thus, also the complete trimethylsilylation is possible for calix[6]- and calix[8]arenes with chloromethylsilane[38] and for calix[4]arenes with the very reactive *N,O-bis*-(trimethylsilyl)-acetamide[61].

Interesting derivatives which may possess new complexing properties for metal cations are obtained by reaction with diethylene glycol mono-tosylate in the presence of potassium tert-butoxide[62], e. g.

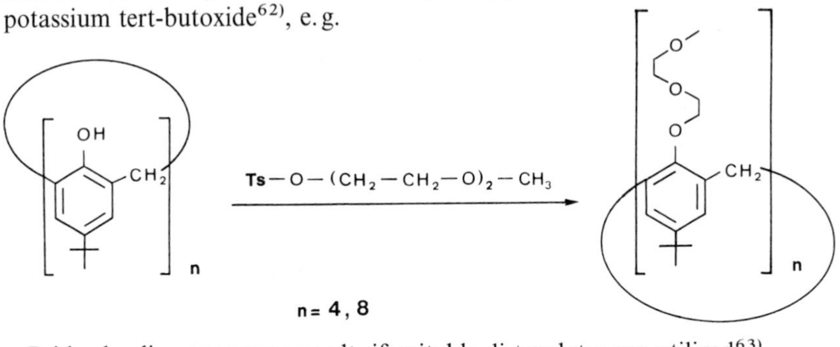

(4.18)

n = 4, 8

Bridged calixarenes may result, if suitable di-tosylates are utilized[63].

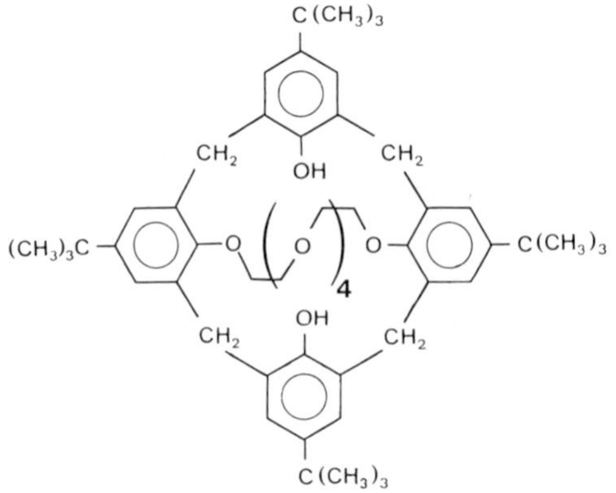

(4.19)

Complete esterification generally proceeds smoothly. Acetates (and other esters) have provided still another method of corroborating the structure of linear oligomers[64] as well as more recently the detailed multi-ring structure of calixarenes[65, 66]. The introduction of bulky ester groups is also possible[66, 67] and reactive esters like acrylates[68] or methacrylates[69] of oligo[(2-hydroxy-1,3-phenylene)-methylene]s have also been prepared. They may be used in copolymerization with monomers like styrene or acrylonitrile[70].

If these esters are treated with an excess of free radical catalyst in very dilute solution, the "polymerization" can be controlled to single molecules. Thus, after cleavage of the ester bonds oligoacrylates or methacrylates (or suitable derivatives) are obtained, and the degree of oligomerization is determined by the number of phenolic units in the starting molecule. Therefore this reaction sequence has been called a "synthetic matrix reaction"[71–74].

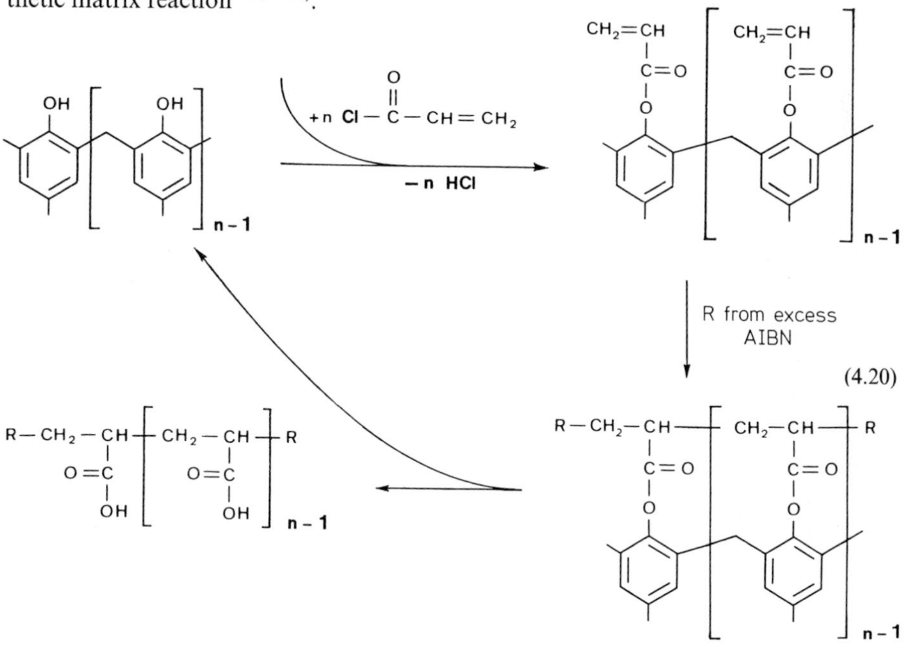

(4.20)

The transalkylation of tert-butyl groups and/or tert-octyl groups has been used to prepare calixarenes with free, reactive *para*-positions[46, 61, 62, 75, 76], e.g.:

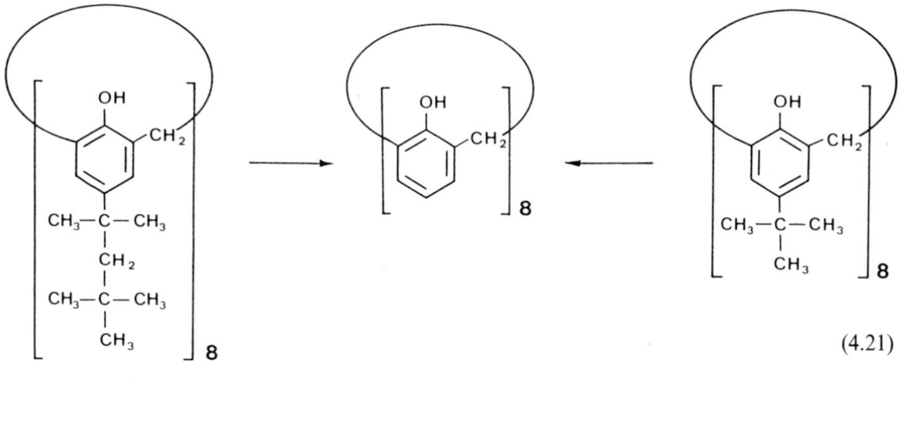

(4.21)

These are now available for the introduction of other functional groups by substitution reactions. With the tetramethylether of calix[4]arene the bromination as well as the Friedel-Crafts-acylation have been carried out. Complete conversion of all *para*-positions is achieved. Both derivatives are stable intermediates for further transformation into a variety of functionalized compounds[77] (Scheme 4.1).

Another possible reaction sequence commences with the tetraallylether of calix-[4]arene which undergoes a fourfold Claisen rearrangement when heated in diethylaniline[61]. From the tetratosylester of *p*-allylcalix[4]arene a variety of functionalized compounds has been prepared, from which the tosyl group may be removed under mildly

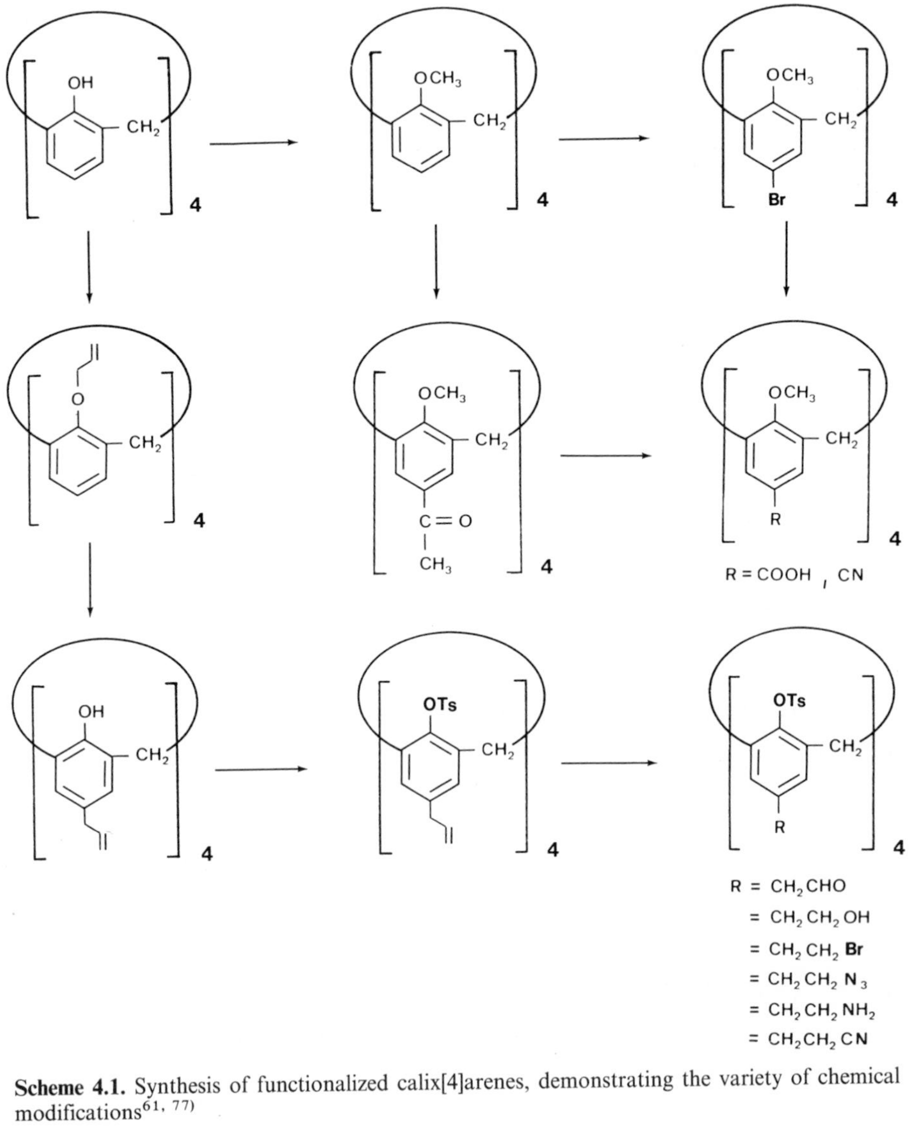

Scheme 4.1. Synthesis of functionalized calix[4]arenes, demonstrating the variety of chemical modifications[61, 77]

basic conditions[77]. A summary of all these reactions is shown in scheme 4.1 and illustrates the vast array of synthetic possibilities.

Although many attempts of nitration have been reported, isolable reaction products could only be obtained from dinuclear compounds[78]. Oligomers containing nitrophenol units therefore must be prepared by an initial stepwise synthesis, starting with nitrophenol[14, 79]. Coupling with diazotized aniline was successful even with a trinuclear compound[15]. Yet the more reactive p-nitro diazonium cation cleaves the methylene bridge under similar conditions[80].

The methylene bridges between the phenolic units can be oxidized with Cr(VI) in acetic acid to the carbonyl function, if the hydroxyl group is acetylated, and if no oxidizable functional groups are present. Oligomers with p-chlorophenol[10] units and p-tert-butylcalix[8]arene[81] are examples of successful oxidation. However, uniform reaction products are not obtained in the latter case.

Oligomers containing hydroquinone units undergo reversible oxidation to the corresponding quinone compounds under mild conditions[17, 34].

Quantitative hydrogenation of the phenolic units to cyclohexanol units is possible without degradation of the oligomeric structure, although high pressure ($1.5 \cdot 10^7$ Pa) and high temperature (150 °C) must be applied.

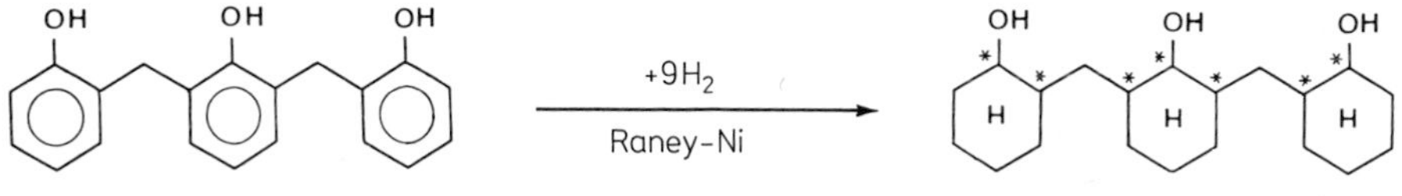

$$(4.22)$$

The resulting mixtures contain the expected number of diastereomeric compounds[82]. Interesting compounds for preparation of cycloaliphatic epoxide resins or polyurethanes can be obtained in that way.

4.2 Properties

4.2.1 Physical Properties

Despite the large number of synthesized oligomers with definite structure, systematic studies of physical properties like solubility characteristics or of thermodynamic data, e.g. heat of solution etc., are not currently available[83].

The melting points of several series of oligo-[(2-hydroxy-5-methyl-1,3-phenylene)-methylene]s were compared by Kämmerer and Niemann[84]. A melting point minimum for $n = 5$ is observed; however, this is not necessarily the case for oligomers with other phenolic substituents.

In contrast to linear oligomers the corresponding calixarenes exhibit remarkably high melting points, in some cases in excess of 400 °C. Examples are given in Table 4.3. In general, calixarene derivatives like esters or ethers have lower melting points, but extremely high values are observed such as 383–386 °C for the tetraacetate of p-tert-butylcalix[4]arene or 410–412 °C for the hexa-trimethylsilyl ether of p-tert-butylcalix-[6]arene[38].

Table 4.3. Melting points ($^\circ$C) of linear and cyclic phenolic oligomers

R	n	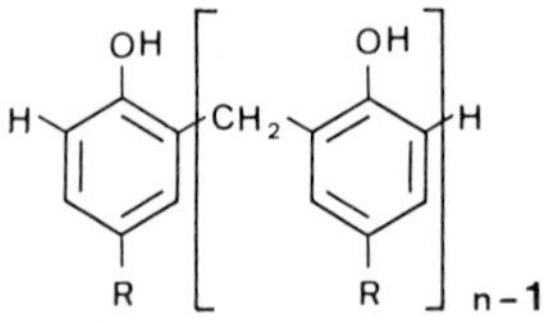	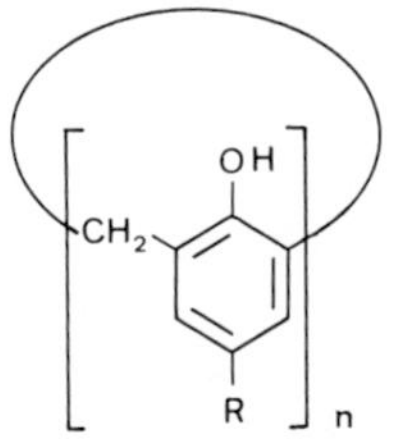
H	4	162–63	315–18
	5	151	295–98
	6	203–4	
	8	240	360
CH_3	4	214–15	320
	5	173	304
	6	215–17	370
$C(CH_3)_3$	4	211	344–46
	5	217–18	310
	6	250–52	380–81
	7	252–54	214
	8	260–61	411–12
C_6H_5	4	182–84	407–9

4.2.2 Acidity

Early potentiometric titration studies of Sprengling in anhydrous solvents (ethylene diamine or benzene/isopropyl alcohol) showed that o,o'-dihydroxy-DPMs have an extremely high acidity in comparison with similar isomers[85]. This effect which was even more pronounced for linear *ortho*-linked tri- and tetranuclear compounds, was designated as "hyperacidity." It is easily explained by a strong intramolecular hydrogen bond which stabilizes the monoanion. Similar observations are reported for maleic acid[86], salicylic acid[87], dihydroxynaphthalenes[88], dihydroxybiphenyls[89],etc.

An extensive study of these effects was performed[90], using "well designed" oligomers consisting of alkyl phenol units and one nitrophenol unit[79]. By its higher acidity and its spectral properties this nitrophenol unit acts as the focal point to monitor various structural features during dissociation of the phenolic hydroxyl group.

A decreasing value of pK_1 (K_1 = equilibrium constant of the first dissociation step) with increasing chain length is found for linear compounds with the p-nitrophenol unit at the terminus. It indicates increased stabilization of the monoanion in comparison to the undissociated compound. Even for tetranuclear oligomers a bulky substituent in *ortho*-position to the hydroxyl group at one end of the molecule causes a higher acidity at the other end of the molecule (Fig. 4.3). This prompts one to assume a chain of intramolecular hydrogen bonds between the phenolic hydroxyl groups which may be assisted by bulky *ortho*-substituents. Stabilization by an intramolecular hydrogen bond can also be deduced from the UV-spectra of the monoanions. Here the absorption coefficient as well as the wavelength of the absorption maximum related to

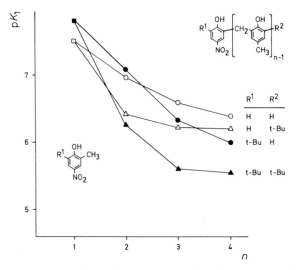

Fig. 4.3. Dependence of pK_1 to the number of phenolic units for a series of differently substituted oligomers with a p-nitrophenol terminus (methanol/water 50:50, 25 °C)[90]

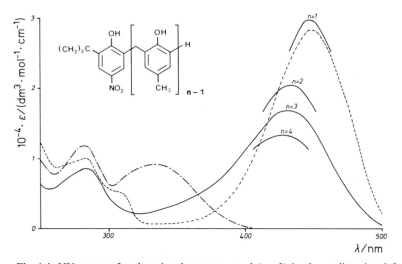

Fig. 4.4. UV-spectra for the trinuclear compound ($n=3$) in the undissociated form (– · –), as monoanion (——), and completely dissociated (– – –). The long-wavelength maximum for the monoanion of the compounds with $n=1$, $n=2$, and $n=4$ indicated shows the increasing strength of the intramolecular hydrogen bond for increasing chain length[90]

the nitrophenol unit decreases, if the strength of the intramolecular hydrogen bond increases (Fig. 4.4).

The pK_1-values summarized in Table 4.4 show (in comparison with those in Fig. 4.3) that an anionic substituent in the interior of a linear chain is better stabilized than at an exterior position. Thus, the monoanion of a trinuclear compound is better

Table 4.4. pK$_1$ values (dissociation of the nitrophenol unit) of different oligomers (water/methanol 50:50, 25 °C), showing the influence of intramolecular hydrogen bonds[90, 91]

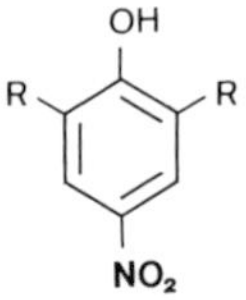

R = CH$_3$ 7.36

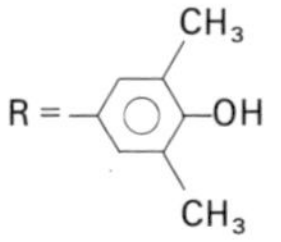

R =   7.38

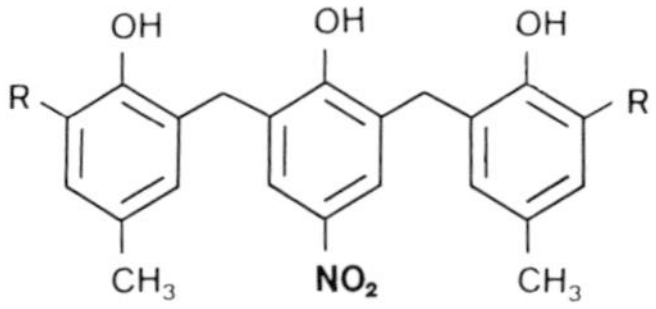

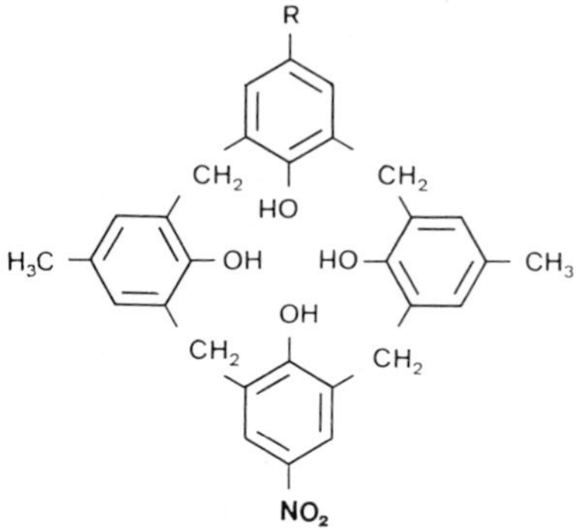

R = H	5.85
R = C(CH$_3$)$_3$	3.92

R = CH$_3$	6.0
R = C(CH$_3$)	4.3

represented by Formula (4.23 a) than by (4.23 b).

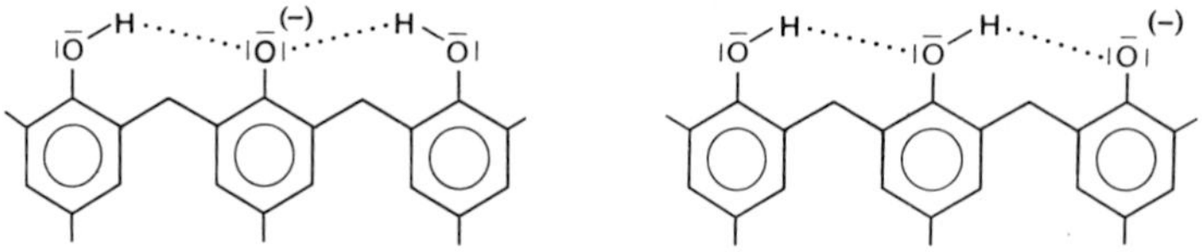

                   (4.23)

a   b

In calix[4]arenes the first dissociation constant obviously is very sensitive to small conformational changes of the cyclic array of hydrogen bonds. Slightly different conformations may be caused already by the exchange of substituents in *p*-position of the opposite phenolic unit[91] (see Table 4.4).

4.2.3 Complexation of Cations

Phenols and oligomeric phenolic compounds are weak acids and are able to form metal salts or complexes with metal ions. The reaction with Fe^{3+}-ions has been studied for analytical purposes with oligo [(2-hydroxy-1,3-phenylene)methylene]s[92]. The red colour is most intense with *o*-hydroxymethyl derivatives[93].

The ability of *p*-tert-butylcalixarenes to transport metal ions through hydrophobic liquid membranes has been reported recently by Izatt and coworkers[94]. In contrast to 18-crown-6, they are ineffective in neutral solution, but show remarkable cation transport ability in basic solution. As demonstrated by the data in Table 4.5, a high selectivity for Cs$^+$-ions is observed for the cyclic tetramer, while the cyclic octamer shows the highest absolute transport ability for Cs$^+$.

Beside low solubility in water, the main advantage of calixarenes in this connection is their ability to form neutral cation complexes through loss of (a) proton(s). Thus, the anion does not accompany the cation through the membrane and the transport

Table 4.5. Cation transport from basic solution by several potential complexing agents through a liquid membrane (0.001 M carrier in CH_2Cl_2/CCl_4 25/75). Data are given as $10^8 \cdot$ flux/mol $\cdot s^{-1} \cdot m^{-2}$ [94]

Source	p-tert-butylcalix[n]arene			p-tert-butylphenol or 18-crown-6
	$n=4$	$n=6$	$n=8$	
1 M LiOH	< 0.9	10 ± 1	2.0 ± 0.2	<0.9
1 M NaOH	1.5 ± 0.4	13 ± 2	9 ± 2	<0.9
1 M KOH	< 0.9	22 ± 3	10 ± 0.4	<0.9
1 M RbOH	5.6 ± 0.7	71 ± 8	340 ± 20	<0.9
1 M CsOH	260 ± 90	810 ± 80	1000 ± 260	<0.9

of cations may be coupled to the reverse flux of protons. Although the exact structure of the complex is still subject to debate (the cation may be surrounded by the oxygen atoms or it may be situated in the cavity after loss of its hydration shell), fascinating results will no doubt be reported in the foreseeable future as a consequence of the large variability of these systems. An example of an intriguing and unexpected observation is the "crowned" calixarene (4.19). It can transfer Na^+- and K^+-ions from a basic to an acid solution while neither is possible with p-tert-butylcalix[4]arene nor dibenzo-18-crown-6[63].

4.2.4 Kinetic Studies

Initial efforts to correlate the reactivity of linear oligomers with the chain length have been reported by Imoto et al.[95]. They studied the acid catalyzed ($HClO_4$) reaction with formaldehyde (in dioxane/water) and found that the reaction rate for $n=5, 6$, and 8 was half of the reaction rate for $n=1$–4. This was explained by cyclic conformations which are stabilized by intramolecular hydrogen bonds and cause one end of the molecule to be unavailable to the attacking reagent.

More detailed studies are reported for the electrophilic bromination in acetic acid. For several isomeric[96] or similar[97] dinuclear compounds containing the same reacting p-cresol (or o-cresol) unit it was found that the reactivity is decreased by an intramolecular hydrogen bond between the phenolic hydroxyl groups. The effect is especially pronounced if this hydrogen bond is assisted by bulky substituents of the adjacent unit toward the reacting phenolic unit. It can be explained by a smaller electron donating effect of a hydroxyl group accepting an intramolecular hydrogen bond in comparison to a "free" hydroxyl group. The directing effect of an ortho-substituent is transmitted from one end of the molecule to the other end in linear, ortho-linked oligomers even up to $n=6$[98] (Fig. 4.5). This is again strong evidence for a concerted intramolecular hydrogen bond phenomenum between all the phenolic hydroxyl groups. And again the strongest effect is observed when the interior phenolic unit is influenced by the adjacent units (Table 4.6)[99]. Surely the bromination reaction has no industrial relevance, but similar effects can be expected for other electrophilic substitutions at least under similar conditions.

Unique differences were found for the reactivity of chloromethyl groups in methanolysis and aminolysis reactions for two series of isomeric dinuclear compounds

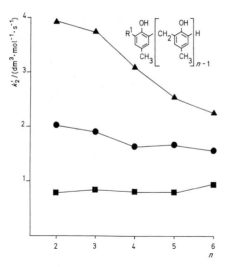

Fig. 4.5. Rate of bromination for three series of oligo[(2-hydroxy-5-methyl-1,3-phenylene)meth-ylene]s as a function of the chain length (acetic acid, bromine, 22 °C, ▲ R^1=H, ● R^1=CH$_3$, ■ R^1=C(CH$_3$)$_3$)[98]. For 2,4-dimethylphenol, as a model of the reacting unit k_2^1=7.6 dm^3 · mol^{-1} · s^{-1} is found under these conditions[96]

Table 4.6. Relative rate of bromination (molecular bromine, acetic acid, 22 °C) of trinuclear compounds with the reacting unit in the middle of the molecule[99]

Compound		Relative rate
	R=CH$_3$	1
	R=C(CH$_3$)$_3$	0.63
		0.61
	R'=R''=CH$_3$	0.34
	R'=C(CH$_3$)$_3$ R''=CH$_3$	0.07
	R'=CH$_3$ R''=C(CH$_3$)$_3$	0.36

Table 4.7. Relative rates for the methanolysis and the first step of the aminolysis with p-nitroaniline in DMSO for two groups of isomeric chloromethylated methylenediphenols in relation to the corresponding model compounds[100]

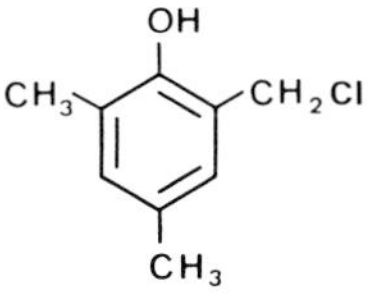

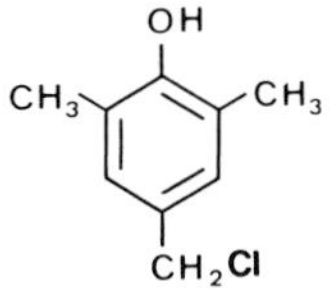

Compound	Position of OH-groups	Methanolysis	Aminolysis
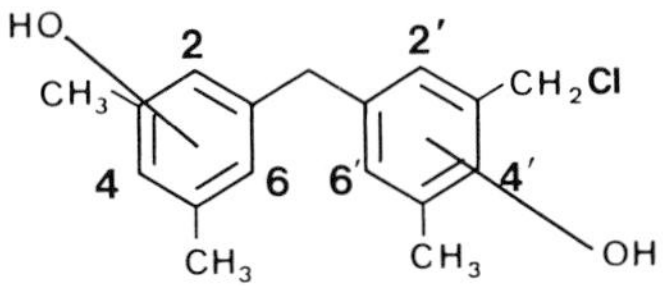	2,2′	0.65	305
	2,4′	1.51	1.06
	4,2′	0.92	0.97
	4,4′	1.10	0.98
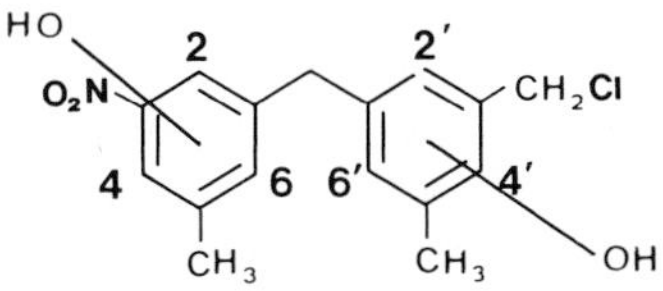	2,2′	0.36	906
	2,4′	0.53	0.72
	2,6′	0.28	143
	6,2′	0.37	2775
	6,4′	0.55	0.74
	6,6′	0.29	303
	4,2′	0.26	0.72
	4,4′	0.38	0.71

(Table 4.7)[100]. The rate of the methanolysis, leading to hydroxybenzyl methyl ethers, is nearly independent of the structure, while the aminolysis is strongly accelerated (by factors greater than 10^3), if the hydroxyl groups of the reacting and the adjacent phenolic unit are both in *ortho* position to the methylene bridge. This is caused by different reaction mechanisms. The methanolysis of the non-dissociated chloromethyl phenols is a S_N1 reaction[101]. The aminolysis proceeds via the reversible elimination of HCl and the subsequent addition of p-nitroaniline to the intermediate quinone methide to form the hydroxybenzyl amine[102]. Obviously this first elimination step is intramolecularly assisted by the adjacent hydroxyl group.

Again similar effects can be expected for those reactions where quinone methides are postulated as intermediates.

In contrast, electrophilic substitutions under alkaline conditions which require the phenolate ion as substrate may be retarded for o,o′-dihydroxy-DPMs. Obviously the intramolecular interaction of the hydroxyl groups, leading to a decrease of pK_1 and an increase of pK_2 makes a complete anionic structure of the reacting phenolic unit impossible. The lowest reactivity which o,o′-dihydroxy-DPM shows in comparison with p,p′-dihydroxy- and o,p′-dihydroxy-DPM in the alkali-catalyzed addition of formaldehyde (in spite of the presence of two more reactive p-positions) was explained by such a hydrogen bonded monoanion[103].

This interpretation is confirmed by the coupling reaction of three isomeric dihydroxy-DPMs with phenyldiazoniumchloride. In these compounds (4.24)

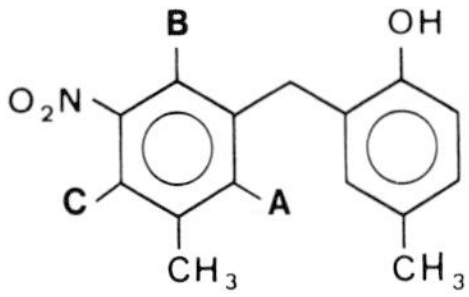

A = OH   B,C = H

B = OH   A,C = H   (4.24)

C = OH   A,B = H

clearly the nitrophenol unit adjacent to the reacting unit is dissociated in the monoanion and the compound with C=OH shows the highest reactivity[104].

4.2.5 Crystal Structure

Single crystal X-ray diffraction analysis is a method which necessarily requires highly purified crystalline compounds and is not applicable to mixtures. In this way the structure of a molecule can be unambigously determined or confirmed. Thus, the first studies in the field of phenolic oligomers were reported for p-tert-butylcalix[4]arene[105] and for the octaacetate of p-tert-butyl[8]arene[65]. This was the "coup de grace" that a mixture of cyclic compounds is produced during the alkaline condensation of p-substituted phenols with formaldehyde, and it was the first definitive proof of the ring size of two oligomers in these mixtures.

X-ray analysis is also a very important tool for the determination of molecular conformations. Obviously a specific conformation is influenced by interaction with other molecules in the crystal lattice (packing effects), if the molecular structure is not completely rigid. But even for flexible molecules it can be assumed that the conformation in the crystalline state is similar to the preferred conformation in solution, if there are no special solvation effects.

For calix[4]arenes and derivatives four conformations are possible which according to Gutsche are designated as "cone," "partial cone," "1,2-alternate," and "1,3-alternate"[35]. All molecular and crystal structures of calix[4]arenes thus far reported show that the molecules are in the cone conformation (Fig. 4.6). This means that the conformation of the molecule is nearly exclusively determined by the cyclic array of intramolecular hydrogen bonds. The substituent in p-position (R=H[106], tert-butyl[105], tert-octyl[107]) as well as occluded solvent molecules have only a slight effect. For all compounds the distances between adjacent oxygen atoms are in the range of 2.63–2.67 Å. Slight differences in the conformation are reflected by the inclination of the

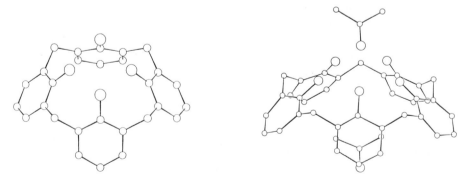

Fig. 4.6. Perspective view of calix[4]- and calix[5]arene as determined by X-ray analysis[106, 108]. Acetone which is incorporated in the crystal lattice in different ways is omitted for calix[4]arene

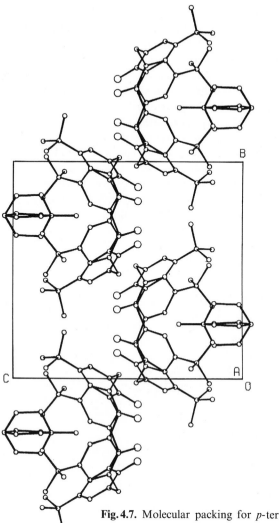

Fig. 4.7. Molecular packing for *p*-tert-butylcalix[4]arene/toluene (1:1)[105]

phenyl rings with respect to the normal of the plane of the oxygen atoms or the methylene carbon atoms. It varies normally between 121 ° and 129 °, but may be deformed to 115 ° or 137 ° by special packing effects[107]. A similar cone conformation is found for calix[5]arene[108]. However, longer O-O-distances of 2.82–2.86 Å indicate weaker intramolecular hydrogen bonds, in accordance with the results obtained in solution (Sect. 4.2.6).

Solvent entrapment in the crystal lattice seems to be a normal characteristic of calix-arenes, although solvent free crystals have been obtained from *p*-tert-octylcalix-[4]arene[107]. However, the guest molecules are not always situated inside the cavity of the host molecules ("cryptato-cavitate") as for instance found for *p*-tert-butylcalix-[4]arene/toluene (1:1) clathrate (Fig. 4.7)[105]. A so called "tubulato-clathrate" has been observed for *p*-tert-octylcalix[4]arene/toluene (1:1)[107] and an "intercalato-clath-rate" for calix[4]arene/acetone (1:1)[106]. It should be mentioned that the ratio host/guest may be different from one[106, 108, 109] and that different ratios have been found for the same host/guest pair[106]. Interesting calixarene contributions to the rapidly developing field of host/guest chemistry can be expected in the future.

Crystal structures for linear oligomers were first reported by Casiraghi et al.[2, 7, 18]. All results show that conformation in the solid state again is determined by intramolecular hydrogen bonds between hydroxyl groups of adjacent phenolic units[110]. In the crystal lattice the molecules are ordered by further intermolecular hydrogen bonds either to indefinite chains (compare Fig. 4.8 a) or to cyclic dimers (compare Fig. 4.8 b). In this arrangement each hydroxyl acts alternately as a donor and an acceptor concurrently ("isodromic hydrogen bonds") and never as a double acceptor. The intramolecular O-O-distances for all structures are in the range of 2.62–2.71 Å for all methylene bridged compounds (2.71–2.79 Å was found for ethylidene bridged trimers[18]) and

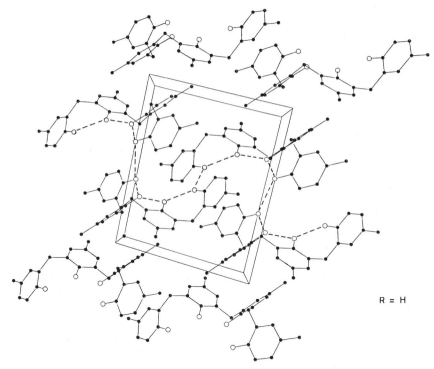

R = H

Fig. 4.8. Molecular and crystal structure of tetranuclear compounds

Conformation is *anti/anti* for R=H and *syn/anti* for R=CH₃. Hydrogen bonds are indicated by dotted lines

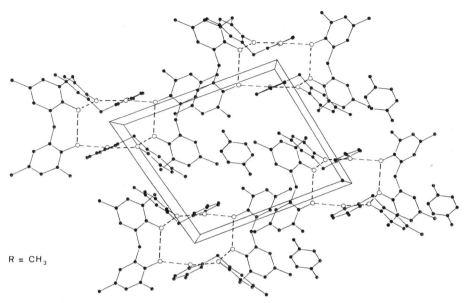

R = CH$_3$

Fig. 4.8 (continued)

only slightly longer (2.68–2.77 Å) are the intermolecular O-O-distances. This indicates that all hydrogen bonds have the same strength. While no special direction of the hydrogen bonds is found for dimers, trimers, and tetramers with unsubstituted *ortho*-positions[2, 7, 18], a methyl or tert-butyl group in one *ortho*-position obviously directs the hydrogen bonds in the opposite direction, as found for tetranuclear compounds[110].

Bond distances, bond angles, and torsion angles are practically equal in all compounds and comparable with literature values for similar groups. The conformation of the molecules is determined by the torsion angles for the rotation around the σ-bonds to the methylene (or ethylidene) bridge. They are in the range of 80–100 °. The resulting dihedral angles between the aromatic planes are found between 105 ° and 130 °.

As shown in Fig. 4.9 three phenolic units may be placed in a chairlike *anti-* or *trans*-position or in a boatlike *syn-* or *cis*-position. In the case of the ethylidene bridged oligomers this sequence is determined by the relationship of the methyl substituent to the O–H⋯O-bond system that occurs at the opposite side[18]. This means, *anti*-conformations correspond to racemic dyads, while *syn*-conformations correspond to meso dyads.

While two subsequent *anti*-conformations are possible for the next higher homologue, the two *syn*-conformations related to the conelike structure of calix[4]arenes are not obtained for tetranuclear compounds. The preference of *anti*-conformations over *syn*-conformations is assumed for higher oligomers. This is mute since single crystals for compounds with 5 or more phenolic units have not as yet been obtained.

It would be intriguing to compare the single crystal X-ray studies with two dimensional NMR analysis. The latter technique is becoming quite popular in comparing large biological molecules with X-ray. Relaxation times and line shapes (2D-NMR) could provide additional information on dynamics of calixarenes and *syn-anti* linear

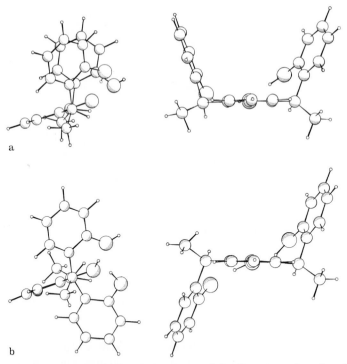

Fig. 4.9 a and b. Molecular conformation of the diastereomeric *ortho*-ethylidene linked phenol trimers seen from different directions. The meso-compound (**a**) has the *syn*-conformation, the racemic compound (**b**) the *anti*-conformation[18]

oligomers. Surely it would be also interesting to study oligomers with *o-p* or *p-p* linked phenolic units.

Crystallographic characterization of 2,4,6-*tris*(hydroxymethyl)phenol (THMP) and 3,3′,5,5′-*tetra*(hydroxymethyl)-4,4′-dihydroxydiphenylmethane (THMDPM) was performed by Perrin et al.[111], and these are the first studies reported for resol model compounds. THMP crystals, grown from a methanol/chloroform mixture, are allied to the monoclinic system, space group P2$_1$/c; THMDPM crystallized from acetone to the triclinic system. Bond distances are conform with literature values, selected bond angles for THMP are cited in Table 4.8. Figure 4.10 shows the stereoscopic view of both compounds.

Table 4.8. Some characteristic bond angles (°) for 2,4,6-*tris*(hydroxymethyl)phenol

C2−C1−O1	115.9 (2)	C6−C1−O1	123.3 (2)
C1−C2−C7	119.2 (2)	C3−C2−C7	121.8 (2)
C3−C4−C8	120.1 (2)	C5−C4−C8	121.5 (2)
C5−C6−C9	120.1 (2)	C1−C6−C9	121.6 (2)
C2−C7−O2	110.0 (2)	C4−C8−O3	112.4 (2)
C6−C9−O4	112.4 (2)		

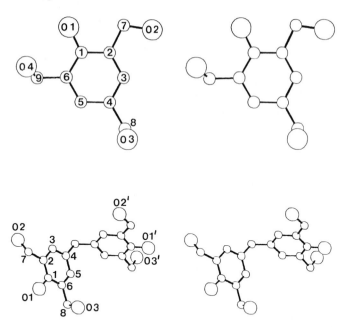

Fig. 4.10. Molecular structure of hydroxymethylated phenolic compounds[111]

Considering THMP the phenolic oxygen O1 is in the ring plane as well as the oxygen O2 of one methylol group in *ortho* position. The two remaining methylol oxygen atoms O3 and O4 are out of plane on the same side with distances of 1.217 Å for O3 and 1.463 Å for O4. Dihedral angles between the benzene ring plane and the C–C–O planes of methylol groups are: O2–C7–C2-ring 8.75° O3–C8–C4-ring 80.68° O4–C9–C5-ring 67.57°. THMDPM shows a more complicated geometry. As for THMP, the oxygen of one methylol group (in *ortho* position) is in the ring plane. Both ring planes form an angle of 82°. Dihedral angles between C–C–O planes and benzene rings are given below:

C6–C8–O3 – ring 1   3,26°   C6'–C8'–O3' – ring 2   105,93°
C2–C7–O2 – ring 1   88,75°   C2'–C7'–O2' – ring 2   143,31°

A very strong intermolecular hydrogen bonding in a tridimensional network is found leading to a very compact arrangement of the molecules in the crystalline state. This is clearly a consequence of the large number of hydroxyl groups in the molecule. Surprisingly no intramolecular bonds were identified.

4.2.6 Conformation in Solution

The "molecular dynamics" of calixarene ring systems range from moderately rigid or slowly interconverting cyclic materials to molecules with fluxional mobility.

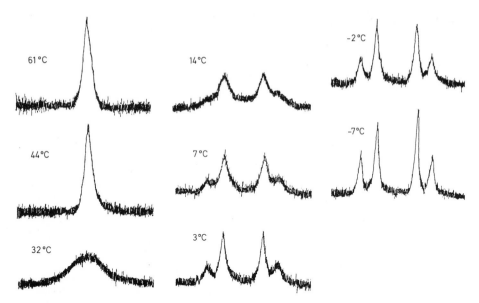

Fig. 4.11. Temperature dependence of the ^{1}H-NMR spectrum of a cyclic tetranuclear compound consisting of two p-cresol and two p-tert-butylphenol units in alternating sequence in the region of the methylene protons (solvent CDCl$_3$)[112]

Calixarene conformational analysis or the interconversion of various calixarene conformers is conveniently followed by ^{1}H-NMR variable temperature conditions. In the methylene proton region (3.1–4.5 ppm) cyclic tetranuclear compounds exhibit a proton pattern (Fig. 4.11) of a sharp singlet at high temperatures which broadens with decreasing temperature to a broad doublet and ultimately a pair of doublets (AB system) at low temperatures.

The coupling constant J = 12–14 Hz is typical for geminal protons. Hence the methylene protons are magnetically equivalent at high temperatures, while at low temperatures a "rigid" conformation exists in which the two protons of a methylene bridge are different in spite of equivalence of the four methylene bridges. The NMR data best describes the cone conformation in which 4 "axial" and 4 "equatorial" hydrogen atoms can be distinguished. Other conformations do not adequately satisfy the spectral data[113].

From those temperature dependent NMR-spectra which are similar for larger cyclic compounds the Gibbs-energy of activation ($\Delta G^{\pm}$) for the interconversion of both protons can be calculated. Typical values are listed in Table 4.9. They show that the role of the p-substituent is indifferent. As is expected, cyclic penta- and heptanuclear compounds are more flexible ($\Delta G^{\pm} = 51$–54 kJ $\cdot$ mol^{-1}) than tetranuclear compounds ($\Delta G^{\pm} = 62$–67 kJ $\cdot$ mol^{-1}). But surprisingly p-tert-butylcalix[8]arene shows in apolar solvents (CDCl$_3$) exactly the same temperature dependence of the ^{1}H-NMR-spectrum as p-tert-butylcalix[4]arene[114], although the flexibility should be even higher than for calix[5]- or calix[7]arenes. This is explained by assuming a "pinched" conformation consisting of two cyclic tetranuclear "subunits" which are stabilized by intramolecular hydrogen bonds. Indeed, with pyridine as solvent, the cyclic octanuclear compound

Table 4.9. Free energies of activation for the conformational inversion of several cyclic oligo-nuclear phenolic compounds ("calixarenes") in different solvents. (t-Bu $= C(CH_3)_3$, t-Oct $= C(CH_3)_2-CH_2-C-(CH_3)_3$)

n	R^1	R^2	R^3	R^4	R^5	R^6	R^7	R^8	Solvent	$\Delta G^{\ddagger}$ kJ·mol^{-1}
4	t-Bu	CH_3	CH_3	CH_3	–	–	–	–	CDCl$_3$	66.5 [44]
4	t-Bu	t-Bu	CH_3	CH_3	–	–	–	–	CDCl$_3$	64.2
4	t-Bu	CH_3	t-Bu	CH_3	–	–	–	–	CDCl$_3$	61.5 [113]
4	t-Bu	t-Bu	t-Bu	CH_3	–	–	–	–	CDCl$_3$	64.5 [48]
4	t-Bu	t-Bu	t-Bu	t-Bu	–	–	–	–	CDCl$_3$	65.6 [114]
									d$_5$-Bromobenzene	63.5
									d$_5$-Pyridine	56.0
5	CH_3	CH_3	CH_3	CH_3	CH_3	–	–	–	CDCl$_3$	50.5 [48]
5	t-Bu	CH_3	CH_3	CH_3	CH_3	–	–	–	CDCl$_3$	50.7
5	t-Bu	t-Bu	t-Bu	CH_3	CH_3	–	–	–	CDCl$_3$	53.4
7	t-Bu	CH_3	CH_3	CH_3	CH_3	CH_3	CH_3	–	CDCl$_3$	51.2 [47]
8	t-Bu	t-Bu	t-Bu	t-Bu	t-Bu	t-Bu	t-Bu	t-Bu	CDCl$_3$	65.6 [114]
									d$_5$-Bromobenzene	63.5
									d$_5$-Pyridine	36
8[a]	t-Oct	t-Oct	t-Oct	t-Oct	t-Oct	t-Oct	t-Oct	t-Oct	CDCl$_3$	67.1 [116]

[a] This compound was described to be the cyclic tetramer, but most probably it was the cyclic octamer

shows the expected high flexibility in comparison to the tetranuclear compound, since all hydroxyl groups form hydrogen bonds with the solvent. On the other hand, it was shown very recently by X-ray analysis that a "pleated loop" conformation is present in the crystalline state in which the eight hydroxyl groups are arranged in a slightly undulating, almost planar, intramolecularly hydrogen bonded cyclic array[115].

If the hydrogen atoms in hydroxyl groups are replaced by larger groups, which are unable to pass the center of the macrocyclic ring, certain conformations are "fixed"[113]. As shown by NMR-spectra[62, 113], in one case also by X-ray analysis[117], these are mostly the "cone" or "partial cone" conformation, seldom the "1,3-alternate." Derivatives of the "1,2-alternate" are not observed.

Less is known about the conformation of linear oligomers in solution, but most probably intramolecular hydrogen bonding also seems to be an important factor (compare 4.2.2 and 4.2.4). Infrared spectroscopic studies show the presence of intra- as well as intermolecular hydrogen bonds between phenolic hydroxy groups[8,118–120] and in some cases also intramolecular OH-π-hydrogen bonds. The dipole moment and its temperature dependence allows also some conclusions on the conformation of the molecules. Dipole moment data for a series of well defined oligomers (di-, tri-, and tetranuclear compounds) have been reported by Tobiason et al.[121]. An interesting approach was recently disclosed by Casiraghi, Kaptein et al.[122]. They used chemically induced dynamic nuclear polarization technique (CIDNP), to gain some insight in the accessibility of hydroxyl groups. For *ortho*-ethylidene bridged oligomers of phenol, they observed a strong effect for the dimer, but only for the phenolic unit in the middle of the trimer. For the tetra- and pentamer also no effect is found for the phenolic units at the ends of the molecule. However, further studies seem necessary for a complete interpretation, taking into account the steric accessibility of hydroxyls and their role in intramolecular hydrogen bonds.

4.3 References

1. Megson, N. J. L.: Chemiker-Ztg. 96, 17 (1972)
2. Casiraghi, G., Cornia, M., Sartori, G., Casnati, G., Bocchi, V., Andreetti, G. D.: Makromol. Chem. 183, 2611 (1982)
3. Kämmerer, H., Lenz, H.: Makromol. Chem. 27, 162 (1958)
4. Kämmerer, H., Rausch, W., Schweikert, H.: Makromol. Chem. 56, 123 (1962)
5. Kämmerer, H., Haberer, K.: Monatsh. Chem. 95, 1589 (1964)
6. Zinke, A., Kretz, R., Leggewie, E., Hössinger, K.: Monatsh. Chem. 83, 1213 (1952)
7. Casiraghi, G., Cornia, M., Ricci, G., Balduzzi, G., Casnati, G., Andreetti, G. D.: Makromol. Chem. 184, 1363 (1983)
8. Kämmerer, H., Eberle, K., Böhmer, V., Großmann, M.: Makromol. Chem. 176, 3295 (1975)
9. Gutsche, C. D., No, K. H.: J. Org. Chem. 47, 2708 (1982)
10. Hakimelahi, G. H., Moshfegh, A. A.: Helv. Chim. Acta 64, 599 (1981)
11. Kämmerer, H., Gros, G.: Makromol. Chem. 149, 85 (1971)
12. Kämmerer, H., Lotz, W.: Makromol. Chem. 145, 1 (1971)
13. Niemann, W., Böhmer, V., Evers, H., Kämmerer, H.: Makromol. Chem. 158, 123 (1972)
14. Böhmer, V., Deveaux, J., Kämmerer, H.: Makromol. Chem. 177, 1745 (1976)
15. Kämmerer, H., Lenz, H.: Chem. Ber. 94, 229 (1961)
16. Manecke, G., Panoch, H. J.: Makromol. Chem. 96, 1 (1966)
17. Manecke, G., Zerpner, D.: Makromol. Chem. 129, 183 (1969)
18. Casiraghi, G., Cornia, M., Ricci, G., Casnati G., Andreetti, G. D., Zetta, L.: Macromolecules 17, 19 (1984)
19. Kämmerer, H., Büsing, G., Haub, H. G.: Makromol. Chem. 66, 82 (1963)
20. Kämmerer, H., Harris, M.: J. Polym. Sci., A 2, 4003 (1964)
21. Kämmerer, H., Happel, G.: Makromol. Chem., Rapid Commun. 1, 461 (1980)
22. Rodia, J. S.: J. Org. Chem. 26, 2967 (1961)
23. Kämmerer, H., Großmann, M.: Chem. Ber. 86, 1492 (1953)
24. Kämmerer, H., Happel, G., Böhmer, V.: Org. Prep. Proced. Int. 8, 245 (1976)
25. Tashiro, M.: Synthesis, 1979, 921 and references cited there
26. Böhmer, V., Rathay, D., Kämmerer, H.: Org. Prep. Proced. Int. 10, 113 (1978)
27. Moshfegh, A. A., Mazandarani, B., Nahid, A., Hakimelahi, G. H.: Helv. Chim. Acta 65, 1229 (1982)
28. Casnati, G., Casiraghi, G., Pochini, A., Sartori, G., Ungaro, R.: Pure and Appl. Chem. 55, 1677 (1983)
29. Casiraghi, G., Casnati, G., Cornia, M., Pochini, A., Puglia, G., Sartori, G., Ungaro, R.: J. Chem. Soc. Perkin Trans. I, 1978, 318
30. Pochini, A., Ungaro, R.: J. Chem. Sci., Chem. Commun. 1976, 309
31. Casiraghi, G., Cornia, M., Balduzzi, G., Casnati, G.: Ind. Eng. Chem. Prod. Res. and Develop 23, 366 (1984)
32. Bigi, F., Casiraghi, G., Casnati, G., Sartori, G., Zetta, L.: J. Chem. Soc., Chem. Comm. 1983, 1210

33. Casiraghi, G., Cornia, M., Casnati, G., Zetta, L.: Macromolecules 17, 2933 (1984)
34. Förster, H.J., Manecke, G.: Makromol. Chem. 129, 165 (1969) and references cited there
35. Gutsche, C.D.: Acc. Chem. Res. 16, 161 (1983); Topics in Current Chemistry 123, 1 (1984)
36. Zinke, A., Ziegler, E.: Chem. Ber. 77, 264 (1944)
37. Cornforth, J.W., Hart, P.D.A., Nicholls, G.A., Rees, R.J.W., Stock, J.A.: Br. J. Pharmacol. 10, 73 (1955)
38. Gutsche, C.D., Dhawan, B., No, K.H., Muthukrishnan, R.: J. Am. Chem. Soc. 103, 3782 (1981)
39. Ninagawa, A., Matsuda, H.: Makromol. Chem., Rapid Commun. 3, 65 (1982)
40. Nakamoto, Y., Ishida, S.: Makromol. Chem., Rapid Commun. 3, 705 (1982)
41. Coruzzi, M., Andreetti, G.D., Bocchi, V., Pochini, A., Ungaro, R.: J. Chem. Soc., Perkin Trans. II 1982, 1133
42. Hayes, B.T., Hunter, R.F.: Chem. Ind. 1956, 193, J. Appl. Chem. 8, 743 (1958)
43. Kämmerer, H., Happel, G., Caesar, F.: Makromol. Chem. 162, 179 (1972)
44. Happel, G., Mathiasch, B., Kämmerer, H.: Makromol. Chem. 176, 3317 (1975)
45. Kämmerer, H., Happel, G.: Makromol. Chem. 179, 1199 (1978)
46. Kämmerer, H., Happel, G., Böhmer, V., Rathay, D.: Monatsh. Chem. 109, 767 (1978)
47. Kämmerer, H., Happel, G.: Makromol. Chem. 181, 2049 (1980)
48. Kämmerer, H., Happel, G., Mathiasch, B.: Makromol. Chem. 182, 1685 (1981)
49. Kämmerer, H., Happel, G.: Monatsh. Chem. 112, 759 (1981)
50. Böhmer, V., Chhim, P., Kämmerer, H.: Makromol. Chem. 180, 2503 (1979)
51. Dhawan, B., Gutsche, C.D.: J. Org. Chem. 48, 1536 (1983)
52. Kämmerer, H., Dahm, M.: Kunststoff Plastics 6, 1 (1959)
53. Hultzsch, K.: Kunststoffe 52, 19 (1962)
54. Moshfegh, A.A., Beladi, E., Radnia, L., Hosseini, A.S., Tofigh, S., Hakimelahi, G.H.: Helv. Chim. Acta 65, 1264 (1982)
55. Högberg, A.G.S.: J. Am. Chem. Soc. 102, 6046 (1980)
56. Högberg, A.G.S.: J. Org. Chem. 45, 4498 (1980)
57. Erdtman, E., Haglid, F., Ryhage, R.: Act. Chem. Scand. 18, 1249 (1964); compare also 123
58. Kämmerer, H., Schweikert, H., Haub, H.G.: Makromol. Chem. 67, 173 (1963)
59. Kämmerer, H., Gros, G.: Monatsh. Chem. 101, 1617 (1970)
60. Gutsche, C.D., Dhawan, B., Levine, J.A., No, K.H., Bauer, L.J.: Tetrahedron 39, 409 (1983)
61. Gutsche, C.D., Levine, J.A.: J. Am. Chem. Soc. 104, 2652 (1982)
62. Bocchi, V., Foina, D., Pochini A., Ungaro, R., Andreetti, G.D.: Tetrahedron 38, 373 (1962)
63. Alfieri, C., Dradi, E., Pochini, A., Ungaro, R., Andreetti, G.D.: J. Chem. Soc., Chem. Commun. 1983, 1075
64. Kämmerer, H., Schweikert, H.: Makromol. Chem. 60, 155 (1963)
65. Andreetti, G.D., Ungaro, R., Pochini, A.: J. Chem. Soc., Chem. Commun. 1981, 533
66. Muthukrishnan, R., Gutsche, C.D.: J. Org. Chem. 44, 3962 (1979)
67. Kämmerer, H., Schweikert, H.: Makromol. Chem. 36, 40 (1959)
68. Kämmerer, H., Hegemann, G., Önder, N.: Makromol. Chem. 183, 1435 (1982)
69. Kämmerer, H., Pachta, J., Ritz, J.: Makromol. Chem. 178, 1229 (1977)
70. Polowinski, S.: Eur. Polym. J. 14, 563 (1979), 11, 183 (1975)
71. Kämmerer, H.: Chemiker-Ztg. 96, 7 (1972)
72. Kämmerer, H.: Proceedings of the Weyerhaeuser Science Symposium "Phenolic Resins, Chemistry and Application," Weyerhaeuser Company, Tacoma, 1981, P. 143
73. Kämmerer, H., Pachta, J.: Colloid. Polym. Sci. 255, 656 (1977)
74. Kämmerer, H., Pachta, J.: Makromol. Chem. 178, 1659 (1977)
75. No, K.H., Gutsche, C.D.: J. Org. Chem. 47, 2713 (1982)
76. Coruzzi, M., Andreetti, G.D., Bocchi, V., Pochini, A., Ungaro, R.: J. Chem. Soc. Perkin Trans. II 1982, 1133
77. Gutsche, C.D., Lin, L.G., Pagoria, P.F.: Unpublished results; see also 35
78. Kämmerer, H., Böhmer, V.: Makromol. Chem. 135, 97 (1970)
79. Böhmer, V., Lotz, W., Pachta, J., Tütüncü, S.: Makromol. Chem. 182, 2671 (1981)

80. Ziegler, E., Zigeuner, G.: Monatsh. Chem. 79, 42, 363 (1948)
81. Ninagawa, A., Cho, K., Matsuda, H.: Polymer Preprints, ACS, Division of Polymer Chemistry, Vol. 24, No. 2, P. 207, ACS 1983
82. Kämmerer, H., Niemann, W.: Unpublished results
83. For the mole refraction of two series of linear oligomers see: Kämmerer, H., Jäger, R.: Makromol. Chem. 81, 78 (1965)
84. Kämmerer, H., Niemann, W.: Makromol. Chem. 169, 1 (1973)
85. Sprengling, G. R.: J. Am. Chem. Soc. 76, 1190 (1954)
86. Cardwell, H. M. E., Dunitz, J. D., Orgel, L. E.: J. Chem. Soc. 1953, 3740
87. Musso, H.: Chem. Ber. 88, 1915 (1955)
88. Zollinger, H.: Helv. Chim. Acta 34, 600 (1951)
89. Musso, H., Matthies, H.-G.: Chem. Ber. 94, 356 (1961)
90. Böhmer, V., Schade, E., Antes, C., Pachta, J., Vogt, W., Kämmerer, H.: Makromol. Chem. 184, 2361 (1983)
91. Böhmer, V., Schade, E., Vogt, W.: Makromol. Chem., Rapid Commun. 5, 221 (1984)
92. Kämmerer, H., Gölzer, E.: Makromol. Chem. 44/45, 53 (1961)
93. Kämmerer, H., Gölzer, E., Kratz, A.: Makromol. Chem. 44/45, 37 (1961)
94. Izatt, R. M., Lamb, J. D., Hawkins, R. T., Brown, P. R., Izatt, S. R., Christensen, J. J.: J. Am. Chem. Soc. 105, 1782 (1983)
95. Imoto, M., Ijiichi, J., Tanaka, C., Kinoshita, M.: Makromol. Chem. 113, 117 (1965)
96. Böhmer, V., Niemann, W.: Makromol. Chem. 177, 787 (1976)
97. Böhmer, V., Stotz, D., Beismann, K., Niemann, W.: Monatsh. Chem. 114, 411 (1983)
98. Böhmer, V., Beismann, K., Stotz, D., Niemann, W., Vogt, W.: Makromol. Chem. 184, 1793 (1983)
99. Böhmer, V., Stotz, D., Beismann, K, Vogt, W.: Monatsh. Chem. 115, 65 (1984)
100. Böhmer, V., Stein, G.: Makromol. Chem. 185, 263 (1984)
101. Stein, G., Böhmer, V., Lotz, W., Kämmerer, H.: Z. Naturforsch. 36 b, 231 (1981)
102. Stein, G., Kämmerer, H., Böhmer, V.: J. Chem. Soc., Perkin Trans. II, 1285, 1984
103. Francis, D. J., Yeddanapalli, L. M.: Makromol. Chem. 119, 17 (1968)
104. Böhmer, V., Schalla, H., Vogt, W.: Unpublished
105. Andreetti, G. D., Ungaro, R., Pochini, A.: J. Chem. Soc., Chem. Commun. 1979, 1005
106. Ungaro, R., Pochini, A., Andreetti, G. D., Sangermano, V.: J. Chem. Soc., Perkin Trans. II 1984, 1979
107. Andreetti, G. D., Pochini, A., Ungaro, R.: J. Chem. Soc., Perkin Trans. II, 1983, 1773
108. Coruzzi, M., Andreetti, G. D., Bocchi, V., Pochini, A., Ungaro, R.: J. Chem. Soc., Perkin Trans. II, 1982, 1133
109. Ungaro, R., Pochini, A., Andreetti, G. D., Domiano, P.: J. Chem. Soc., Perkin Trans. II 1985, 197
110. Paulus, E., Böhmer, V.: Makromol. Chem. 185, 1921 (1984)
111. Perrin, M., Perrin, R., Thozet, A., Hanton, D.: ACS Polymer Preprints, Vol. 24, No. 2, August 1983, P. 163
112. Kämmerer, H., Happel, G.: Unpublished results
113. Gutsche, C. D., Dhawan, B., Levine, J. A., No, K. H., Bauer, L. J.: Tetrahedron 39, 409 (1983)
114. Gutsche, C. D., Bauer, L. J.: Tetrahedron Letters 22, 4763 (1981)
115. Gutsche, C. D., Gutsche, A. E., Karaulov, A. I.: To be published, private communication
116. Munch, J. H.: Makromol. Chem. 178, 69 (1977)
117. Rizzoli, C., Andreetti, G. D., Ungaro, R., Pochini, A.: J. Mol. Struct. 82, 133 (1982)
118. Cairns, T., Eglinton, G.: Nature 162, 535 (1962)
119. Kovac, S., Eglinton, G.: Tetrahedron 25, 3599 (1969)
120. Cairns, T., Eglinton, G.: J. Chem. Soc. 1965, 5906
121. Tobiason, F. L., Houglum, K., Shanafelt, A., Böhmer, V.: ACS Polymer Preprints, Vol. 24, No. 2, August 1983, P. 131
122. Zetta, L., Casiraghi, G., Cornia, M., Kaptein, R.: Macromolecules, in press
123. Canceill, J., Gabard, J., Collet, A.: J. Chem. Soc., Chem. Commun. 1983, 122

5 Resin Production

The versatility of phenolic resins in a broad spectrum of market areas requires the production of phenolic resins in a variety of physical states such as powder or flaked solid resins, solvent based or aqueous solution resins. More recently "in situ dispersions" of phenolics are available. Aqueous, solid, or solvent based resins which are required to meet a number of special criteria are produced from prime raw materials, phenol and formaldehyde, by varying reaction conditions:
– molar ratio of phenol to formaldehyde;
– reaction time and temperature;
– water content and residual phenol;
– modification with other aldehydes and/or substituted phenols;
– etherification and/or dissolution in organic solvents;
– modification with other compounds;
– efficient utilization of phenol;
– environmental, ecological, and toxicological safeguards.
All of these process variables are important depending on product diversity. Besides streamlined resin production, environmental, ecological, and toxicological safeguards are constantly assessed and maintained to provide optimum customer, plant, and community safety.

Resin production is conducted by a discontinuous or batch process. Many continuous processes have been described in the literature[1-5], but are not entirely in operation with the exception of the continuous process for standard novolaks[3] and perhaps particle board and laminate resol[6,7] production. The multitude of resin specifications required by the market and irregular production would render the continuous process uneconomical except in those instances where a single large volume resin can be absorbed into the market. A continuous process is particularly susceptible to the formation of a cured resin coating on the hot reactor surfaces.

Of paramount importance in the batch production of phenolics is the high exothermic reaction which limits batch volume. As resin manufacturers have gained confidence and experience through process R & D fundamentals, batch reactor scale has increased from 3–5 m^3 during the infancy of the industry to the present reactors with capacity up to 60 m^3, primarily for novolaks.

The total reaction enthalpy of the phenol-formaldehyde substitution and the subsequent condensation reaction in acidic medium at a molar ratio of 1:1 was determined experimentally by Manegold and Petzold[8] and Jones[9] among others to be $H_o = 81.1$ kJ/mol and 82.3 kJ/mol respectively. Individual values of 20.1 kJ/mol for the substitution reaction, 78.1 kJ/mol for the condensation reaction and 98.2 kJ/mol for the total reaction have been calculated via combustion enthalpies[10]. The heat gen-

Fig. 5.1. First phenolic resin kettle. (Photo: Bakelite GmbH, D-5860 Iserlohn-Letmathe)

Fig. 5.2. Production plant of phenolic resins. (Photo: Bakelite GmbH, D-5860 Iserlohn-Letmathe)

eration per unit of time and peak temperature depend upon production conditions, the molar ratio of reactants, and catalyst concentration.

The use of aqueous formaldehyde, which is generally the norm, has the advantage that the amount of heat formed by the exothermic reaction is absorbed by the water and can be utilized to heat the contents within the reactor and later be consumed as evaporation heat. This procedure prevents an uncontrolled and sometimes a potentially explosive reaction. The danger of such a reaction can be serious, if the aqueous formaldehyde solution is partly substituted by paraformaldehyde to either increase batch capacity or reduce energy consumption or shorten the distillation step.

Reactors are double walled, closed vessels which are jacketed into separate heating and cooling sections (Fig. 5.2). Stainless steel alloy reactors and auxiliary equipment are used; sometimes nickel clad reactors are utilized. These particular materials do not promote the discoloration of resins. Copper and copper alloys also exhibit good resistance to phenol but lead to discoloration. Detailed description of polymerization reactors regarding theory and applications to various polymer systems is summarized by Gerrens[11].

There is a notable distinction between pure and impure phenol when corrosive behavior is considered (Table 5.1). Very pure phenol does not attack high-alloy ferritic and austenitic stainless steels even at its boiling point[12]. Structural parts under stress consisting of unalloyed steel are susceptible to stress crack corrosion, especially at the weld seams[13]. Reactor corrosion occurs more readily with liquid phenol than phenol vapor. The corrosion rate is enhanced significantly at low pH. Minor amounts of carboxylic acid and water affect reactor corrosion.

Several production examples of various phenolic resins are provided in the literature[5, 14, 15].

Table 5.1. Corrosion resistance of some materials to pure
and used phenol at 240 °C[9]

Material	Corrosion rate mm/year at 240 °C	
	Pure phenol	Used phenol
Unalloyed steel	0.76	1.09
18/8 Cr-Ni-steel	0.00	0.00
Nickel 99.4	0.00	0.79
Aluminum	56.00	–

5.1 Novolak Resins

Oxalic acid, MP 101 °C, as dihydrate is the predominate catalyst for the manufacture
of novolaks. Oxalic acid sublimes in vacuum at about 100 °C and at normal pressure
at 157 °C without decomposition. At a higher temperature (180 °C) it decomposes to
carbon monoxide, carbon dioxide and water so that a removal process is unnecessary.
Due to its reducing behavior very light resins are obtained. Other less frequently used
acid catalysts are HCl, H_2SO_4, toluene sulfonic acid and H_3PO_4. Historically hydrochloric acid was used because of its low cost but is being used sparingly because
of its high corrosiveness. It should be noted that the use of hydrochloric acid in
phenolic resin manufacture can result in the generation of a hazardous intermediate.
When formaldehyde and hydrochloric acid are present in the gas phase in concentrations of more than 100 ppm, 1,1-dichlorodimethyl ether is obtained, a highly hazardous and carcinogenic compound. Maleic acid has been recommended for the production of high melting novolaks.

The molar ratio of phenol to formaldehyde is normally within the range of 1 : (0.75–
0.85). The influence of the P/F molar ratio and catalyst type on MWD is shown in
Fig. 3.6, Chap. 3. While novolaks with a MP of 70–75 °C are used for the production
of foundry resins for shell cores, those with MP's between 80–100 °C are used for all
other application areas. These novolaks are flaked to an appropriate size, mixed with
HMTA, and ground to powder in special mills and processed in this form.

In the batch process[16] phenol, which is stored in tanks of alloyed steel at approximately 60 °C, is transferred to the reaction vessel via a scale and heated to 95 °C.
After catalyst addition formaldehyde solution is introduced with stirring at a rate so
that the mixture is boiling gently. When all the formaldehyde is added, the temperature is maintained until the formaldehyde is consumed. Then, the water is removed
at normal pressure and, by further heating to 160 °C, with vacuum the unreacted
phenol. Removal of volatiles can be facilitated by introducing steam. As soon as the
desired melting point is reached, the resin is transferred to a heated vessel and then
flaked on a continuous cooling belt (Fig. 5.3).

In the continuous process[3, 16, 17] shown in Fig. 5.4 formaldehyde, phenol, and catalyst are transferred from appropriate storage tanks to the first stage reactor, where
their quantity is automatically measured and controlled. In this stirred reactor, equipped with an external heating jacket, the reaction between phenol and formaldehyde
commences. The reaction is continued and completed in a second stage reactor. The

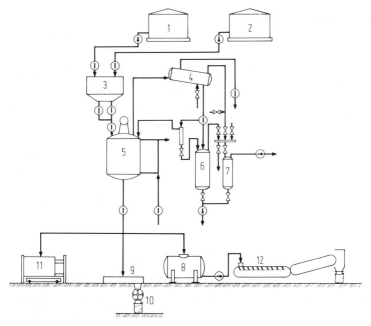

Fig. 5.3. Phenolic resin production plant, batch process. (Drawing: Bakelite GmbH, D-5860 Iser-lohn-Letmathe). *1* Phenol; *2* Formaldehyde; *3* Scale; *4* Condenser; *5* Reactor; *6* Condensate receiver; *7* Vacuum; *8* Resin receiver; *9* Resin trough; *10* Mill; *11* Cooling carriage; *12* Cooling belt

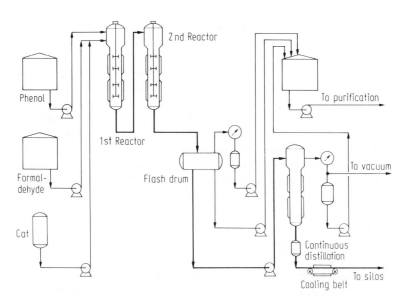

Fig. 5.4. Continuous novolak resin production process[17)] (Drawing: Euteco SPA, I-20161 Milano)

reaction is carried out under pressure to 7 bar at a temperature range between 120 °C and 180 °C to enhance the reaction rate. The reacted mixture leaving the second stage reactor is conveyed into a flash drum, acting also as a vapour-liquid separator. The flashed vapour is condensed, collected in a receiver and sent to the purification section while the liquid phase separating on the bottom of the flash drum settles in two layers. The upper, aqueous layer containing small amounts of phenol, is withdrawn and sent to the purification section, whilst the lower layer, i.e. the resin layer, is pumped for further treatment.

Since the resin still contains small portions of water, it must be removed. This operation is carried out in a vacuum evaporator of special design from which two streams are obtained: the overhead vapor, essentially consisting of water with very small amounts of phenol, is condensed and the other streams combined with other materials for disposal in the purification section, while the bottom stream, i.e. dewatered resin, is fed to a belt flaking machine.

A phenol novolak may be prepared on a laboratory scale according to the following procedure[18]:

"A 1,000 ml resin kettle is charged with 130 g of phenol (1.38 mole), 13 ml of water, 92.4 g of 37% aqueous formaldehyde (1.14 mol) and 1 g of oxalic acid dihydrate. The mixture is stirred and refluxed for 30 min. An additional 1 g of oxalic acid dihydrate is then added, and refluxing is continued for another hour. At this point, 400 ml of water are added and the mixture cooled. The resin settles for 30 min and the upper layer of water decanted or withdrawn through a siphon. Heating is then begun with the condenser modified for vacuum distillation. Water is distilled at 50–100 mm pressure until the pot temperature reaches 120 °C or until a sample of the resin is brittle at room temperature. About 140 g of resin are obtained."

Generally the yield does not exceed 105% based on phenol weight for novolak resin production.

Water has a remarkable influence on the plasticity of novolaks[19]. Only 1% water content can reduce the MP of novolaks with a medium MW between 450 and 700 by an average of 3.4 °C. This behavior was evaluated within the range of 0.1 to 4% water[20]. The effect on the melt viscosity is even more pronounced. It was reduced by 90% by increasing the water content from 0.1 to 3.1% at 120 °C (MW 450). As expected, the effect of water as an internal lubricant in low concentrations is especially high. The addition of 0.5% water (from 0.2 to 0.7%) reduces the melt viscosity by 50%. The reactivity of novolaks to HMTA increases with increasing water (and free phenol) content, so that the flow distance (a measure for the melt viscosity and reactivity) is reduced inspite of decreasing melt viscosity[20]. The influence of free phenol is less severe. The MP is lowered according to MW between 2.1–2.9 °C/1% of free phenol. The influence on the melt viscosity is quite significant at high MW[21].

5.2 Resols

The alkaline PF reaction is more versatile and related to the molar ratios, catalysts, and reaction conditions applied than the production of novolaks. PF molar ratios are typically between 1:1 and 1:3. The type of catalyst has a greater influence on molecular structure and MWD than in the acid/novolak system (Fig. 5.5). Sodium hydroxide, sodium carbonate, alkaline earth oxides and hydroxides, ammonia, HMTA and tertiary amines are the most important catalysts. The catalyst performance/removal, if required, and costs are the main criteria for catalyst selection. Sodium hydroxide and carbonate usually remain in solution. The oxides and hydroxides of calcium and barium can be precipitated as sulfates or carbonates by addition of sulfuric acid or carbon dioxide at the end of the reaction and separated. Tertiary amines, i.e. triethylamine, can be separated by distillation. The catalyst separation is necessary, if dielectric properties, ageing- and humidity resistance are required. The molar catalyst ratio based on phenol as applied in commercial resols is quite diverse. It ranges from 1 : 1 to 1 : 0.01.

Exact temperature and time control are required. Adequate vacuum, at least 50 mbar, and a cooling water temperature as low as possible, are necessary to maintain a low resin temperature (max. 60 °C) and a relatively brief distillation period. Distillation is terminated when the desired resin content is obtained. The viscosity of the resin can be regulated, if necessary, by post-condensing at approximately 70 °C. The resin is then quickly cooled to room temperature. A partly cured resin coat gradually forms on the surface of the reactor wall, reducing the heat transfer and prolonging the time of distillation. The reactor and auxiliary equipment must, therefore, be thoroughly cleaned after a certain number of batches.

The following procedure is recommended for the production of resol on a laboratory scale[18]:

"A 500 ml resin kettle is equipped with a reflux condenser, stirrer, thermometer, and siphoning tube leading to a collecting trap for the removal of samples for testing. To the reaction vessel is added 94 g (1 mol) of distilled phenol, 123 g aqueous formaldehyde 37% by weight (1.5 mol) and 4.7 g barium hydroxide octahydrate. The reaction mixture is stirred and heated in an oil bath at 70 °C for 2 h. Two layers form if stirring is stopped. Sufficient 10% sulfuric acid is added to reduce the pH to 6–7. Vacuum is then applied by means of a water aspirator (pressure regulated at about 30–50 mm) and water is removed through the condenser which is now modified for distillation. The temperature should not exceed 70 °C. Samples are withdrawn every 15 min through the vacuum siphon apparatus and tested for gel time by working with a spatula on a hot plate at 150 °C" (Sect. 7.5.2).

A wide range of resol compositions is obtained depending on the type, amount of catalyst, and molar ratio as well as reaction conditions. A low molecular weight liquid phenol resol is used to impregnate paper for laminates to reduce their water absorption (Resol A, Fig. 5.5). However, different requirements are necessary for plywood glues. The resin must wet very well, but not penetrate too deeply into the porous structure of the wood. This would cause high glue consumption and render the adhesive process uneconomical. The molecular weight must, therefore, be considerably increased by post reaction (Resol B, Fig. 5.5).

High melting resols of phenol and cresols made with ammonia as catalyst are of special importance, such as in the manufacturing of laminates, coatings, and brake linings. The uniform production of solid resols requires a high degree of experience. After the raw materials are charged in the reactor, composition is carefully heated to the boiling point and the temperature is maintained for about 50 min. Then, distillation is carried out under vacuum at approximately 50 °C. As soon as water is removed, the temperature of the highly fluid resin is gradually increased under vacuum to 95 °C,

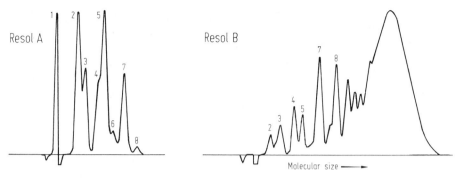

Fig. 5.5. Molecular weight distribution of resols produced under different reaction conditions. *1*=Ph; *2*=*o*-HMP; *3*=*p*-HMP; *4, 5*=2,4- and 2,6-BHMP; *6*=DPM Der.; *7*=THMP; *8*=DPM derivative

whereby the MP of the resin gradually rises due to the condensation reaction. As soon as the desired MP is reached, which is limited by the reactivity of the resin, the resin is quickly discharged and cooled. Since the MP-determination takes considerable time, the end point is determined by experience. If the exact point of time is missed, the temperature and MP of the resin continues to increase due to condensation reaction. It becomes a very tough and hard amorphous material which can only be removed from the reactor by the use of a pneumatic hammer.

5.3 High-*Ortho* Resins

Phenolic resins with a significantly high amount of *ortho-ortho'*-orientation are prepared at an intermediate pH range of 4–7 (see Chap. 3). The unique properties of a wholly *ortho* substituted phenolic resin are primarily high cure rate with HMTA and attractive rheological behavior. Both features are desirable in selected applications such as foundry (Chap. 16), molding materials (Chap. 12) and RIM (Chap.10).

5.3.1 Novolak

Conventional novolak equipment is utilized to manufacture high-*ortho* novolaks except mildly acidic catalysts such as divalent metal acetates (Ca, Mg, Zn, etc.) are used with a molar ratio of phenol to formaldehyde in the range of 1:0.8. Formaldehyde is added during the course of the reaction. Usually the reaction/condensation and high temperature distillation (to 145 °C) is conducted in two stages to minimize any exotherm or gelation that can occur due to the use of higher concentration of formaldehyde solutions (50%). A lower effective utilization of phenol occurs or 90–100% as compared to 105–110% for typical novolaks.

Newer processing techniques have been developed for the preparation of high-*ortho* novolaks and liquid resols (see following section). A common innovation[22] for either materials is the use of toluene or xylene as an azeotropic solvent. Constant removal of water provides more favourable control of reaction exotherm and rate enhancement. Other features include dual catalyst system and split feed of formaldehyde.

Table 5.2. Isomers ratios in novolak resins

	o-o'	p-p'	o-p
Conventional novolak	10%	45	45
Hybrid	33%	25%	42
High-*ortho*	60–67	–	40–33

Other novel methods for high-*ortho* novolaks have been reported and consist of conducting the reaction of phenol and paraformaldehyde in an aprotic non polar media (xylene) at 170–220 °C for 12 h[23] or by the reaction of selected metal phenolates with paraformaldehyde[24].

More recently high-*ortho* novolaks based on phenol and acetaldehyde via non-transition metal and "hydrogen bond assisted regio and stereo controlled conditions" have been described[25]. The metal phenolate reacts in a nonpolar aprotic solvent with acetaldehyde to yield the high-*ortho* resin. Molecular chelation with high-*ortho* specificity is proposed.

"Hybrid" novolak resins have also been described[26]. These materials are prepared in a two step manner whereby the first step is a normal (low pH) acid catalyzed novolak preparation followed by adjustment of the pH to 5–6. Additional phenol and formaldehyde are then introduced along with a catalyst selected from Mn, Mg, Cd or Co salts and heated until reaction is completed. Removal of volatiles (water, phenol) yields a solid hybrid resin whose methylene substitution pattern is intermediate between conventional novolak and high-*ortho* content (Table 5.2). High-*ortho* resins approaching 100% o-o can be prepared.

5.3.2 Liquid Resol

Similar divalent metal salt catalyst systems are used[27] in the preparation of high-*ortho* liquid resols. Phenol to formaldehyde ratio varies from 1:1.5–1.8. Aqueous formaldehyde (or paraformaldehyde) is incrementally added while water is concurrently removed by the azeotropic solvent. The slower reaction (as compared to typical strong base catalyzed resols) requires 6–8 h within a temperature range of 100–125 °C for reaction/dehydration to resols with water content of below 2%. The high-solids resols are lower in bulk viscosity with values reported between 1,000–10,000 mPa · s. These resins contain hemiformal groups (Sect. 3.2) which are attributable to formaldehyde reaction with phenolic hydroxyl group and methylol groups. The neutral resins are more stable under ambient storage conditions than the corresponding caustic catalyzed resols.

5.4 In Situ Dispersions

A recent technical advance[28] of considerable importance is the preparation of PF resins as stable dispersions or discrete particles. Normally during the early stages of a typical base catalyzed phenol formaldehyde reaction, the low MW PF resin is water soluble, and as reaction proceeds, the resin becomes insoluble due to increased MW leading to phase separation. The "in situ dispersion" technique provides a method to manufacture high MW PF resol resins as stable dispersion. Depending on formulation and manufacturing conditions, the resulting phenolic compositions can be obtained as stable aqueous dispersions or free flowing discrete PF particles. The relatively high free-phenol content of these resins (up to 6%) can be a disadvantage.

5.4.1 Aqueous Dispersions

The manufacture[29] of stable aqueous resol dispersion of phenol or substituted phenols with formaldehyde requires the use of e.g. polysaccharide-type protective colloid with a base catalyst. It is primarily a two stage reaction with the first stage consisting of phenol formaldehyde condensation to a specific MW followed by transformation of an aqueous phenolic solution to a phenolic-in-water dispersion. The protective colloid is added during phase change and provides the necessary stabilization of the phenolic particles without agglomeration. Reaction continues to the desired gel time. Solids content are 40–50% with dispersion viscosity ranging between 5–8,000 mPa·s. Selective protective colloids like gum arabic, gum ghatti, and hydroxyalkylated gum are effective. Fine particle size and narrow size distribution dispersions are obtained with gum arabic and ghatti. The effect of pH is noteworthy. Intermediate or neutral pH allows optimum dispersion stability or minimal sedimentation.

5.4.2 Solid Resols

The type and concentration of the protective colloid coupled with a rapid sedimentation rate allows the isolation of discrete, uniform spherical particles of solid resol from the HMTA catalyzed reaction of P and F in water[30]. These particles have been named "phenolic thermospheres" or PTS. By conducting the PF condensation in a dispersion medium, the polymerization is conducted in a low viscosity medium (water) that provides excellent heat transfer, control of gel time and uniform product composition. The solid resol particles (10 μm average) are easily filtered, have reduced moisture absorption, improved sinter resistance and higher Tg. Table 5.3 illustrates a comparison between PTS, typical bulk solid resol, and a novolak.

Particularly efficient colloids are gum arabic and 88% hydrolyzed polyvinyl alcohol. The latter is more effective in providing uniform particle size and distribution. Physical characterization of PTS by GPC indicates that polymer MWD is comparable to conventional resols in spite of the disparity in reaction conditions and product iso-

Table 5.3. Moisture behavior and Tg of phenolic resins[28]

Resin	%-Wgt gain[a]	Tg · °C[b]	Appearance
PTS	3	48–54	Slightly sintered
Resol	8	33–43	Fused
Novolak	6	58–74	Fused

[a] 24 h saturated moisture exposure at 22 °C
[b] Dynamic mechanical analysis

lation. Polymer microstructure characterization by ^{13}C-NMR shows that the particulate resol contains less methylol and methylene ether structures but increased amount of benzylamine and methylene bridges. These subtle microstructure differences are responsible for reduced moisture absorption and high sinter resistance or Tg (Table 5.3).

Both types of dispersion products are being actively examined in a variety of applications such as polymer alloys, friction elements, fiber bonding, wood bonding, coated abrasives, "in situ" method of phenolic molding compounds preparation, and friction granules. The latter application is particularly remarkable in that a totally synthetic friction granule[31] can be prepared without the traditional skin irritating cashew nut shell liquid mixture.

5.5 Dust Explosions

For most solid resin applications, the solid, flaked or lump phenolic resins (resol or novolak) must be pulverized into fine particulate materials with a desirable size distribution, e.g. for phenolic molding compounds, fiber bonding, grinding materials, and brake linings. Most finely divided combustible materials suspended in air are hazardous. These extremely fine materials below 500 mesh (less than 75 μm) or dusts are susceptible to dust explosions. The hazard of any dust is related to ease of ignition, maximum rate of pressure rise and maximum explosion pressure of the ensuing explosion[32]. US Bureau of Mines has conducted many studies which are quite useful in measuring the explosive hazard. Terms such as ignition sensitivity (IS), explosion severity (ES) and explosivity index (EI) (= product of IS · ES) have been identified and are related to the type of explosion (Table 5.4).

Table 5.4. Characterization of explosion hazards

Type of explosion	IS	ES	EI
Weak	0.2	0.5	0.1
Moderate	0.2–1.0	0.5–1.0	0.1– 1.0
Strong	1.0–5.0	1.0–2.0	1.0–10
Severe	5.0	2.0	10

Table 5.5. Explosion characteristics of phenolic and related materials

Type of dust	EI	IS	ES	Minimum explosion concentration g/m³
Resol	10	7.9	5.3	40
Novolak	10	13.9	4.0	25
PF molding compound/wood flour	10	8.9	4.7	30
UF molding compound/grade II fine	1.0	0.6	1.7	85
Phenol/furfural resin	10	15.2	3.9	25
MF laminating resin	0.1	0.1	0.2	85
HMTA	>10	32.7	5.6	15

There are further distinctions which add to the peril of dust explosions. These include reduced particle size (below 325 mesh), concentration (weight per unit volume), and maximum oxygen concentration. Besides US Bureau of Mines, Fire and Explosion Hazards Testing Services are available[33] to identify a variety of evaluation facilities and capabilities.

Dusts of many thermosetting resins, molding compounds, and HMTA (Chap. 2) have been examined and are summarized in Table 5.5.

It is apparent by inspection of the above data that dusts of PF-resins and compounds represent severe explosion hazards and necessary precautions must be enforced to avoid severe dust explosions[34, 35].

5.6 Spray Dried Resins

Resol resins in general are too reactive to be transferred to the solid state by distillation. The removal of water from a heat sensitive material (resol) can be accomplished by a spray drying operation. However, the scope of resins capable of spray drying is limited. At present, NaOH catalyzed resols are spray dried[36, 37] to be used in waferboard and plywood production. A spray drier is a large, usually vertical chamber through which a hot gas is introduced and into which a liquid, slurry or paste is sprayed by a suitable atomizer. All emerging droplets depending on size (10 μm– 1 mm) must be completely dried with no tack before they strike the chamber wall. Special flat bottom spray driers with a pneumatic powder collector specially designed to handle heat sensitive materials are available. These resins are easily redispersible in water. Considerable savings in transportation cost can be achieved.

Additives are used[36] during resin production to avoid sticking to caul plates causing cure and cleanup problems for waferboard production, and to improve handling and storage properties of the resin.

Features of spray dried resols resins are better stability, redilutability, shorter press time and lower temperature, and transportation economics. Phenolic microballoons, a spray dried and completely crosslinked phenolic material, has been available for many decades.

5.7 References

1. Biesterfeld & Stolting: DE-PS 758 436 (1939)
2. Rütgerswerke AG: DE-PS 965 923 (1954)
3. Societa Italiana Resine S.P.A.: US-PS 3,687,896 (1972)
4. Perstorp AB: US-PS 1,460,029 (1976) Deutsche Texaco AG: DE-OS 25 38 100 (1977)
5. Hesse, W.: Phenolharze. In: Ullmanns Encyclopädie d. techn. Chem. Vol. 18, 4. Ed., Verlag Chemie, Weinheim 1979
6. Formica Ltd.: US-PS 4,413,113 (1983)
7. VEB Sprela Werke: DE-OS 31 13 554 A1 (1982)
   VEB Sprela Werke: DE-OS 32 26 114 A1 (1983)
   VEB Sprela Werke: DE-OS 32 26 115 A1 (1983)
8. Manegold, E., Petzold, W.: Kolloid.-Z. 94, 284 (1941)
9. Jones, T. T.: J. Soc. Chem. Ind. (Lond.) 65, 264 (1946)
10. Vlk, O.: Plaste u. Kautschuk 4, 127 (1957)

11. Gerrens, H.: Chem. Tech. 380, 434 (1982)
12. Donndorf, R. et al.: Werkstoffeinsatz und Korrosionsschutz in der chemischen Industrie, Leipzig: VEB Deutscher Verlag für Grundstoffindustrie, 1973
13. Dunn, C. L., Liedholm, G. E.: Oil Gas J. 51, 560 (1952)
14. Sandler, S. R., Caro, W.: Polymer Syntheses, Vol. 2, Chap. 2. Phenol-Aldehyde Condensations, New York: Academic Press 1977
15. Wegler, R., Herlinger, H.: Polyaddition u. -condensation v. Carbonyl- u. Thiocarbonylverbindungen. In: Houben-Weyl: Methoden d. Org. Chem. Vol. XIV/2 P. 272–291; Chem. Abs.: Phenol-Condensation Products, 8th Collect. Vol. 1967–1971, 9th Collect. Vol. 1972–1976
16. N. N.: Modern Plastics Int., April 1981, p. 45
17. Euteco S. P. A.: Euteco Continuous Process, Technical Bulletin
18. Sorenson, W. R., Campbell, T. W.: Preparative Methods of Polymer Chemistry, 2nd Ed. New York: Interscience 1968
19. Jones, T. T.: J. Appl. Chem. 2, March 1952, 134
20. Čiernik, J.: Chemický prûmysl, 25/50, 260 (1975)
21. Čiernik, J., Fiala, Z.: Chemický prûmysl, 25/50, 374 (1975)
22. Monsanto: US-PS 4,112,700 (1978)
23. Casiraghi, G., et al.: Makromol. Chem. 182, 2973 (1981)
24. Casiraghi, G., et al.: Makromol. Chem. 182, 2151 (1981)
25. Casiraghi, G.: Polym. Preprints 24 (2) 183 (1983)
26. Pacific Resins and Chemicals Inc.: US-PS 4,403,076 (1983)
27. Union Carbide Corp.: US-PS 4,403,066 (1983)
28. Brode, G. L., Jones, T. R., Chow, S. W.: Chem. Tech. 676 (1983)
29. Union Carbide Corp.: US-PS 3,823,103 (1974)
    Union Carbide Corp.: US-PS 4,026,848 (1977)
    Union Carbide Corp.: US-PS 4,039,525 (1977)
30. Union Carbide Corp.: US-PS 4,206,095 (1980)
31. Union Carbide Corp.: US-PS 4,316,827 (1982)
    Union Carbide Corp.: US-PS 4,420,571 (1983)
32. Schwab, R. F.: "Dust Explosions" In: McKetta, J. J. Ed.: Encyclopedia of Chemical Processing and Design, Vol. 17, p. 67, Marcel Dekker Inc., New York 1983
33. N. N.: Chemical Processing, March 1984, p. 56.
34. Clark, S.: Chem. Eng., May 1984, p. 153
35. N. N.: Manufacturing Chemist, March 1984, p. 67
36. Reichhold Chemicals Ltd.: US-PS 4,098,770 (1978)
37. Borden: US-PS 4,424,300 (1983)

6 Toxicology and Environmental Protection

There are potential risks associated with handling or use of chemical starting materials that are transformed into resinous products and ultimately cured products. The burgeoning use of enormous amounts of natural and chemical raw materials for energy, transportation, agriculture, and industry has had a debilitating effect on air/water quality, the environment, and even inhabitants. Most highly developed nations require vast amounts of energy due to technological advancements. These factors were instrumental in many countries adopting necessary legislation to improve the health, well-being of the citizens as well as the environment.

In the USA, the Environmental Protection Agency (EPA) was established to protect individual quality of life and the environment (air and water quality). Within the EPA, the U.S. Occupational Safety and Health Administration (OSHA) monitors worker safety; National Institute for Occupational Safety and Health (NIOSH) provides standards. The responsible authorities in West-Germany are the Umweltbundesamt (UBA), Bundesgesundheitsamt (BGA), and Bundesanstalt für Arbeitsschutz und Unfallforschung (BAU).

The US Congress enacted several related series of legislative acts that dealt with many of these areas: Clean Air Act (air quality and laws and regulations), Clean Water Act for control and reduction of water pollutants; Toxic Substances Control Act (TSCA) for premarket notification of new chemicals introduction; Superfund Act authorizing cleanups and remedial actions to protect public health and the environment from the harm that can result from release of hazardous wastes; Consumer Product Safety Commission-agency determining consumer product safety or hazards.

Similar actions have occurred on a global basis primarily in those highly developed countries faced with similar ecological, toxicological, and environmental problems.

To judge the risks connected with the handling of phenolic resins (during manufacture, formulation or fabrication into finished products), a clear distinction must be between starting materials (P, F), phenolic resin oligomers, and cured phenolic resin. Aside from compositional characteristics, the MW of the resin is a convenient or necessary parameter that aids in categorizing the physiological effects of resins. Low to medium MW resins possess varying content of free phenol and formaldehyde. High MW resins, primarily novolaks, are formaldehyde free and low in free phenol whereas solid resols possess small amounts of formaldehyde and phenol. Cured phenolic resins are completely innocuous. FDA permits phenolic coatings and phenolic molded articles to come in contact with food.

Toxicological characteristics of phenols and formaldehyde raw materials are discussed below. Many other chemicals are reacted or combined with P, F in resin manufacture, formulations, coatings, molding compounds, etc. and are beyond the scope of this book in describing their respective toxicological behavior. Reader is referred to Sax[1] or Federal Register 1975 (U.S.). The latter is revised periodically to update changes.

6.1 Toxicological Behavior of Phenols

Mononuclear phenols are highly toxic compounds causing protein degeneration and tissue erosion. The oral LD_{50}-value (rats) is 530 mg/kg[1]. Phenol has a debilitating effect on human skin. Immediately on contact, skin becomes white, subsequently red with blister like appearance; a strong burning sensation is experienced. Either solid or liquid phenol is quickly absorbed by the skin and causes very severe damage. Contact with large amounts of phenol results in death due to paralysis of the central nervous system. Minor ingestion causes damage mainly to the kidneys, liver, and pancreas. If phenol is inhaled or swallowed, local cauterization occurs with headaches, dizziness, vomiting, irregular breathing, respiratory arrest, and heart failure.

The degenerating effect of phenol on skin is reduced by the introduction of lypophilic groups (methyl-, higher alkyl-, or chloro groups). Neutral molecules are far more toxic than the corresponding ionic species. The biological activity of phenols is the result of their ability to alter biological structures, i.e. attack cell membranes resulting in cytochrome diffusion[2]. Cresols are similar to phenol in activity but less severe (Table 6.2). An excellent review summarizing environmental and health risks of phenol is recommended[3].

6.2 Toxicological Behavior of Formaldehyde

Aqueous formaldehyde is a protoplasmic poison with a cauterizing and protein degenerating effect. It causes irritation on contact with the skin, eyes, nose, throat, or lungs. The use of aqueous formaldehyde to preserve medical or biological specimens is well known. Formaldehyde is believed to attack bacteria by reacting with protein amino groups thus altering the nature and activity of these groups. Formaldehyde in the organism is partly exhaled or rapidly metabolized into carbon dioxide and formic acid and the remainder excreted in the urine.

Gaseous formaldehyde is very irritating to the mucous membranes; the pungent smell is noticeable below 1 ppm. Formaldehyde is mutagenic in bacteria, funghi, and certain insects. Studies performed[4, 5] with rats and mice indicate that nasal tumors developed when exposed to very high concentrations during a long period of time. The final report of the Chemical Industry Institute of Toxicology (CIIT)/Battelle was[6] issued on February 16, 1982. In this study rats and mice were exposed to 15, 6, and 2 ppm of formaldehyde gas six hours per day, five days a week for twenty-four months. Nasal cancer mainly squamous cell carcinoma was seen in 106 rats (of the 240) and 2 mice at the 15 ppm level and 2 rats at 6 ppm. A statistically significant increase was only found with the rats exposed to 15 ppm; all other nasal tumors were not significantly increased. No significant effect upon mortality was seen in the rats at 2 ppm or for all of the exposed mice. In a Formaldehyde Institute (established in 1979 by a group of companies that produce or utilize formaldehyde) sponsored sub-chronic inhalation study[7] at Bio/dynamics, rats, monkeys, and hamsters were exposed to 3, 1, and 0.2 ppm formaldehyde gas 22 h per day for 6 months. The main finding was that squamous cell metaplasia occurred in the nasal cavity of rats and monkeys at 3 ppm. No tissue damage was seen in rats and monkeys at 1 or 0.2 ppm and in hamsters at any of the levels tested. This supplies additional evidence that at 1 ppm and below no significant effects occur. Concentrations above are known to be extremely irritating to the eyes and respiratory tract in humans, and individuals would not tolerate these levels of exposure for more than a few minutes.

Several papers concerning the mechanism of formaldehyde induced carcinogeneity have been published[8-12].

None of the epidemiological studies of persons occupationally exposed to formalde- hyde have shown an excess in nasal or respiratory tract cancer – studies indicate that toxic effects are to be expected only in those tissues directly exposed to formaldehyde – or even a significant increase in any other form of cancer attributable to formal- dehyde. Mortality or epidemiological studies conducted by Du Pont[13] and BASF[14] and engaged in the manufacture of formaldehyde for many decades showed no excess of cancer deaths for those employees who had worked with formaldehyde as compared to other employes indicating the long history of safe use.

In the USA the Formaldehyde Institute jointly with the National Cancer Institute are conducting an epidemiological study of 17,000 workers exposed to formaldehyde to determine if there is a link between formaldehyde exposure and the incidence of cancer in humans. Completion of the study was expected in 1984.

The recent report[15] issued by the European Chemical Industry Ecology and Toxicology Centre (ECETOC) claims that the nasal cancers in experimental animals develop only at concentrations which produce chronic tissue irritation and that there is no relationship between formaldehyde exposure and cancer in humans.

In the USA, OSHA has set a 3 ppm limit for formaldehyde exposure during an 8 h time weighted average (TWA) with a ceiling concentration of 5 ppm. In Western Eu- rope exposure levels in the work facility vary from country to country with a threshold limit value (TLV) of 1 ppm in many European countries (Table 6.1).

However, the discussion concerning the rating of formaldehyde is still controversial and significally lower threshold limit value is expected[16].

The toxicological effects of formaldehyde extend beyond resin manufacture. Con- sumer products based on formaldehyde resins (urea, melamine or phenol) release for- maldehyde into the atmosphere in varying amounts. The US consumer Product Safety Commission banned UF foam insulation 1982; this ban was later rescinded by a fed-

Table 6.1. Formaldehyde exposure levels in the work facility

	Formaldehyde (TLV)
USA	3 (TWA)
UK	2
West Germany	1
Switzerland	1
Scandinavia	1
USSR	0.4

eral court. Composite wood products (particle board, plywood, and panelling), fiberglass insulation materials, and other formaldehyde treated products have been examined by a modification of the JIS desiccator procedure. The highest measurable rates of formaldehyde emissions were noted[17] for urea/formaldehyde resin bonded wood products (see Chap. 11).

6.3 Environmental Protection and Regulation

Aside from the toxic effects and necessary safeguards in handling of phenol/formaldehyde raw materials, air/water quality, worker safety and ecology have been of utmost concern to many nations by implementing controls in eliminating the wanton disposal or emission of hazardous materials. The various "Acts" mentioned previously have provided adequate timetables for industry to implement the required changes. Changes have occurred in resin processes and product modifications by resins manufacturers as well as customers in regard to handling reaction by-products, effluents, and emissions.

Most resin manufacturing preparations require a condensation/distillation step whereby aqueous distillate contains varying amounts of phenol and formaldehyde. With the passage of the Clean Water Air Act (1977) permissible levels of phenols in waste water (distillate) were established in the USA by the EPA, and published in the Federal Register. It provided the guidelines as to the quantity of phenol (0.1 mg/l) permissible for discharge via Best Practical Treatment (BPT)[18]. Lower amounts (0.02 mg/l) are proposed by Best Available Technology (BAT) that is reasonable and affordable to the company in 1984[19].

Even relatively low concentrations of phenols below 10,000 µg water are fatal to several species of fish after 1–3 days (Table 6.2). Lower concentrations are injurous

Table 6.2. Phenols in water[20]: Threshold odor and taste concentration and acute fish toxicity

	Threshold odor concentration µg/1	Threshold taste concentration µg/1	Acute fish toxicity LC_{50} µg/1
Phenol	300	4000	5000
o-Cresol	3	1400	2000
m-Cresol	2	800	6000
p-Cresol	2	200	4000
2,3-Xylenol	30	500	–
2,4-Xylenol	500	400	28000
2,5-Xylenol	500	400	2000
3,4-Xylenol	50	1200	4000
3,5-Xylenol	–	5000	50000
p-tert-Butylphenol	30	800	–
Resorcinol	2000	6000	–
o-Chlorophenol	0.1	10	8000
m-Chlorophenol	0.1	50	3000
p-Chlorophenol	0.1	60	3000

Table 6.3. Comparison of TSCA and 6th Amendment (EEC)

	U.S. TSCA	EEC 6th Amendment
Covers existing and new chemicals	Yes	New chemicals only
New chemical notification	Premanufacturing	Premarketing
Applies to raw materials, intermediates, precursors	Yes	No
Emphasizes packing/labelling with toxic criteria/classification	No	Yes
Inventory of existing substances	Yes	Yes
Applies to imports	Yes	Yes

and adversely affect the taste of fish resulting in unacceptability for human consumption. Phenols in chlorinated water lead to the formation of chlorophenols which impart an objectionable odor and taste in water below 0.01 mg/l.

Under certain circumstances waste water may be discharged into municipal biological water treatment plants together with household sewage[21]. In any case, the municipal requirements must be observed; the composition of the waste water should not affect biological processes or vegetation.

The disposal of solid waste material is subject to local, state, and federal regulations depending whether it is innocuous or hazardous. Recognition of abandoned hazardous waste sites and its unknown perils prompted the enactment of Comprehensive Environmental Response, Liability, and Compensation Act in 1980 or what is commonly referred to as "Superfund Act."

Disposal of waste materials containing hazardous substances must be deposited at official disposal sites[22]. Flammable waste is preferably destroyed in appropriate incinerators at the plant. Studies have shown that detectable quantities of phenol which may elute from resin bonded sands (foundry) are even lower than those amounts determined for household waste under similar circumstances[23].

The U.S. Toxic Substances Control Act of 1976 or TSCA and the European Economic Community's (EEC) version or "the Sixth Amendment" apply to existing and new chemicals[24]. There are some differences and similiarities between both acts (Table 6.3).

The EEC 6th Amendment places' strong emphasis on packaging/labelling with the establishment of toxic criteria/classification procedure resulted in EEC rules and regulation for materials utilized in manufacture of phenolic resins.

Phenols and phenolic resins, formaldehyde, and solvents such as methanol, propanol, toluene and some others which are used to produce resinous solutions, are governed by these regulations.

All phenolic resins containing more than 5% free phenol must be designated "poisonous" by a skull. Phenolic resins containing 1–5% free phenol are considered detrimental to health and are to be marked with a "St. Andrew's cross." Formulations are not considered toxic if the amount of free phenol is below 0.2%. Package or container label should clearly identify the producer, kind of toxic components, warnings of the special dangers involved, and safety precautions to follow.

During handling of phenolic resins, sensitive persons may be affected by dermatitis. To prevent such an adverse reaction, it is advisable to treat exposed portions of the body with an appropriate protective cream and wear rubber or plastic gloves during work. After work, hands and arms are washed with a special soap followed by protective cream treatment.

Furthermore, special attention should be given to clean working conditions and effective ventilation in the plant. The MAK-value (maximum concentration at place of work) is 5 ppm for phenol and 1 ppm for formaldehyde, in West Germany. Additional technical safety data for phenol are given in Chap. 2, Table 2.

6.4 Waste Water and Exhaust Air Treatment Processes

There is no uniform procedure for solid and liquid waste disposal for plants engaged in the manufacture of phenolic resins. The selection of an optimum process requires an individual analysis of the kind and amount of hazardous substances as well as the infrastructure of the plant and laboratory performance tests. Occasionally, a combination of different processes may be feasible. Such processes are microbial degradation, thermal combustion, physical, and physico-chemical scrubbing, chemical oxidation or resinification reactions and adsorption methods.

6.4.1 Microbial Transformation and Degradation

The breakdown of aromatic compounds is an important step in the natural carbon cycle and many micro-organisms: eubacteriales, pseudomonas, actinomycetas, endomicetas, higher funghi – are capable of degrading aromatic substrates[25]. The essential step required for biological degradation is the conversion of the aromatic compound to an *ortho-* or *para*-dihydroxybenzene structure. The enzymes responsible for this hydroxylation have the characteristics of oxidases or dioxygenases. The first steps of the three possible oxidative cleavage reactions of *o-* and *p*-dihydroxy compounds are shown in the formulae (6.1/5). In the case of 1,2-dihydroxybenzene, *ortho-* or *meta*-cleavage (6.1, 6.2) may occur. *Ortho-* and *para*-hydroxybenzoic acids (6.3, 6.5) may be formed as intermediates during the degradation of phenolic resin prepolymers.

ortho cleavage (6.1)

meta cleavage (6.2)

p-dihydroxybenzene cleavage (6.3)

(6.4)

(6.5)

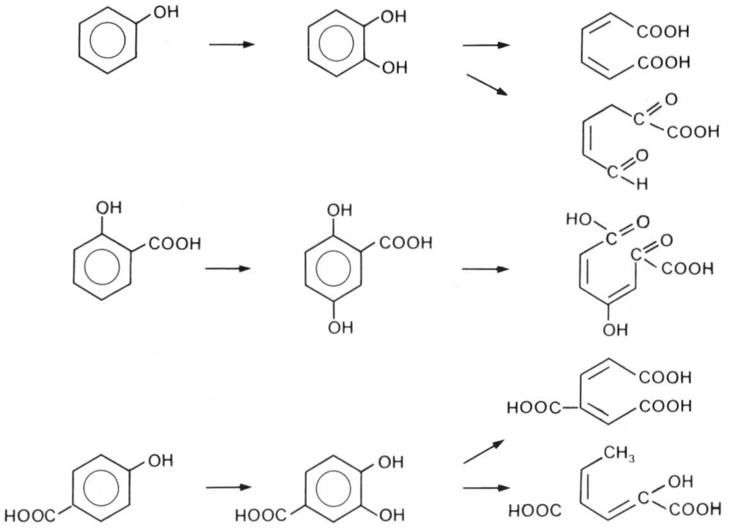

The aliphatic mono- and dicarboxylic acids formed are further converted to 3-oxo-hexanedioic acid, which is taken up in the Krebs cycle, or to fumarate, pyruvate, acetaldehyde, and acetoacetate. After this, the degradation to CO_2 and H_2O follows.

Certain kinds of microorganisms are able to thrive in water containing up to 1,000 mg/l of phenol[27]. They are most active at temperatures between 25–35 °C. Further essential prerequisites are a sufficient nutrient content (N, P) and oxygen, pH between 7.5–8.5 and the absence of heavy metal ions (< 5 mg/l). To provide nutrient, it is advisable to combine and treat the waste water with household sewage. Ammonium phosphate is most frequently used as a nutrient. The effectiveness of the biomass increases with time to an upper limit value because biological selection processes take place and the resistance and degradative ability of some microorganisms increase. The lagoon must have an effective aerating and circulation system so that dissolved oxygen is always in excess. The Unox process[28] uses oxygen instead of air for a higher oxygen concentration.

Biochemical degradation is the most frequently used and most effective process for treating waste waters containing phenol. Final effluents in the range of 0.1 mg/l are reported. To ensure feed and environmental conditions are constant for biomass, properly designed equalization systems are required for optimum efficiency. Particular problems arise if plants are operated discontinuously or 5 days a week[29].

Suitable reductions in phenol are reported for waste water from the manufacture of adhesive resins. The use of rainwater dilution in a balancing lagoon, primary activated sludge with secondary biological treatment in filters or other activated sludge processes facilitated phenol reductions from 123 to ≤ 0.1 mg/l[30].

A novel two-phase method[31] has been proposed to remove and recover phenols from aqueous alkaline waste streams. The method is based on the reaction of phenols with selected acyl or sulfonyl halides (benzoyl chloride or p-toluenesulfonyl chloride) dissolved in a water immiscible solvent (toluene) in the presence of a phase transfer catalyst. In a continuous reactor-settler arrangement, phenol content was reduced from 2,035 ppm to 45 ppm with phenol recovered as phenyl benzoate.

6.4.2 Chemical Oxidation and Resinification Reactions

In chemical oxidation processes phenols are normally transformed into intermediate non-toxic compounds (not CO_2 and H_2O), and so only a certain decrease in COD will result. The removal of phenol may reach final levels of less than 1 mg/l or $>99\%$ according to the ratio of chemicals used[29].

$$\text{C}_6\text{H}_5\text{OH} + 7\,O_2 \longrightarrow 6\,CO_2 + 3\,H_2O \qquad COD = 2.38 \text{ mg } O_2/\text{mg phenol} \qquad (6.6)$$

Hydrogen peroxide[32] in the presence of small amounts of iron-, manganese-, chromium-, or copper salts is an effective oxidizer of phenols (and other organics). The temperature has little effect on reaction rate and conversion, a pH in the range of 3–5 is most effective. Hydrogen peroxide may be used to treat concentrated wastes high in phenols or for pretreating of high-phenol waste before biological treatment to obtain uniform low phenol content.

Ozone is a more effective oxidant than hydrogen peroxide. Lower amounts are normally applied as necessary for complete destruction to carbon dioxide and water. The selectivity of ozone is low, operating at pH values of 11.5–11.8 results in preferential oxidation of phenol[33]. Ozone is often used in the final treatment step leading to very low phenol levels (lower than 0.1 ppm).

Sodium hypochlorite or chlorine dioxide, which is the oxidizing agent, will oxidize phenols to benzoquinones (pH 7–8). At pH above 10 further oxidation to maleic acid and oxalic acid will occur; chlorophenols are not formed. Chlorine is not used because of the formation of chlorophenols which are more toxic and have more objectionable taste and odor than the original phenols. Potassium permanganate or potassium dichromate are also effective oxidants; however, the handling of the precipitated sludge can be a serious problem.

Resinification reactions followed by precipitation of the polymeric material can be used for waste waters which contain phenol, phenol prepolymers and formaldehyde by adding sulfuric acid or ammonia and reacting at higher temperature. Ferric chloride or aluminum sulfate are recommended as precipitants. The deposits are incinerated in most cases. It is customary in the plywood-, particle board-, and fiber board industries to acidify the waste waters with aluminum sulfate to pH 4. By this method, the resinous components precipitate almost completely as a deposit which settles rapidly and is filterable, especially if the precipitation occurs at elevated temperatures. Afterwards, the water must be neutralized with lime (pH 6.5–8.0) and calcium sulfate is filtered.

6.4.3 Thermal and Catalytic Incineration

The treatment of effluent air by oxidation – thermal or catalytic incineration – is accomplished if recovery of the solvents is not feasible or economical.

The organic components are oxidized to CO_2, CO, and water. The catalytic incineration occurs at temperatures between 350–400 °C. Metal oxides are preferable, however, elements of the platinum group on different supports are used as catalysts. Catalysts are very sensitive. Sulfur-, phosphorus-, halogen-, silicon-, arsenic compounds and many other contaminents lead to catalyst poisoning[34, 35].

Catalytic incineration is preferred when the exhaust air contains only minor concentrations of organic substances (≤ 3 g C/m^3), catalyst poisons are absent, and the content of dust in the air is small. It is very rare for these factors to be absent in plants working with phenol or phenolic resins, thus catalytic incineration is not important in these industries. The exhaust air is often rich in combustible materials, due to the solvents used (varnish industry, production of laminates) or has a high solid particles content (production of mineral wool). The phosphoric acid esters which are components of many formulations act as catalyst poisons. They are transformed to phosphorus-pentoxide which may react with the catalyst support (for instance with Al_2O_3 forming $AlPO_4$), thus leading to deactivation[36]. The low molecular phenol alcohols may cause resinification and formation of coke on the surface of the catalyst.

In thermal incinerators, the exhaust gases are burned in metal chambers at 700–800 °C. Gas or oil fired jet or area burners are used. The effectiveness of thermal incinerators, which depends on temperature, is considerably higher (residual content of carbon <5 mg/m^3 at 750 °C) than that of the catalytic process. This process results

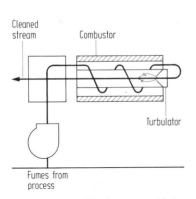

Fig. 6.1 Thermal incinerator with heat exchanger (Drawing: UOP-KAVAG, D-6467 Hasselroth)

Fig. 6.2. Overall view of a thermal incinerator (Photo: UOP-KAVAG, D-6467 Hasselroth)

in a larger consumption of energy so that the utilization of the exhaust gas heat is a necessary economical factor. The principle of thermal incineration and an example of an incinerator are shown in Figs. 6.1 and 6.2[37].

This process is used in most plants producing technical laminates as well as in larger varnish drying equipment. In both cases, phosphoric acid esters and a relatively large quantity of solvents are contained in the exhaust air. The combustion heat is utilized to preheat the exhaust air as well as to heat the process air, occasionally also for hot water. The exhaust air, after appropriate adjustment, may be led directly into the dryer.

Combustion is not limited to exhaust air. Combustion of waste (distillate, etc.) offers the only economical solution in many instances. The liquid waste which must be atomized at temperatures below 200 °C is sprayed into the combustion chamber either immediately or, in the case that the liquid is not able to sustain combustion, after the addition of supporting fuel, and burned. Theoretically, a mixture of about 18% phenol and 82% water would be self-sustaining when burned at 760 °C[29]. However, the solubility of phenol in water is limited to about 10% at ambient conditions, and therefore effective mechanical mixing is necessary at higher phenol concentration.

6.4.4 Extraction Processes and Recovery

The coking plants in the Ruhr area have been confronted with the problem of the recovery of phenol from waste waters since their inception (Chap. 2). During the intervening years, different extraction processes have been developed. Some processes are used extensively for phenol recovery in coking plants today[27, 38, 39]. The effectiveness of the benzene/caustic process according to Pott and Hilgenstock, which uses benzene as a solvent and effects the recovery with an aqueous sodium hydroxide solution, can be increased to 98% by the use of multi-stage Podbielniak countercurrent extractors. However, it is not possible to lower the content of phenol below 20 mg/l. Using the Phenosolvan process developed by Lurgi[40], which employs diisopropyl ether as extractive solvent in a 10-stage counter-current extractor, the phenol content can be reduced to values around 10 mg/l. Neither process is sufficiently attractive to achieve the limit of 0.5 mg of phenol per liter. Therefore, these extraction processes are only important for the recovery of phenol in coking plants, coal liquefication plants and installations producing phenol and phenolic resins.

6.4.5 Activated Carbon Process

Activated carbons have been developed for the treatment of waste water and exhaust gas. They have a high phenol adsorption capacity, high stability, and abrasion resistance[41]. A polymeric adsorbent, also highly effective in removing phenol and similar compounds from aqueous waste yielding an effluent with generally less than 1 ppm of phenol has been developed[42]. Other examples of adsorbents are aluminum oxide, silica gel, and zeolites. Generally, these processes consist of three process steps. During the pretreatment, the waste water or exhaust gas is free of solid substances. In the adsorption step, the organic substances are adsorbed by an activated carbon bed by a purely physical process based on electrostatic interaction. The adsorption process may be accompanied by chemisorption and capillary condensation.

In the regeneration step, the desorption and reactivation of the activated carbon take place. Either dilute caustic, solvents (methanol, acetone) or inert gas mixed with steam are used to desorb phenols. A sulfuric acid treatment is recommended to recover phenol from the sodium salt. If solvents are used, the treatment is done by distillation. Regeneration with inert gas/steam requires a later thermal combustion of the exhaust gas.

Recently the biogeneration of granular activated carbon under conditions typical of water treatment plant operation was demonstrated[43]. Influent consisted of a pH 7.5, mineral ions containing medium, 2 mg/l phenol, and dissolved oxygen was at air saturation value of 9 mg/l. The authors defined bioregeneration as an increase in adsorptive capacity resulting from bacterial removal of material absorbed on carbon.

6.4.6 Gas Scrubbing Processes

Gaseous components are absorbed in a liquid either by a physical solution process or by chemical reaction. Both mechanisms may be applied to exhaust gases containing phenol. The exhaust gases may also contain solids and droplets[44]. The principle of a continuous absorption plant is that the finely distributed washing agent and the exhaust gas pass each other in the absorber in a countercurrent operation. The contaminated water can either be regenerated or, in some cases, returned to the resin plant in a closed loop thus omitting further addition of water. Otherwise, the exhaust gas problem would become a waste water problem as well.

This process is extremely suitable in cases where relatively small amounts of hazardous substances are to be removed from relatively large amounts of air. For example, this is the case in the production of mineral wool products. Here, this process would be the only practicable solution. The essential parameters for the absorption are the mass transfer constant, the diameter of the drops, the exchange area and the time of direct contact. Common absorbers are spray tower washers, filter washers and plate columns.

6.5 Smoke and Gas Evolution from the Combustion of Phenolics

In recent years the effects of smoke and toxic gases have been singled out as being one of the most important causes of injury and death in fires. An increasing amount of attention and work is now being directed to studies of polymer combustion products and to the development of flame retardant polymer formulations. Improved polymer resistance to ignition and reduced rate of burning are key properties to delay or lessen the onset of total obscuration or flammability for escape and/or rescue.

The initiative has been mainly undertaken by the aircraft industry to define the permissible or maximum values of flammability, smoke density, and toxic gases evolution. A multitude of testing methods was developed to evaluate flammability data and ratings. The commonly utilized procedures in phenolic resin technology are briefly mentioned. Additional information is contained in "Flammability Handbook for Plastics"[45] as well as recent publications[46-49].

Thickness and shape of test specimens usually have a significant effect on flammability and arc/track data. For this reason flammability and arc/track test values and ratings must always be accompanied by specimen dimensions. The Underwriters' Laboratories flammability series of tests includes a standard burning test (Bulletin 94) applied to vertical and horizontal test bars, from which a general flammability rating is derived. There are additional tests that provide an estimate of the susceptibility of plastics to ignition from electrical sources (hot wire ignition and high current arc ignition tests) because arc/track behavior is frequently important in electrical insulation applications.

The ASTM flammability and arc/track tests are the ones most commonly utilized for a broad spectrum of plastics. These were the only methods other than UL Bulletin 94 for which a compendium of data exists to permit comparisons. Details of the oxygen index, D 2863-77, and D 635 horizontal bar flammability tests as well as other ASTM flammability and arc/track tests can be found in Vol. 35, Plastics, ASTM Standards published by the American Society for Testing & Materials, 1916 Race St., Philadelphia, PA.

Table 6.4. UL Burning rating code

---

94 V-0 = Extinguishment of flaming combustion in a vertical bar test within 5 s avg. of 10 ignitions on five specimens, and within 10 s for each individual ignition; no glowing combustion after 30 s each specimen; no flaming drips that ignite cotton.

94 V-1 = Extinguishment of flaming combustion in a vertical bar test within 25 s avg. of 10 ignitions on five specimens, and within 30 s for each individual ignition; no glowing combustion after 60 s each specimen; no flaming drips that ignite cotton.

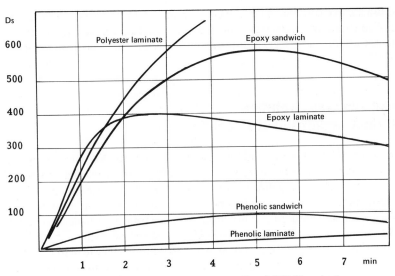

Fig. 6.3. Smoke density[50] of thermosetting composites (NBS Chamber)

The Epiradiateur test after the French norm AFNOR/P 92-507 is of similar importance in Europe. The test-specimen is irradiated by a heat radiator source of 3 W/cm², the nascent combustible gas can be burnt by pilot flames. In West Germany testing methods after DIN 4102 are used among others. Railroad companies Deutsche Bundesbahn (West Germany) and SNCF (France) have developed their own particular methods and standards[50].

The Limiting Oxygen Index (LOI), ASTM-D 2863-77, or AFNOR NFT 51071 test was designed to overcome certain inherent limitations in "Bunsen burner tests." Principally it provides a graduated numerical measure of flammability rather than categorical rating with improved reproducibility of flammability testing. The latter is a serious problem in "Bunsen burner tests," particularly in comparison between laboratories. The LOI is the % oxygen in an oxygen-nitrogen mixture that will barely sustain combustion of a vertically mounted specimen (bar) that has been ignited by application of an external gas flame to its upper end.

Smoke properties are measured according to National Bureau of Standards in a smoke density chamber by means of an integrating optical system. Smoke toxicity standards are being considered[51, 52]. Various groups (US National Bureau of Standards along with several universities) are developing protocols or preliminary test procedures to serve as a final ASTM test for determining the toxicity of combustion products emitted by different polymeric materials and the magnitude of their threat to human life.

Most polymeric materials are inherently flammable and must be modified with a flame retardant chemical for either a V-1 or V-0 classification. Phenolic resins are exemplary among all other polymeric materials. Phenolics are fire resistant materials with low smoke emission and low toxicity, hence they exhibit very favorable flame retardant characteristics under fire conditions. Since phenolics are mainly composed of carbon, hydrogen, and oxygen, their combustion products are water vapour, carbon dioxide, carbon char, and moderate amounts of carbon monoxide depending on combustion conditions. The toxicity of the corresponding combustion products is, therefore, relatively low. In an extensive thermal and flammability study[53], modified phenolic, bismaleimide, and polyimide composites were studied as alternatives to the current flame retardant epoxy used as face-sheet/adhesive resin in aircraft interior composite panels. In addition to low smoke release, the modified phenolic composite had the lowest toxicity of all the materials tested using the NASA-Ames pyrolysis toxicity apparatus. A comparison with thermally stable thermoplastics was included.

Analysis of the molecular structure and its high crosslink density of the completely cured resin is helpful in gaining an understanding of the high flame resistance of phenolic resins and compounds. The rigid three-dimensionally crosslinked structure resists thermal stress without softening or melting. These morphological characteristics favor decomposition and volatilization of flammable low MW species. As the phenolic material is heated to ignition temperature, it is transformed into a char forming material. The phenolic structure facilitates the formation of a high carbon foam structure (char) that radiates heat and functions as an excellent heat insulator (Chap. 10).

Returning to oxygen index and how it relates to flammability, correlations have been developed between the structure of polymers and their char formation. Hence char yield becomes an important parameter in flammability studies. A linear relationship has been established between char yields and oxygen indices of halogen-free polymers[54, 55]. The systematic design of a variety of substituted and unsubstituted phenolic resins resulted in a correlation between oxygen index and char yield. The study showed that meta and para cresol resins exhibited slightly lower oxygen indices than the unsubstituted phenolics. Highest char yields (Table 6.5) were observed for unsubstituted phenolics[56].

The oxidative decomposition of phenolic resins is initiated by peroxide formation at the methylene bridge of two phenolic units [see Eq. (8.1), Chap. 8] under certain conditions. This highly exothermic reaction becomes self-sustaining, leading to a phenomenon known as punking (Chap. 13). Smoldering combustion or afterglow appears and continues until the mass is consumed, provided a large surface area is available for oxidation and the rate of heat removal is low. Such observations were identifiable with early phenolic foams. Newly developed phenolic foams no longer exhibit the punking phenomenon. The punking tendency can be decreased by inhibiting peroxide formation through acetylation of the phenol group. The methylene bridge is stabilized through steric and inductive effects and, therefore, punking tendency diminished; however, flame resistance is reduced.

Table 6.5. Oxygen index and char yield of various phenolic resins[56]

	Oxygen index (%)		Char yield (%)	
	Novolak	Resol	Novolak	Resol
Phenol	34–35	36	56–57	54
m-Cresol	33	–	51	–
m-Chlorophenol	75	74	50	50
m-Bromophenol	75	76	41	46

Flame retardant additives are used in phenolics[48] mainly if moderately flammable fillers or reinforcing fibers, e.g. cellulose, wood particles, textile fibers, are employed. Satisfactory additives include Tetrabromobisphenol-A (TBBA) and other brominated phenols, *p*-bromobenzaldehyde, organic and inorganic phosphorous compounds [e. g. *tris*(2-chloroethyl) phosphate, diphenylcresylphosphate, ammoniumphosphate], red phosphorus, melamine, and melamine resins, urea, dicyandiamide, boric acid, and borates and other inorganic materials.

6.6 References

1. Sax, N.J.: Dangerous Properties of Industrial Materials van Nostrand Rheinhold Co.: 1975
2. Albert, A.: Selective Toxicology. London: Methuen & Co.: 1968
3. Babich, H., Davies, D.L.: Regul. Toxicol. Pharmacol. 1 (1), 90 (1981)
4. N.N.: Chem. Eng. News, July 9, 1979, p. 19
5. "Formaldehyde and other Aldehydes": National Research Council, National Academy Press, pp 183–185 (1981)
6. Battelle Columbus Laboratories: Final Report on "A chronic inhalation toxicology study in rates and mice exposed to formaldehyde." CIIT, Dec. 31 (1981)
7. Bio/dynamics; Formaldehyde Institute: A 26-week inhalation toxicity study of formaldehyde in the monkey, rat and hamster. Jan. 21 (1982)
8. Ragan, D.L., Boreiko, C.-J.: Initiation of C3H/10T1/2 cell transformation by formaldehyde. Cancer Lett., 13, 325 (1981)
9. Ross, W.E., McMillan, D.R., Ross, C.F.: Comparison of DNA damage by methylamines and formaldehyde. J. Natl. Cancer Inst., 67, 217 (1981)
10. Moerman, D.J., Baillie, D.L.: Formaldehyde mutagenesis in the nematode Caenorhabditis elegans. Mutat. Res. 80, 273 (1981)
11. Bedford, P., Fox, B.W.: The role of formaldehyde in methylene dimethylsulphonate – induced DNA crosslinks and its relevance to cytotoxicity. Chem.-Biol. Interact. 38, 119 (1981)
12. d'A. Heck, H.: Biochemical toxicology of inhaled formaldehyde. CIIT Activities, 2, No. 3, 3 (1982)
13. Fayerweather, W.E., Pell, S., Bender, J.R.: Case-control study of cancer deaths in Du Pont workers with potential exposure to formaldehyde, May 17 (1982)
14. Goldmann, P. et al.: Formaldehyd – Morbiditäts Studie. Zbl. Arbeitsmed. 32, 250 (1982)
15. European Chemical Industry Ecology and Toxicology Centre: Technical Report No. 6: Formaldehyde Toxicology, Sept. 4 (1982)
16. N.N.: C & EN, May 28 (1984), p. 8
17. Pickrell, J.A., Mokler, B.V., Griffis, L.C., Hobbs, C.H., Bathija, A.: Environ. Sci. Technol. 17, 753 (1983)
18. EPA 440/1-74-0414a, April 1977
19. Chem. Eng. News p. 14 (August 19, 1982)
20. Dietz, F., Traud, J.: gwf-Wasser/Abwasser 119 (6) 318 (1978)
21. Hinweise für das Einleiten von Abwasser aus gewerbl. und industr. Betrieben in eine öffentliche Abwasseranlage, ATV-BDI-KfK 1970, BDI-Drucksache Nr. 90, BDI, 5 Köln 51
22. Gesetz über die Beseitigung von Abfällen (Abfallbeseitigungsgesetz-AbfG) vom 5. Jan. 1977, BG-Bl. I Nr. 2, S. 41 (1977)
23. Bradke, H.-J., Klein, T.: Deponierverhalten und Verwertung von Gießereisanden, Teil I und II: Laborauslaugungen von Gießereisanden zur Ermittlung der eluierbaren Stoffe und des voraussichtlichen Deponieverhaltens. IWL, 5 Köln 51, 1975
24. N.N.: Chem. Week, Nov. 18, 1981, P. 64
25. Kieslich, K.: Microbial Transformations of Non-Stereoid Cyclic Compounds. Stuttgart: Thieme 1976

26. Rozovskaya, T. I., Lazareva, M. F.: Mikrobiologiya 40, 370 (1971)
27. Sierp, F.: Die gewerblichen und industriellen Abwässer, Berlin/Heidelberg/New York: Springer 1967
28. Martin, P., Schönfelder, H., Tischer, W.: gwf, Wasser/Abwasser 116, 272 (1975)
29. Lanuette, K.: Chem. Engng.-Deskbook issue, Oct. 17, 99 (1977)
30. Housden, A. J.: Water. Pollut. Control 80 (4), 490 (1981)
31. Krishnakumar, V. K., Sharma, M. M.: Ind. Eng. Chem. Process Des. Dev. 23, 410 (1984)
32. Kibbel, W. H. Jr., Raleigh, C. W., Shepherd, J. A.: Hydrogen Peroxide for Industrial Pollution Control, 27th Ann. Purdue Indl. Waste Conf. Part 2, West Lafayette, Ind., 1972.
33. Nebel, D., Gottschling, R. D., Holmes, J. L., Urangst, P. C.: Ozone Oxidation of Phenolic Effluents. Welsbach Ozone Systems Corp. 1976
34. Quillmann, H.: Chem. Ind. 25, 497 (1973)
35. Rüb, F.: Wasser, Luft u. Betrieb 19, 702 (1975)
36. Kirchner, K., Angele, B.: Chem.-Ing.-Techn. 49, 243 (1977)
37. UOP-KAVAG, UOP Thermal Incineration Systems, Techn. Bulletin
38. Wurm, H. J.: Chem.-Ing.-Techn. 48, 840 (1976)
39. Hahn, E., Herpes, E. Th., Neff, I.: Chem. Ind. 28, 591 (1976)
40. Lurgi GmbH: Dephenolization of Effluents by the Phenosolvan Process. Techn. Bulletin
41. Jüntgen, H., Klein, J.: Energy Sources 2 (4) 311 (1976)
42. Fox, Ch. R.: Hydrocarbon Proc., July 1976, P. 109
43. Chudyk, W. A., Snveyiak, V. L.: Environ. Sci., Technol. 18 (1), 1 (1984)
44. Reither, K.: Wasser, Luft u. Betrieb 10, 3 (1974)
45. Hilado, C. J.: "Flammability Handbook for Plastics," 2nd Ed., Technomic, Westport, Conn. 1974
46. Cullis, C. F., Hirschler, M. M.: "The Combustion of Organic Polymers." Clarendon Press, Oxford, 1981
47. Landrock, G. H.: "Plastics Flammability and Combustion Toxicology." Noyes Publications, Park Ridge, N.J. 1983
48. Sunshine, N. B.: "Flame Retardancy of Phenolic Resins and Urea- and Melamine-Formaldehyde Resins." In: Flame Retardancy of Polymeric Materials. Vol. 2, Chap. 4, Marcel Dekker, New York (1973)
49. Troitzsch, J.: Brandverhalten von Kunststoffen, Carl Hanser Verlag, München (1982)
50. Schick, J.-P., Schönrogge, B., Perrier, A.: Kunstharz-Nachrichten Hoechst, 20, Sept. 1983, p. 26
51. Modern Plastics, March 1982, p. 48
52. Plastics World, August 1983, p. 36
53. Kourtides, D. A., Gilwee, W. J., Parker, J. A.: Polymer Engineering and Science, 19 (1), 24 (1979) and 19 (3) 226 (1979)
54. Van Krevelan, D.W.: Chimia, 28, 504 (1974)
55. Van Krevelan, D.W.: Polymer, 16, 615 (1975)
56. Zaks, Y., Lo, J., Raucher, D., Pearce, E. M.: J. Appl. Polym. Sci. 27, 930 (1982)

7 Analytical Methods

In the pursuit of phenolic resin chemistry, a multitude of analytical methods[1, 2] and testing procedures are used to determine product quality, batch consistency, monomer content, resin microstructure, resin macrostructure, and structural configuration of cured resins. Various analytical techniques as well as structural and configurational characteristics of phenolic prepolymers were reviewed by Tsuge[3]. The summary to 1980 included in excess of 100 references and placed special emphasis on spectroscopic methods. Principles of spectroscopic methods applied to polymers were reviewed by Klopfer[4]. More modern monomer and polymer analytical procedures are being developed as "systems protocol" by coupling several methods in tandem. Tandem systems such as GC/MS, GC/FTIR, GC/FTIR/MS, GC/LC/MS, NMR/FTIR are rapid, convenient methods that provide a moderately comprehensive assay of phenolic components under investigation.

7.1 Monomers

7.1.1 Phenols

Phenols in general undergo a number of reactions leading to colored products which were formerly utilized for the qualitative identification of phenolic resins. The ferric chloride test is probably the most widely known identification reaction[5, 6]. Qualitative color tests are no longer used.

Koppeschaar Method

Free phenol is mainly determined by the Koppeschaar method[7] as described in ASTM D 1312. Phenol is steam distilled and then converted to tribromophenol by use of a 0.1 N potassium bromide-bromate solution as bromine source. Excess of bromine is determined by sodium thiosulfate procedure. The selectivity of the method is limited; other bromine reactive compounds such as furfural or furfuryl alcohol, cannot be distinguished.

Colorimetric Determination

The quantitative determination of low phenol levels is conducted colorimetrically. The phenol obtained by steam distillation of an aqueous solution is converted into an azo dye by the addition of *p*-nitroaniline and a sodium nitrite solution, which is then determined colorimetrically either directly or after extraction with butanol.

The 4-aminoantipyrine(1-phenyl-2,3-dimethyl-4-aminopyrazolone-5) method (ASTM D 1783–70) is widely used, e.g. for the determination of phenols in industrial waste waters, despite some shortcomings of the method[2, 8]. 4-Aminoantipyrine reacts with phenols in the presence of ferricyanide ions in alkaline solution resulting in quinoid-type compounds (7.1). *para*-Alkylphenols give negative or weak results.

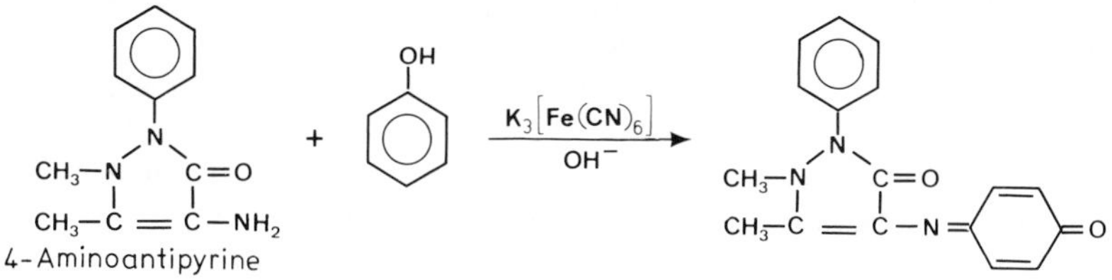

(7.1)

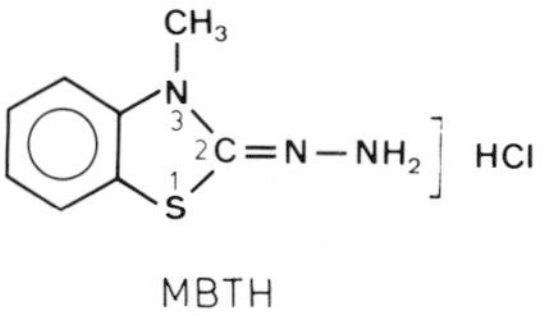

4-Aminoantipyrine

(7.2)

MBTH

Some disadvantages of 4-aminoantipyrine can be avoided by use of MBTH (2-hydrazono-2,3-dihydro-3-methylbenzthiazol hydrochloride).

Gas-Chromatographic Determination

Gas-chromatographic determination is being used to an increasing extent for the determination of free phenols[9] in phenolic resins as well as phenol raw material purity. This method is very sensitive and has the advantage that cresol and xylenol isomers can be determined quantitatively and other bromine sensitive monomers are unaffected. An internal standard compound is introduced for quantitative determination. GC is discussed in greater detail in the chromatography section.

7.1.2 Formaldehyde

The (free) formaldehyde[10] is converted quantitatively into the corresponding oxime by the addition of hydroxylamine hydrochloride. During reaction an equivalent amount of HCl is released and determined by acid/base titration or pH meter.

$$CH_2{=}O + [NH_2{-}OH]\,HCl \rightarrow H_2C{=}N{-}OH + H_2O + HCl \qquad (7.3)$$

A simple and highly sensitive colorimetric identification test is based on the formation of a specific blue-violet color when formaldehyde reacts with Schiff's fuchsine-bisulfite reagent in the presence of strong acids.

7.2 Polymers

7.2.1 Infrared Spectroscopy

The infrared spectra (IR) of phenol formaldehyde resins have been studied by many authors. Hummel[11, 12] has collated a multitude of typical infrared spectra. *Ortho-*

ortho linked novolaks show a strong band at 13.3 μm whereas *ortho-para* novolaks exhibit bands of almost equal intensity[13] at 13.3 and 12.2 μm. In the same vicinity, however, the typical hydroxymethyl absorption appears at 9.9 μm. The resol structure is further identified by strong absorption at 3 μm (hydroxyl group) and 6.9 μm. The dibenzyl ether peak appears at 9.5 μm. Etherified resols show an absorption at 9.2 μm. IR spectra are also very effective for the identification of additives, however, identification requires an extensive library of IR spectra for comparison.

The use of Fourier transform infrared spectroscopy (FTIR) is becoming the modern method of IR analyses of polymers[14]. Sojka[15] used FTIR to study the reaction of phenol with HMTA at 100 °C. Identification was problematic apparently due to peak overlap. Pearce et al.[16] analyzed cured novolak and resol samples.

7.2.2 Nuclear Magnetic Resonance Spectroscopy

Nuclear magnetic resonance spectroscopy (NMR) is unique among spectroscopic methods[17] in distinguishing the structure of transient molecules involved in resin formation[18], species involved at HMTA cure, and final structure of phenolic oligomers[19].

Early NMR studies consisted of proton ^{1}H-NMR and resulted in detailed structural determination of a variety of intermediate methylol components within a resol[19-22]. Chemical shifts of ^{1}H-NMR of the various types of protons contained in phenolic resins are summarized in Table 7.1. Besides structural characterization of resols, NMR can distinguish between resol and novolak resins and identify isomer distribution[23, 24], e.g. patterns of novolak resins (*o-p*, *p-p'*, or *o-o'*).

There are inherent limitations in the use of proton NMR. Chemical shifts often overlap and, in some cases, peak assignments are difficult without model compounds.

Table 7.1. Phenolic resin ^{1}H-NMR chemical shifts[3, 19, 21-24]

General ^{1}H-NMR characterization

	δ (ppm)
Aromatic proton (C−H)	6.6–7.1
Methylene ether (−CH$_2$OCH$_2$−)	4.6–4.8
Aryl methylene (−CH$_2$−Ar)	3.7–4.2
Methylol (−CH$_2$OH)	4.3–4.5

Novolak methylene substitution pattern

	δ (ppm)	
	Quinoline	Pyridine
p-p'	4.0	3.8
o-p	4.3	4.1
o-o'	5.0	4.5

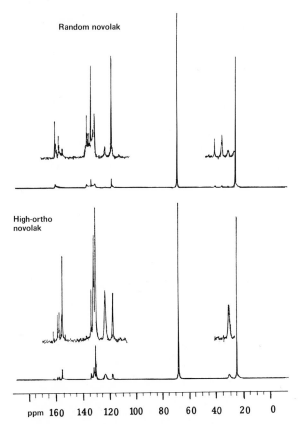

Random novolak

High-ortho
novolak

ppm 160   140   120   100   80   60   40   20   0

Fig. 7.1. ^{13}C-NMR spectrum of random novolak and high-*ortho* novolak

Solvent effects can occur. Methylol groups are usually derivatized by acetylation (dry sample) with acetic anhydride.

The use of ^{13}C-NMR initially complemented ^{1}H-NMR spectra of resols and novolaks. Today, it is the preferred NMR method of analysis for PF resins. The many advantages of ^{13}C-NMR include greater resolution; all nonequivalent carbon atoms are identifiable and derivatization is not required. ^{13}C-NMR is primarily used for detailed microstructure characterization of novolaks. Based largely on the efforts of Sojka[25], Kim[26], and Casiraghi[27] random and high-*ortho* novolaks are extensively characterized. ^{13}C-NMR spectra of random and high-*ortho* novolak are shown in Fig. 7.1.

Sojka and coworkers[28] have prepared a large variety of linear trimers and tetramers with definite structure (Chap. 4) and in combination with dimers have developed a computerized "structures library" of these novolak oligomers (Table 7.2). Further they have shown that the analyses of ^{13}C-NMR peaks of the phenolic end group provide a wealth of valuable structural information; i.e. free phenol content of the novolak, type and concentration of phenolic end and middle groups and concentration of branching centers. Resol ^{13}C-NMR assignments included in Table 7.2 are attributable to Mechin[29] and Pasch[30]. The condensation reaction between phenol, melamine and formaldehyde in relation to pH was analyzed by means of ^{13}C-NMR and GPC by Braun and Krauße[31]. More detailed information of ^{13}C-NMR for resols would be desirable.

Table 7.2. Compilation of ^{13}C chemical shift regions for various carbon types found in phenolic oligomers[25, 28-30]

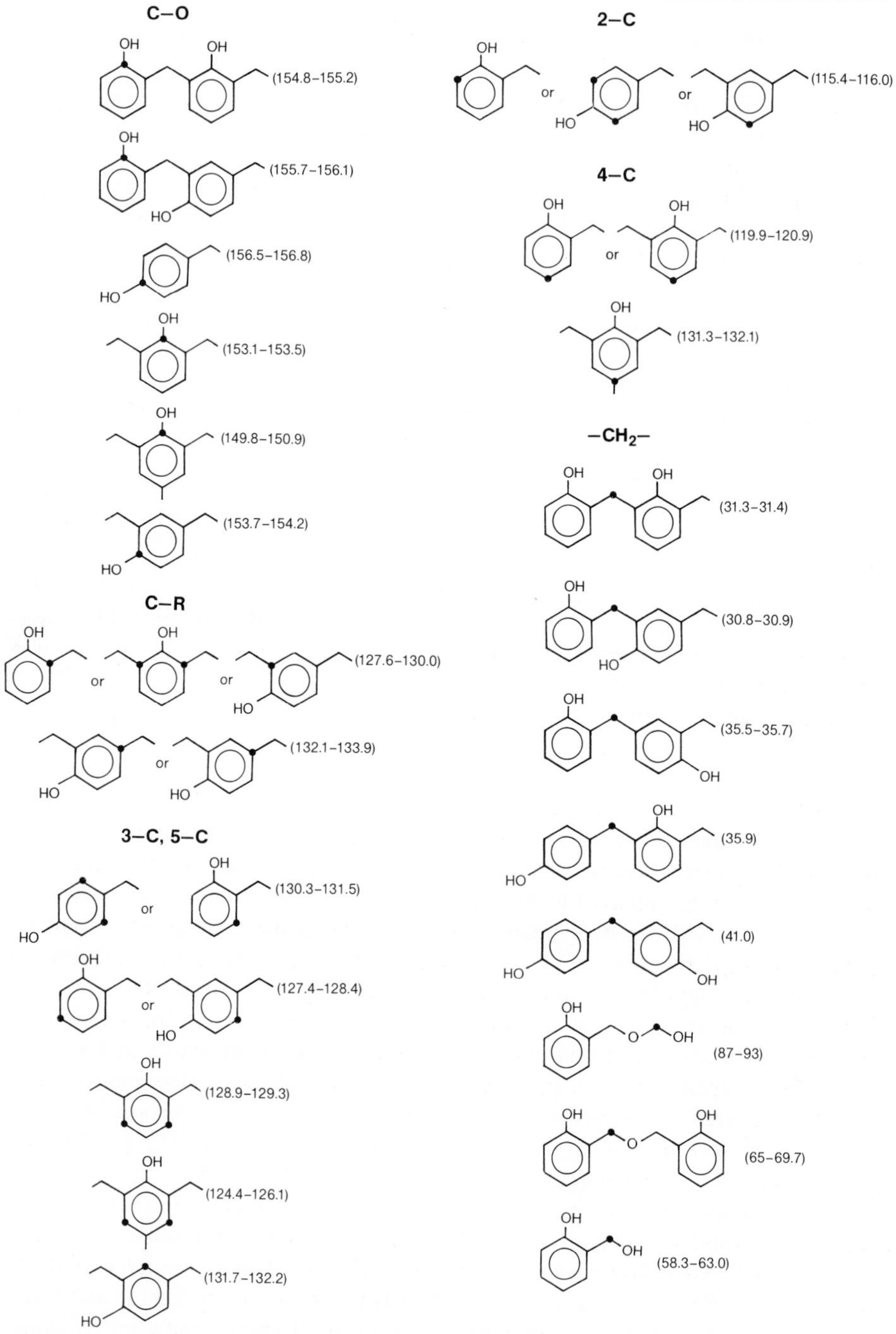

Numbers in parentheses are the ^{13}C chemical shift regions for the dotted carbons in the adjacent structure

Branching in novolak is identifiable with an upfield value because of increased alkyl-ation. A branching segment in an "all *ortho*"-novolak is observed[28] at 151.1–151.2 ppm.

Computer fitting of ^{13}C-NMR spectra of novolaks has recently been reported[32]. By incorporating the basic features of known phenolic oligomers and conducting the analyses via computer graphics in color, one high-*ortho* and two random resins were characterized. Various substitution patterns (*o-p*, *p-p'*, *o-o'*) and chain branching were described and attested to the significance of the method. It is our opinion that the com-puter fitting technique can be utilized to characterize various novolak resins and iden-tify action of catalyst, process conditions (continuous or batch production) and other subtle features such as branching, and, e.g. raw material origin. It is very likely that the use of computer fitting of novolak structure and computer simulation (Chap. 3) will allow structure and process optimization with a minimum of experimental work.

Improvement in NMR instrumentation has allowed the study of molecular mo-tion of solid polymers[33, 34]. The use of solid state ^{13}C-NMR with cross polarization (CP) and "magic angle" spinning (MAS) has provided structural information relevant to HMTA cure of novolaks[35] as well as thermal decomposition of solid resols in pres-ence and absence of air[36]. In a recent publication analyzing the cure of resols the ex-tensive involvement of the phenolic hydroxyl group and condensation/crosslinking of the methylene bridge with formaldehyde evolved or with the hydroxymethyl group has been suggested[37]. Additional fundamental information is expected from solid state ^{13}C-NMR regarding resin microstructure, two phase systems, impact strength modification, interpenetrating polymer networks and ion exchange resins (Chap. 10).

7.2.3 Electron Spectroscopy

Electron spectroscopy for chemical analysis (ESCA) or X-ray photoelectron spectros-copy (XPS) is a recent analytical technique that is used to identify elements distributed in a polymer and to analyze surface structure and composition of (solid) chemical com-pounds[38–49]. ESCA supplies information regarding binding energies or ionisation energies of the electron belonging to the surface of the polymer ($\delta = 1$ to 2 μm) which is influenced by the "chemical environment". Surface effects to be investigated include chemical treatment, photo-oxidation, corona discharge, carbonization and plasma/ion beam technology. Several studies have characterized new materials/structures pro-duced via plasma/ion beam technology. Plasmas are used in the fabrication of inte-grated circuits and pattern generation from resists (Chap. 10). ESCA studies of plasma treated novolak resists have identified the interaction of inert gas (CF_4) or oxygen plasmas with the novolak resist[41, 42].

ESCA has also been used[43] to distinguish the location of sodium ions of a conven-tional caustic catalyzed solid resol with a resol prepared by dispersion technology (Chap. 5). Analyses of phenolic composites by ESCA would be quite beneficial in many critical applications such as carbonized surfaces in brake linings, alumina or SiC-phenolic interface in grinding wheels, copper clad laminates, and other surface ef-fects in adhesive applications.

7.2.4 Mass Spectrometry

Field desorption mass spectrometry (FDMS) has been used for characterization of individual polymeric components and mixtures of commercial polymers. Furthermore the determination of MWD's of various kinds of oligomers and polymers can be ascertained. The advantage of the method compared to other pyrolytic techniques is "in situ" thermal cleavage in the absorbed polymer layer with coincident soft ionization and yields larger ionized subunits of the polymer. FDMS was applied to the identification of molecular species in several novolaks[44] prepared by the condensation of formaldehyde and phenol, *p-tert*-butylphenol, *p*-octylphenol and *p*-phenylphenol. The same technique (FDMS) was utilized[45] to characterize commercial phenolic tackifying resins with *tert*-octylphenol and *tert*-dodecyl-phenol substituents.

7.2.5 High Performance Liquid Chromatography

High performance liquid chromatography (HPLC), sometimes called high pressure liquid chromatography, offers advantages of high speed reuseable columns, automatic and continuous solvent addition, reproducible programmed gradients of solvents, and automatic and continuous monitoring of samples eluted. New HPLC's are available with microprocessors which control chromatographs, detectors, and selective accessories like fraction collectors. Program entries are displayed by LED's, LCD's or CRT's. Besides sensitivity, selectivity and speed, solvent consumption is gaining in importance. The recent comprehensive review by Tesarova and Pacakova[46] of GC and HPLC of phenols contains over 290 references. It lists many satisfactory stationary phases, mobile phase composition, and retention data. A review emphasizing HPLC of resols and novolaks was published by Werner and Barber[47] in 1982.

Inspection of the highly polar structure and thermal instability of hydroxy methyl phenols suggests that HPLC may be the best technique to utilize for the separation of low molecular weight prepolymers up to a MW of 1,000. HPLC has, in fact, developed into an extremely facile and powerful tool to completely characterize resols without the fear of decomposition or the necessity to derivatize the phenol alcohols. HPLC is preferably combined with UV, ^{13}C-NMR or mass spectrometry.

Resol and novolak type prepolymers have been analyzed by Šebenik and Lapanje on Mercosorb SI-60, 10 μ[48] and Li Chromosorb SI-60, 10 μ[49]. The gradient elution technique was an important step towards higher resolution[50]. Newer publications include those of Hitzer et al.[51] (cresol resols and their tung oil reaction products on Li Chromosorb RP 8, 10 μ); Casiraghi et al.[52] (high-*ortho* novolak resins on Bondapack C 18) and Much and Pasch[53].

Current status in resolution and identification is best illustrated by showing the impressive results (Fig. 7.2) obtained by Mechin, Hanton et al.[29, 54].

For a NaOH catalyzed resol (P/F = 1 : 1) reacted at 70 °C, about 30 mono-, di-, and trinuclear hydroxymethyl derivatives were separated by a semipreparative HPLC technique and identified by ^{13}C- and ^{1}H-NMR (Fig. 7.2 and Table 7.3). Surprisingly, no *ortho*-bridged compounds or hemiformals could be identified in this resin. On the other hand, a barium hydroxide catalyzed resol with a high formaldehyde ratio (P/F = 1 : 3.3) showed a much more complex mixture including hemiformals. A comparison of both resins is available by examining GC spectra in Fig. 7.4).

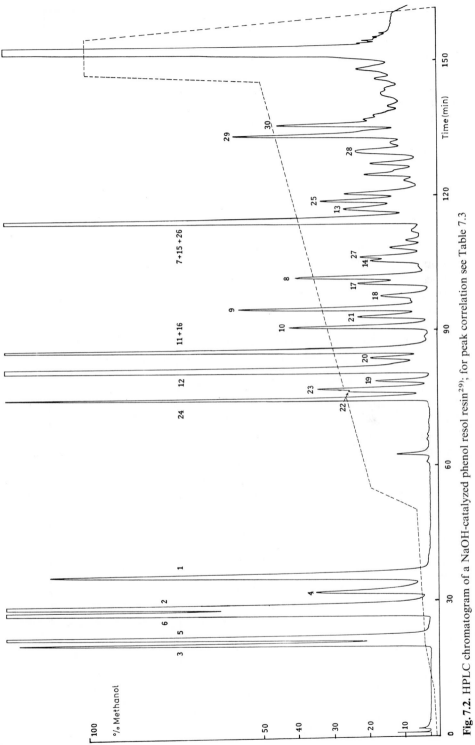

Fig. 7.2. HPLC chromatogram of a NaOH-catalyzed phenol resol resin[29]; for peak correlation see Table 7.3

Table 7.3. Peak correlation to HPLC and GC spectra Figs. 7.2 and 7.4

Formula	Formula	Formula	Formula	Formula
1	7	13	19	25
2	8	14	20	26
3	9	15	21	27

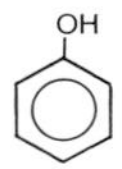

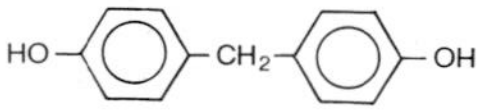

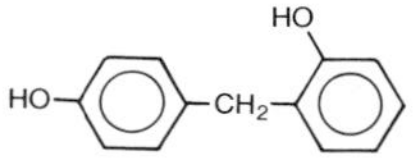

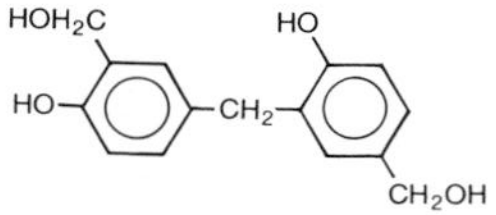

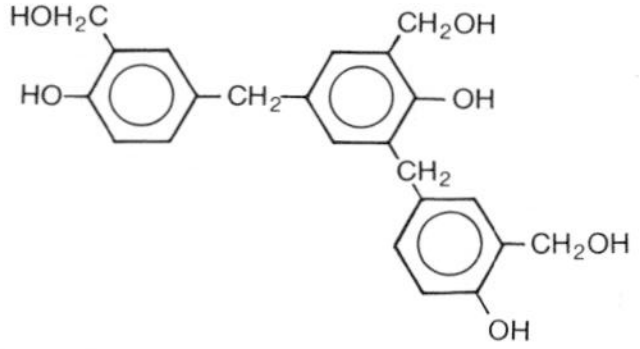

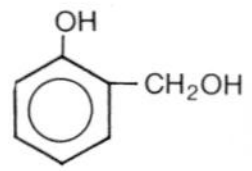

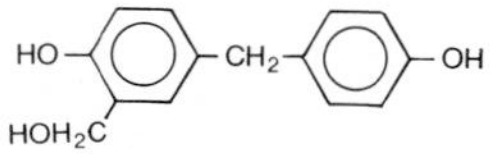

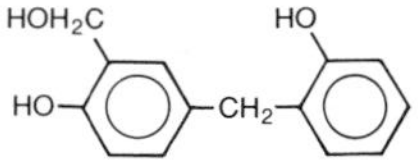

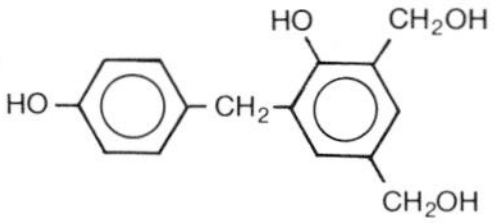

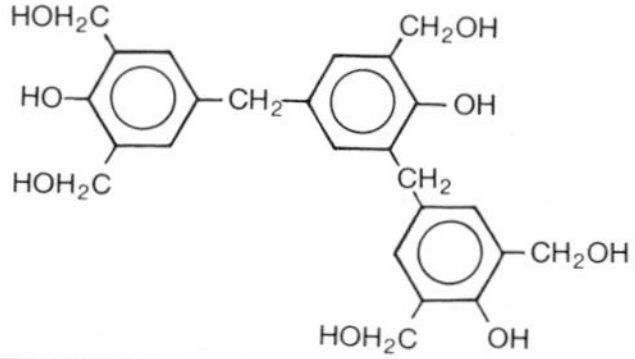

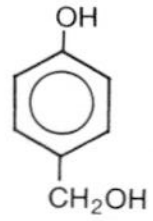

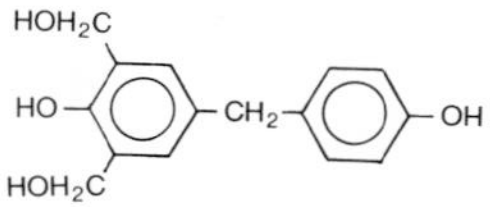

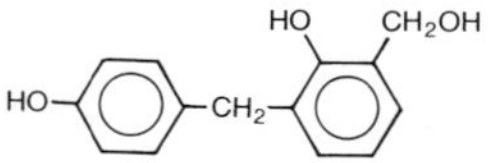

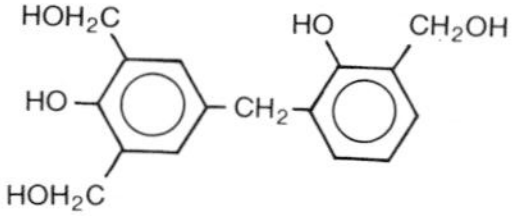

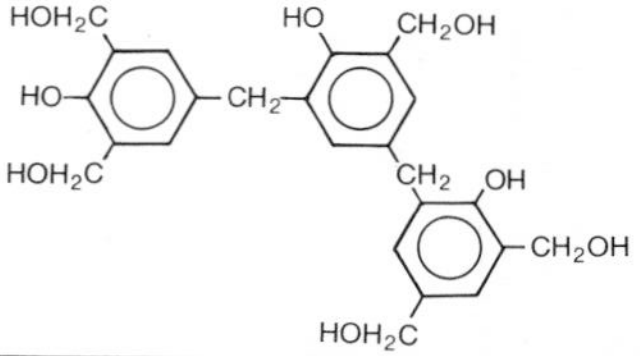

4 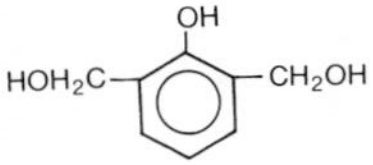	10 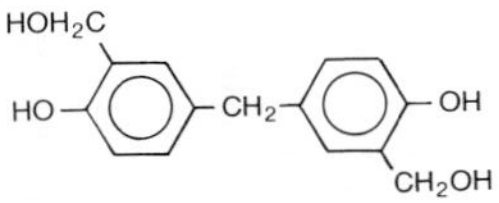	16 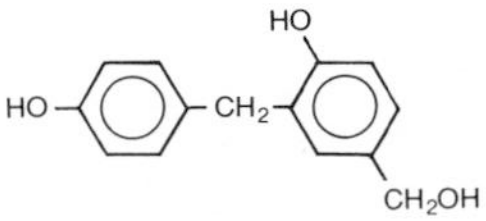	22 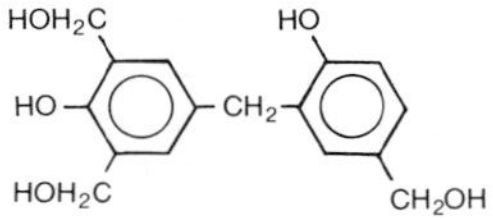	28 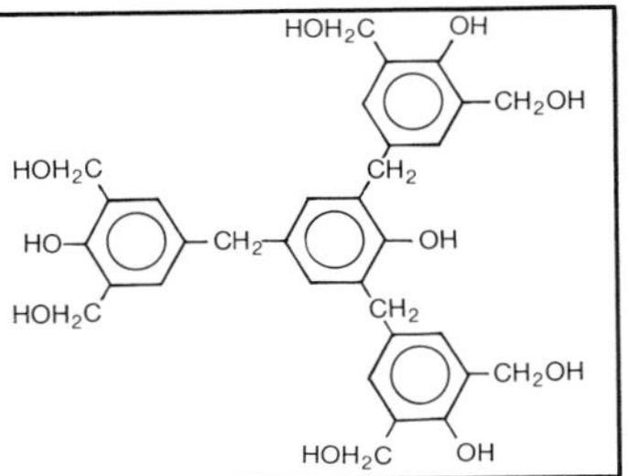
5 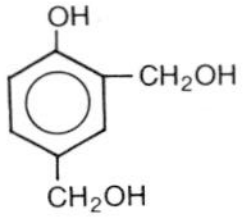	11 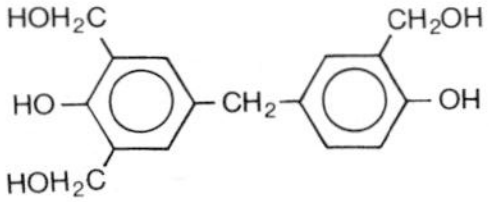	17 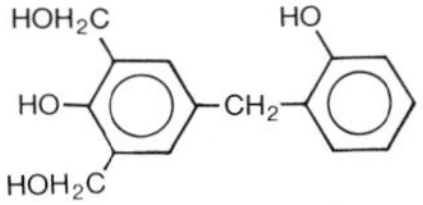	23 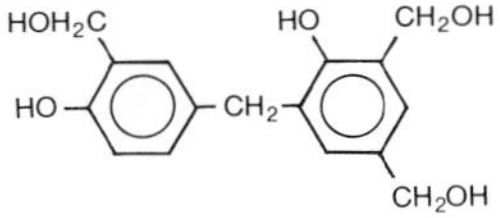	29 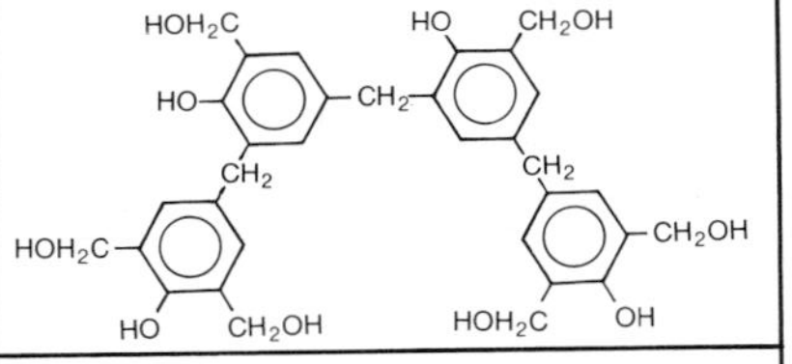
6 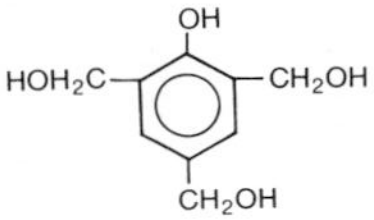	12 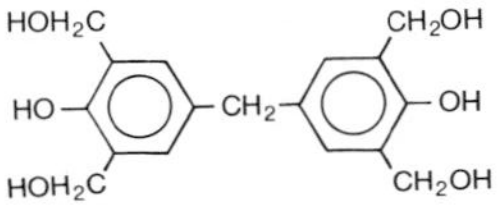	18 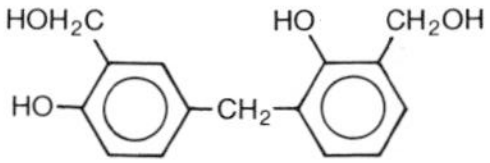	24 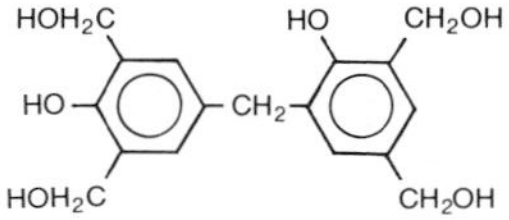	30 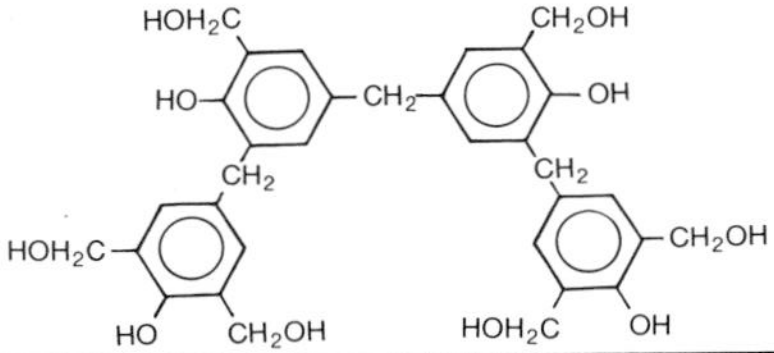
Formula	Formula	Formula	Formula	Formula

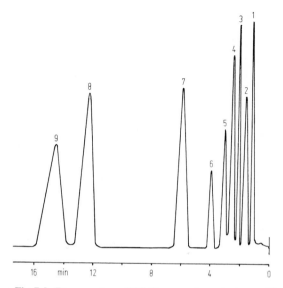

Fig. 7.3. Reverse-phase HPLC separation of phenols[56]. Stationary phase: Si-100-C_4. Eluent: water/methanol. Samples: 1 = methanol; 2 = hydroquinone; 3 = resorcinol; 4 = catechol; 5 = orcinol; 6 = phenol; 7 = p-cresol; 8 = m-xylenol; 9 = o-xylenol

Reverse-phase HPLC has been used to separate phenols, alkylphenols and dihydroxybenzenes by Engelhardt et al.[55, 56] (Fig. 7.3). This process is recognized as the premier and most expedient method for the determination of phenols and phenol alcohols.

Casiraghi et al.[52] demonstrated the elegance of the reverse-phase HPLC procedure with *ortho* linked novolak oligomers. Separation of dimer to octamer was accomplished on Bondapack C 18 column with isocratic ambient elution and 280 nm monitoring.

7.2.6 Paper and Thin Layer Chromatography

Two dimensional paper and thin layer chromatography (TLC) are quite satisfactory for qualitative identification of phenols, phenol alcohols, and DPM prepolymers[57-59]. Reproducibility is within the range of 5–15%. Solvents of differing polarity are utilized as eluents with color development performed by spraying with diazotized p-nitroaniline.

A new modification of TLC is called high pressure thin layer chromatography (HPTLC). Smaller plates are used with uniform absorbent coating and narrow adsorbent particle size. The technique is faster than high performance liquid chromatography (HPLC) because many different samples can be examined simultaneously on the same plate.

7.2.7 Gas Chromatography

A comprehensive review[46] of gas chromatography (GC) and HPLC of phenols was mentioned in the HPLC section. Strong hydrogen bonding and thermal instability of

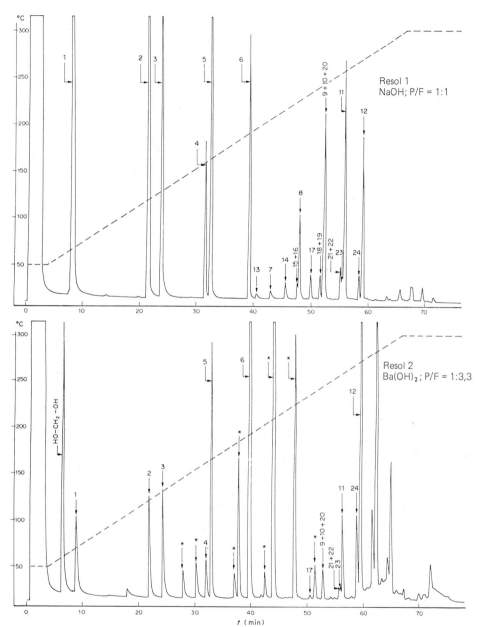

Fig. 7.4. GC chromatogrammes[29] of a NaOH (resol 1, P/F = 1 : 1) and Ba(OH)$_2$ catalyzed resol (resol 2, P/F = 1 : 3.3). Peak identification in Table 7.3 (* = hemiformal compounds)

phenol prepolymers require etherification of the hydroxyl group, e.g. with N,O-*bis*-(trimethylsilyl-trifluoro-acetamide) (BSTFA) for effective GC separation. Haub[60] and Lindner[61] separated mono-, di-, and trinuclear methylolphenols by silylation with BSTFA using e.g. Chromosorb W AW DMCS + 10% OV-101. Significantly improved resolution was obtained by Gnauk et al.[62]. Combined GC/MS analyses of resols and etherified resols led to a complete characterization of the individual components of NaOH catalyzed PF resins[63]. Capillary GC facilitated separation of the trimethylsilyl derivatives.

Figure 7.4 shows the GC separation[29] of a NaOH-catalyzed phenol resol (resol 1, $P/F = 1:1$) and a barium hydroxide catalyzed resol (resol 2, $P/F = 1:3.3$). Both resols are identical with the resins described in the previous section, peak identification is given in Table 7.3. Additional peaks appear in the barium hydroxide catalyzed resin and are attributed to hemiformal species.

7.2.8 Gel Permeation Chromatography

Gel permeation chromatography (GPC) is a special type of liquid-solid chromatography that separates molecules according to molecular size. Polystyrene, PVA or silica gels of very narrow sieve fractions of approximately 5 μm are used as stationary phase. GPC chromatograms[64] are shown in many chapters to illustrate MWD (Figs. 3.1, 3.6, 5.5, 11.3, and 19.1). Molecular weight, size and distribution were recently investigated by Braun and Arndt[65, 66] among others. The studies encompassed solvent and concentration effects and monomer content[67]. Rudin and Fyfe[68] point out, at least in the case of GPC of resols which differ in ability to H bond formation with solvents, the separation of resol components is incomplete due to aggregation. High-alkali resols, e.g. wood binder resins, are favourably analyzed[69] by solubilizing these resins with trichloroacetic acid such that MWD can be measured using conventional instrumentation and techniques.

7.2.9 Thermogravimetric Analysis

Thermogravimetric analysis[70] (TGA) is a method whereby weight loss or gain is measured as a function of temperature or time. The TGA technique is generally used to monitor the thermal stability (or lack of) of polymeric materials. Thermal degradation of phenolics is conveniently followed by TGA, providing information regarding the sequenced steps occuring during different temperature intervals (Chap. 8). Linear correlations have been developed[71] between oxygen index or a specific aspect of polymer flammability and TGA char formation for a large number of char-forming polymers or phenolics (Chap. 6). TGA is of particular importance in the development of brake lining compositions and polymer carbon precursors.

7.2.10 Differential Thermal Analysis

Differential thermal analysis (DTA) technique[70] provides precise temperature measurements during which heat related phenomena occur. DTA and TGA methods were used by Orrel and Burns[72] and others[73] to investigate phenol HMTA and novolak HMTA reactions.

7.2.11 Differential Scanning Calorimetry

Differential scanning calorimetry (DSC) measures temperatures and heats of reaction or transitions and has been an effective method to determine reaction kinetics of resol

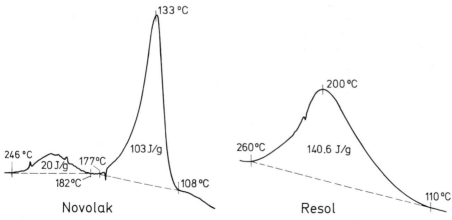

Fig. 7.5. DSC analysis of novolak/HMTA and resol resin curing reaction[78]

and novolak compositions (Fig. 7.5) as well as degree of cure, cure rate and activation energy[74–77]. Data related to phenolic resin glass transition temperatures (Tg) are obtainable by DSC.

Furthermore Erä[79] used DSC (sealed sample cells) to examine the oxalic acid catalyzed PF reaction and reported activation energies of 155–165 kJ/mol. Heat of reaction of 94–100 kJ/mol was determined and is in accord with Vlk's value of 98.2 kJ/mol (see Chap. 5). Similarly resol curing was followed by DSC with sealed cells. The study examined several PF resols with PF ratios of 1:1.5 to 1:2.1. Heat of reaction correlated with PF mol ratio with lower value for higher PF ratio. Experimentally determined heat of reaction of resol resin impregnated paper for high pressure laminates resulted in predicting the cure behavior of practical paper laminates.

7.2.12 Dynamic Mechanical Analysis

The dynamic mechanical analysis (DMA) method is a rapid technique for obtaining the temperature/modulus relationship of a polymeric material. It allows a rapid determination of the glass transition temperature (Tg), especially in the high temperature range. It is applicable to a wide variety of materials from soft elastomers to very rigid thermoset products. DMA has been effective in establishing an interrelationship between chemical properties of phenolic resins, cure rate, and ultimate mechanical properties of wood composites[80]. Besides modulus-temperature characteristics, phase separation of a thermoplastic component during the cure of a filled epoxy-phenolic coating can be determined by DMA[81, 82].

7.2.13 Torsional Braid Analysis

The characterization of the cure phenomena of a thermosetting resin requires the measurement of the elastic and loss moduli as the resin is transformed from liquid to rubbery and ultimately to a fully cured system. These phenomena can be monitored using the torsional braid analysis technique (TBA) in which, e.g., a resin impregnated glass braid is the specimen in a torsion pendulum[83, 84].

The effect of structure on cure and properties of *ortho-ortho'* and random novolak resins cured with HTMA was measured by TBA. A significantly lower activation energy for gelation (105 kJ/mol) was determined for *ortho-ortho'* cured system as compared to 147 kJ/mol for the random material[85]. A comparison of the curing behavior of *ortho* rich and *para* rich resols was studied by TBA[86]. Dynamic mechanical properties showed that the curing rate of the high-*ortho* resol was higher with a corresponding lower rigidity than that of the high-*para* resol.

Both DMA and TBA are routinely employed in the characterization of thermoplastic and thermoset coatings consisting of single or multiphase systems. Gillham and co-workers[83] have developed a time-temperature-transformation (TTT) cure diagram which fully described gelation and vitrification of an epoxy systems. A similar study describing the multiple transitions of phenolics (liquid → rubbery → glassy state) is warranted.

7.3 X-Ray Diffraction Analysis

X-ray diffraction analysis is a very important tool for the determination of molecular conformation. Unfortunately, only a limited number of low molecular weight species forms crystals which are necessary for X-ray investigation. X-ray analysis data for structurally uniform oligomers are reported in Chap. 4.

It is our opinion that additional X-ray diffraction studies and theoretical calculations will supply the basic figures and coordinates for the software required for computerized modelling and a three-dimensional insight even into the extremely complex structures of phenolic polymers. Computer aided molecular design will open new horizons in phenolic resin chemistry.

7.4 Other Characterization

Although phenolic resins are the earliest synthetic resins, many chemical and physical properties are rather complex and as yet are not well understood or fully investigated because of the wide assortment of conformations and components in the resin or limited instrument capability.

7.4.1 Thermodynamic Properties

Thermodynamic properties of phenolic resins are not widely studied. Thermal characteristics related to linear thermal expansion coefficient, specific heat capacity and elastic constants or Young's modulus are reported[87] for a cured resol type resin in the low temperature range (6 to 100 K). The temperature relationship of the specific heat of a resol resin in a wide temperature range (30 to 370 K) was estimated to be $C_p = 0.0042$ T J/K/g[89, 90]. This kind of proportionality has also been observed in a large number of (linear) thermoplastic resins by utilizing an automatic adiabatic calorimeter. Specific heat of wood flour and asbestos PF compounds has been determined (223 K to 473 K) by Wilski[88]. The moisture or water content of PF resins and composites has a pronounced effect on the observed properties.

Table 7.4. Novolak – Mark-Houwink-Sakúrada Expressions[91-94]

$[\eta]$	Novolak	Solvent	Temperature (°C)	$\bar{M}n$	Ref.
$0.019\,\bar{M}n^{0.47}$	Random	Acetone	25	1,000– 8,050	Tobiason
$0.631\,\bar{M}n^{0.28}$	Random	Acetone	30	370–28,000	Kamide
$0.730\,\bar{M}n^{0.28}$	Random	THF	30	540– 4,700	Kamide
$0.813\,\bar{M}n^{0.5}$	High-*ortho*	Acetone	30	690– 2,600	Kamide
$1.075\,\bar{M}vd^{0.20}$	Random	Acetone	20	–	Ishida

7.4.2 Solution Properties

Dilute solution properties of phenolic resins are quite dependent upon solvent type, solvent interaction and molecular configuration. Most dilute solution properties have been confined to novolak resins. Tobiason[91] reported Mark-Houwink-Sakurada (MHS) expression for novolak resin fractions of $\bar{M}n$ from 1,000 to 8,050 (Table 7.4). Kamide and Miyakawa[92] examined random and high-*ortho* novolak resins in acetone and THF at 30 °C and described MHS expressions for these materials. A dilute solution relationship has also been developed between viscosity-diffusion molecular weight in acetone at 20 °C by Ishida and coworkers[93, 94].

Examination of an intermediate molecular weight (MW = 1,154) random novolak by Kamide[92] indicated that the 11 phenolic ring molecule investigated had two branches, one with one phenolic ring and one with two phenolic rings per molecule. The theoretical (computer simulation, Sect. 3.10) and actual (^{13}C-NMR) branching aspects in novolaks have already been mentioned.

The discrepancy in MHS values between random and high-*ortho* resins relates to a consideration of exponent "a" of the MHS equation or $[\eta] = K_m M^a$ and can be interpreted as the difference in molecular configurations of both resins. The high-*ortho* resin is a highly linear low MW polymer whose excluded volume effect in dilute solution is quite small and the exponent "a" can be approximated to 0.5. Thus the random resin is somewhat coiled with two branches while the high-*ortho* material is completely linear within a low MW regime. In the absence of X-ray diffraction data due to the amorphous nature of the novolak resins, dilute solution properties studies give some understanding of novolak resin linearity or branching superimposed with random or high-*ortho* structural configurations. Application of dilute solution properties to liquid resols particularly high-*ortho* (Chap. 5) would be interesting due to the unusually low bulk viscosities reported for these high solids resols (Chap. 10, RIM).

7.4.3 Dipole Moments

Additional fundamental aspects of phenolic resin molecular chain configuration have been examined by Tobiason and coworkers[95, 96] by determining the dipole moments of several substituted PF novolak resins. A systematic listing of dipole moments, temperature coefficients and dipole moment ratios for several dimers, tetramers, and various oligomers ($n = 1$–6) of *p*-cresol have been reported. Extended data calculations have suggested that the conformational angle (methylene linkage) is $\pm 85°$ and relates to the dihedral angle of $85°–90°$ determined by X-ray crystallography for *ortho* linked PF resins (see X-ray section, Chap. 4).

7.5 Resin Macrostructure

Resin macrostructure analysis consists of many conventional analytical methods used to characterize resins for quality control during manufacture or resin properties conveyed to customers. Some of these procedures have been in existence for many decades and are still the preferred method (gel time, B-stage time and flow distance). Most test methods are DIN methods; similar methods are reviewed by Riley[97] and Urbanski[98].

7.5.1 Nitrogen and Water

Total Nitrogen

The Kjeldahl method facilitates the decomposition of nitrogen containing compounds by heating with concentrated sulfuric acid. Carbon is oxidized to carbon dioxide while organically bound nitrogen is converted quantitatively into ammonium sulfate[99]. The decomposition is facilitated by the addition of dehydrating substances like potassium sulfate or catalysts like copper(II)-sulfate. The amount of ammonia displaced by sodium hydroxide addition is determined quantitatively by absorption in a measured amount of acid of known concentration. A rapid total Kjeldahl nitrogen (TKN) method apparatus is available from Brinkmann Instruments and provides data in less than an hour.

HMTA in Powdered Resins

Two methods are used. In the first method, the powder resin is dissolved in a butanol-ethylene glycol mixture and titrated either potentiometrically or by addition of thymol blue as indicator with 0.1 N hydrochloric acid. According to the second method, the determination is performed in acetone solution with 0.1 N perchloric acid or potentiometrically.

Water

The water content of liquid resin is determined by the Karl Fischer titration method according to DIN 51 777. This method is based on the oxidation of iodine. The Karl Fischer solution contains an oxidizing agent J_2 as well as a reducing agent SO_2. The redox reaction occurs only when water is added to the dry system:

$$3\,Py + SO_2 + J_2 + H_2O + CH_3OH \rightleftharpoons 2\,[PyH^+ + J^-] + PyH^+ + CH_3{-}O{-}SO_3^- \qquad (7.4)$$

The equation shows that 1 mol of iodine is equivalent to 1 mol of H_2O. The titration is performed according to the Dead-Stop-technique.

Recently a stable pyridine free Karl Fischer reagent useful in determining water was reported[100]. The newly developed system based on an alkali metal or alkaline earth metal salt of benzoic acid in alcohol and/or glycol allows clearer end point determination, less toxic reagent and less expensive material.

7.5.2 Resin Properties and Quality Control

Resin Content

Two to three grams of resin in a porcelain cup or aluminum metal lid are heated to 150 °C (30 min) in an oven to constant weight. The cup is cooled in a desiccator and weighed.

Viscosity

The most important methods are:

The determination with the falling ball viscosimeter (DIN 53015); the principle of measuring is the rolling and sliding movement of a ball in a sloping, cylindrical tube which is filled with the liquid to be tested.

The use of a cone-plate-rotational viscosimeter (DIN 53229, Paragraph 8.1); the measuring device consists of a plane plate on which a cone is rotating. The substance fills the space between the plate and cone. The time necessary for one measurement is 1–2 min.

The viscosity cup (Ford cup, DIN 59211), the Ubbelohde capillary viscosimeter (DIN 53177) and the rotational viscosimeter (DIN 53229, Paragraph 8.2) are also used.

Water Dilutability

Water dilutability plays a role in the processing of many phenolic resins. When diluting with water, two considerations are noted: the first is turbidity (opalescence) and the second is heavy turbidity (milky or precipitation). It is the heavy turbidity, above all, which is important in dilutability determination.

A measured quantity of liquid resin is diluted with water drop by drop and shaken until a permanent turbidity is achieved. The water dilutability, which is effected by the content of alkali and solvents, depends mainly upon the extent of resin condensation and represents a good criterion for molecular weight.

Specific Gravity

The SG is determined with the pycnometer et 20 °C according to DIN 53217.

Refractive Index

The resin content of a liquid resin can be determined very easily and quickly by the determination of the refractive index according to DIN 53491 with an Abbé-refractometer. The conversion of phenol with formaldehyde during resin production is followed by the refractive index determination.

Melting Point

The MP is determined by the capillary method according to DIN 53181. Since phenolic resins are not uniform compounds with a well-defined MP, it is more reasonable to consider a melting range which is generally within 3 °C.

Bulk Weight and Stamp Volume of Powder Resins

The bulk weight is determined according to DIN 53468. A 100 ml measuring cylinder, which is cut off at the 100 ml mark is used. The powder is poured in from a height of 2 cm above the rim of the cylinder. The liter weight of the powder is an approximate criterion of the fineness of grain and should fluctuate only within very narrow limits.

For the determination of the stamp volume (DIN 53194), the reproducibility is somewhat better. Special stamp volumeters with a design fixed by standards serve for the determination.

Sieve Analysis

A complete sieve analysis to determine the particle size distribution, which in accordance to DIN 53734 is to be performed with an air blast sieving equipment, is too expensive for quality control. Therefore, in most cases only the residue on the 0.06 mm sieve (10,000 mesh per cm^2) is determined according to DIN 1171. For example, for powdered resins to be used for the manufacturing of grinding wheels, this residue should not exceed 1–2%.

Determination of B-Stage Time

B-stage time represents the time when the resin changes from the liquid (A-stage) to the rubber-like B-stage at a distinct temperature. The determination is performed on an iron plate with a diameter of 200 mm and a height of 20 mm. Round cavities with a diameter of 20 mm and 5 mm deep are arranged on a concentric circle with a diameter of 130 mm. A hole for the thermometer is drilled in the center of the plate. The iron plate is put on an electrically heated plate with exact temperature control. 0.5 g of liquid phenol resin or 0.15 g powdered resin are filled into the cavity where the temperature is held at 130 or 150 °C and slightly stirred with a thin glass rod. As soon as a certain degree of three-dimensional crosslinking is reached, the resin shows a rubber-like condition and is unable to flow. The time in seconds until this stage is reached (B-stage) is the measurement for the curing rate.

Gel-Time

A weighed amount of resin is filled into a test tube and placed into an oil bath at 130 °C. The increase in viscosity is determined with a glass rod formed like a piston. As soon as the gel point of the resin is reached, the liquid resin becomes stringy or almost rigid. Elapsed time is automatically recorded (DIN 16945).

Flow Distance

Information about flow and reactivity of solid resin-HMTA mixtures is obtained by determining the flow distance in an oven at 125 °C. Half gram of resin is pressed into a pellet with a diameter of 12.5 mm and 4.8–5.2 mm height on a pellet press. The flow distance is determined on a hot, inclined glass plate. A tilting device along with the glass plate is preheated to the required temperature in an oven which is set at a temperature of 125 °C. The pellet is placed on the horizontal glass plate which is slanted after three minutes to form an angle of 60 ° with the base plate of the device. The whole device is left in the oven in this position for 30 min. The length of the flow in mm is the measurement for the melt viscosity and curing rate. Resins of high molecular weight and low reactivity flow very slightly and only an expansion of the pellet is observed. In this case, the diameter of the pellet is expressed as flow distance.

7.6 Structural Analysis of Cured Resins

The intractibility of cured PF resins makes it exceedingly difficult to adequately identify the resulting cured material. In the cured state the resins are infusible, covalently crosslinked, thermally stable network polymer structures. The technique of pyrolysis/gas chromatography has probed the structure of cured resins as well as the corresponding behavior at elevated temperature. Analysis of the gaseous degradation products[101, 102] by GC provides some understanding of the cure mechanism. Elaborate analyses consisting of GC in conjunction with HPLC, GPC, and PC were recently reported for several compounds evolved during thermal degradation of cured resol[103]. These included phenol, cresols, xylenols, and other non-phenolic aromatic compounds thermally generated in the presence of tetralin, a H donor.

Quite recently the use of solid state NMR[35] has been applied in the examination of HMTA cured random and high-*ortho* novolaks[35] and resol cure[37]. Structural assignments by solid state NMR[36] were developed during the thermal decomposition of resol at elevated temperatures (Chap. 8).

7.7 References

1. Haslem, J., Willis, H. A., Squirell, D. C. M.: Identification and Analysis of Plastics, 2. ed. London: Iliffe Books 1972
2. Ettre, L. S., Obermüller, E.: Phenols. In: F. D. Snell, L. S. Ettre (ed.): Encyclopedia of Industrial Chemical Analysis, Vol. 17, New York: Interscience
3. Tsuge, M.: Progress in Organic Coatings, 9, 107 (1981)
4. Klopfer, W.: Introductions to Polymer Spectroscopy. Springer-Verlag, Heidelberg, New York, 1984
5. Wesp, E. F., Brode, W. R.: J. Am. Chem. Soc. 56, 1037 (1934)
6. Soloway, S., Wilen, S. H.: Anal. Chem. 24, 970 (1952)
7. Koppeschaar, W. F.: Z. anal. Chem. 15, 233 (1876)
8. Feigl, F., Anger, V.: Analyt. Chem. 33, 89 (1961)
9. Harborne, J. B.: Chromatography of Phenolic Compounds. In: E. Heftmann (ed.): Chromatography, 3. ed. New York: Van Nostrand Reinhold 1975
10. Bleidt, R. A.: Formaldehyde. In: F. D. Snell, L. S. Ettre (ed.): Encyclopedia of Industrial Chemical Analysis, Vol. 13. New York: Interscience
11. Hummel, D. O.: Polymer Spectroscopy. Weinheim: Verlag Chemie 1976
12. Hummel, D. O., Scholl, K.: Infrared Analysis of Polymers, Resins and Additives. Carl Hanser-Verlag Chemie 1973
13. Bender, H. L.: Mod. Plastics 30, 136 (1953)
14. Fanconi, B. M.: J. Test. Eval., 12 (1) 33 (1984)
15. Sojka, S. A., Wolfe, R. A., Guenther, G. S.: Macromolecules, 14, 1539 (1981)
16. Zaks, Y., Lo, J., Raucher, D., Pearce, E. M.: J. Appl. Polym. Sci. 27, 913 (1982)
17. Bovey, F. A.: High Resolution NMR of Macromolecules. New York: Academic Press 1972
18. Kopf, P. W., Wagner, E. R.: J. Polym. Sci.-Polym. Chem. Ed. 11, 939 (1973)
19. Woodbrey, J. C., Higginbottom, H. P., Culbertson, H. M.: J. Polym. Sci. Part. A 3, 1079 (1965)
20. Hirst, R. C., Grant, D. M., Hoff, R. E., Burke, W. J.: J. Polym. Sci. Part. A 3, 2091 (1965)
21. Navratil, M., Pospisil, L., Fiala, Z.: Plaste und Kautschuk, 29, 6 (1982)
22. Tong, S. N., Park, K. Y., Harwood, H. J.: ACS Polymer Preprints, 24 (2), 196 (1983)
23. Yoshikawa, T., Kumanotani, J.: Makromol. Chem. 131, 273 (1970)

24. Yoshikawa, T., Kimura, K.: Makromol. Chem. 175, 1001 (1974)
25. Sojka, S. A., Wolfe, R. A., Dietz, Jr., E. A., Dannels, F.: Macromolecules 12, 767 (1979)
26. Kim, M. G., Tiedeman, G. T., Amons, L. W.: "Weyerhauser Science Symposium", Vol. 2, 1981
27. Dradi, E., Casiraghi, G., Satori, G., Casnati, G.: Macromolecules 11, 1295 (1978)
28. Walker, L. E., Dietz, Jr., E. A., Wolfe, R. A., Dannels, B. F., Sojka, S. A.: ACS Polymer Preprints 24 (2) 177 (1983)
29. Mechin, B., Hanton, D., Le Goff, J., Tanneur, J. P.: Eur. Polym. J. 20, 333 (1984)
30. Pasch, H., Goetzky, P., Grundemann, E., Raubach, H.: Acta Polymerica, 32, 14 (1981)
31. Braun, D., Krauße, W.: Angew. Makromol. Chem. 118, 165 (1983)
32. Mackey, J. H., Tiede, M. L., Sojka, S. A., Wolfe, R. A.: ACS Polymer Preprints 24 (2) 179 (1983)
33. Mehring, M.: Principles of High Resolution NMR in Solids. Springer-Verlag, Heidelberg, New York, 1983
34. Shaefer, J., Stejskal, E. O.: J. Am. Chem. Soc. 98, 1031 (1976)
35. Bryson, R. L., Hatfield, G. R., Early, T. A., Palmer, A. R., Maciel, G. E.: Macromolecules 16, 1669 (1983)
36. Fyfe, C. A. et al.: Macromolecules 16, 1216 (1983)
37. Maciel, G. E., Chuang, I-S., Gollob, L.: Macromolecules 17, 1081 (1984)
38. Briggs, D.: Ed. Handbook of X-Ray & UV Photoelectron Spectroscopy, Heyden & Son, Ltd., Phil, PA 1978
39. Dwight, D. W., Thomas, H. W.: Polymer Preprints 22 (1) 302 (1981)
40. Clark, D. T.: Applications of ESCA to Structure & Bonding in Polymers in characterization of Metal Polymer Surface (L. H. Lee editor)
41. Takacs, M., et al.: ACS Polymer Preprints 21 (1) 141 (1980)
42. Leary, Jr., H. J., et al.: ibid 25 (1) 328 (1984)
43. Brode, G. L.: "Phenolic Resins" Vol. 17 In: Kirk-Othmer, Encyclopedia of Chemical Technology, 3rd Ed., Wiley, NY 1982, p. 401
44. Saito, J., Toda, S., Tanaka, S.: J. Thermosetting Plastics 2, 72 (1982), Japan; C.A. 96, 53108e (1982)
45. Lattimer, R. P., Hooser, E. R., Diem, H. E., Rhee, C. K.: Rubber Chem. Techn., 55 (2) 442 (1982)
46. Tesarova, E., Pacakova, V.: Chromatographia 17 (5) 269 (1983)
47. Werner, W., Barber, O.: Chromatographia 15, 101 (1982)
48. Šebenik, A., Lapanje, S.: J. Chromatogr. 106, 454 (1975)
49. Šebenik, A.: J. Chrom. 160, 205 (1978)
50. Van der Maeden, F. P. B., Biemond, M. E. F., Janssen, P. C.: J. Chromatogr. 149, 539 (1978)
51. Hitzer, H., Bethke, K., Quast, O.: Plaste und Kautschuk 27, 130 (1980)
52. Cornia, M., Sartori, G., Casnati, G., Casiraghi, G.: J. Liquid Chrom. 4, 13 (1981)
53. Much, H., Pasch, H.: Acta Polymerica 33, 366 (1982)
54. Mechin, B., Hanton, D.: in print
55. Engelhardt, H., Weigand, N.: Anal. Chem. 45, 1149 (1973)
56. Karch, K., Sebastian, I., Halasz, I., Engelhardt, H.: J. Chromatogr. 122, 171 (1976)
57. Harborne, J. B.: Chromatography of Phenolic Compounds. In: E. Heftmann (ed.): Chromatography, 3. ed. New York: Van Nostrand Reinhold 1975
58. Pastuska, G., Petrowitz, H. J.: Chemiker-Ztg. 86, 311 (1962)
59. Haub, H. G., Kämmerer, H.: J. Chromatogr. 11, 487 (1963)
60. Haub, H. G.: Unpublished
61. Lindner, W.: J. Chrom. 151, 406 (1978)
62. Gnauck, R., Habisch, D.: Plaste und Kautschuk 27, 485 (1980)
63. Anthony, C. M., Kemp, G.: Ang. Makromol. Chem. 115, 183 (1983)
64. Knop, A., Trapper, W.: Unpublished
65. Braun, D., Arndt, J.: Angew. Makromol. Chem. 73, 133 (1978) and 73, 143 (1978)
66. Braun, D., Arndt, J.: Fresenius Z. Anal. Chem. 294, 130 (1979)
67. Schulz, G., Gnauck, R., Ziehbarth, G.: Plaste und Kautschuk 29, 398 (1982)
68. Rudin, A., Fyfe, C. A., Vines, S. M.: J. Appl. Polym. Sci. 28, 2611 (1983)
69. Bain, D. R., Wagner, J. D.: Polymer 25, 403 (1984)

70. Wendlandt, W. (ed.): Thermal Methods of Analysis, 2. ed. New York: John Wiley, 1974
71. Van Krevelen, D.W.: Polymer 16, 615 (1975)
72. Orrell, E.W., Burns, R.: Plastics & Polymers 36, 469 (1968)
73. Katovic, Z.: J. Appl. Polym. Sci. 11, 85, 95 (1967)
74. Kay, R., Westwood, A.R.: Europ. Polymer J. 11, 25 (1975)
75. Kamal, M.R., Sourour, S.: Polym. Engng. and Sci. 13, 59 (1973)
76. Taylor, L.J, Watson, S.W.: Anal. Chem. 42, 297 (1970)
77. Katovic, A., Stefanic, M.: ACS Polymer Preprints 24/2, 191 (1983), Washington DC Meeting
78. Bakelite GmbH: Unpublished
79. Erä, V.A.: J. Ther. Anal. 25, 79 (1982)
80. Young, R.H., Barnes, E.E., Caster, R.W., Krutscha, W.P.: ACS Polymer Preprints, 24 (2), 199 (1983)
81. Prime, R.P.: NATAS Preprints, Sept. 23-6 (1984)
82. Burns, J.M.: NATAS Preprints, Sept. 23-6 (1984)
83. Gillham, J.K.: Chapter 3, "TBA of Polymers" in Developments in Polymer Characterization 3, Applied Science Publishers, London, 1982
84. Enns, J.F., Gillham, J.K.: J. Appl. Polym. Sci. 28, 2567 (1983)
85. Mackey, J.H., Lester, G., Walker, L.E., Gillham, J.K.: ACS Polymer Preprints 22 (2) 131 (1981)
86. Fukuda, A., Horiuchi, H.: Netsu Kokasei Jushi 2 (3), 125 (1981); C.A. 96 105140 f (1982)
87. Harwood, C., Wostenholm, H.H., Yates, B., Badami, D.V.: J. Polym. Sci. Polym. Phys. Ed., 16, 759 (1978)
88. Wilski, H.: Progr. Colloid. & Polym. Sci. 64, 33 (1978)
89. Chang, S.S.: In "Thermal Analysis in Polymer Characterization", E.A. Turi, Ed., P.98, Heyden (1981)
90. Chang, S.S.: ACS Polymer Preprints 24 (2), 187 (1983)
91. Tobiason, F.L., Chandler, C., Schwarz, F.E.: Macromolecules 5, 321 (1972)
92. Kamide, K., Miyakawa, Y.: Makromol. Chem. 179, 359 (1978)
93. Ishida, S., Hakagawa, M., Suda, H., Kaneko, K.: Kobunshi Kagadu, 28, 250 (1971)
94. Ishida, S., Kitagawa, T., Makamoto, Y., Kaneko, K.: C.A. 100, 52413 q (1984)
95. Tobiason, F.L., Houglum, K., Shanafelt, A., Hunter, P., Böhmer, V.: ACS Polymer Preprints 24 (2) 181 (1983)
96. Tobiason, F.L.: "Phenolic Resins Chemistry and Application", Weyerhaeuser Co., Tacoma, WA (1981), pp. 201–211
97. Riley, H.E.: "Phenolic Resins" In: GM Kline ed. Analytical Chemistry of Polymers Vol. XII Part I Interscience Publishers, Inc., 1959, pp. 239–271
98. Urbanski, J.: "Handbook of Analysis of Synthetic Polymers and Plastics", Halsted Press, a division of John Wiley & Sons, Inc., NY 1976, Chap. 6
99. Pohloudek-Fabini, R., Beyrich, Th.: Organische Analyse. Leipzig: Akademische Verlagsges., 1975
100. Merck Patent GmbH: US-PS 4,416,997 (1983)
101. Conley, R.T.: Thermal Stability of Polymers, Chap. 11, New York: Marcel Dekker, Inc., 1970
102. Gautherot, G.: Contribution a l'étude de la dégradation des résines phénoliques. Office national d'études et de recherches aérospatiales 1969
103. Braun, D., Steffen, R.: Fresenius Z. Anal. Chem. 307, 7 (1981)

8 Degradation of Phenolic Resins by Heat, Oxygen and High-Energy Radiation

8.1 Thermal Degradation

For particular applications, the high and low temperature characteristics of a polymer material are of critical importance in the ultimate use of the polymer. The general term polymer thermal stability is not completely accurate in defining thermal characteristics of a polymer. It mainly refers to changes of one or more selected physical properties that are temperature and time dependent. Another characteristic of polymer stability is the propensity of the polymer to degrade. TGA appears to have only a restricted value, however, in relating the relative performance in an actual environment, which is far different from laboratory conditions. Some phenolic copolymers with relatively better TGA, e. g. PF-resins containing naphthol or p-phenylphenol, show inferior ablative performance, indicating that the mechanical strength of the material is also an important factor[1].

In general, all thermochemical and physical phenomena must be considered. PF-resins are known to be quite temperature resistant polymeric materials and yield high amount of char during pyrolysis. These characteristics are important for their use as ablative materials, friction materials, grinding wheels, HT-resistant molding materials and polymer carbon. The thermal degradation of PF-resins[1, 2] is conveniently segmented into three stages, indicated by weight loss and volume change[3].

In the first stage to 300 °C, the polymer remains virtually intact. The quantity of gaseous components released during this stage is relatively small (1–2%). It consists mainly of water and unreacted monomers, phenol and formaldehyde, which were entrapped during cure.

Decomposition commences at approximately 300 °C. From 300 to 600 °C mainly gaseous components are emitted. The reaction rate reaches a maximum within this range. In this second stage, water, carbon monoxide, carbon dioxide, methane, phenol, cresols and xylenols are released. During this degradation stage random chain scission occurs; no depolymerization takes place. Concurrently, a high concentration of ketone- and carboxyl groups can be observed by IR. Shrinkage, however, is relatively low. The internal porosity increases with a corresponding decrease in density.

In the third stage above 600 °C, CO_2, CH_4, H_2O, benzene, toluene, phenol, cresols and xylenols are liberated. High shrinkage is characteristic; density increases and the permeability decreases considerably. The electrical conductivity increases and is documented at temperatures of 600–700 °C for reinforced glass or cloth filled phenolics[4].

The mechanism of the thermo-chemical degradation of novolaks was thoroughly investigated by Conley[5] and Gautherot[6]. It was stated that there is always a thermo-oxidative process taking place regardless of whether the pyrolysis reaction occurs in

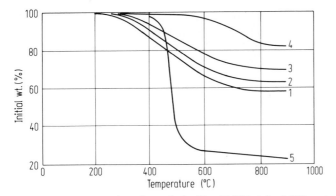

Fig. 8.1. Thermogravimetric analysis (N$_2$, 15 °C/min) of different phenolic resins *1–3*, Poly-*p*-phenylene *4* and Polycarbonate *5*.
1 Phenol-novolak resin, 10% HMTA; *2* Phenol novolak/resol resin 60:40, 6% HMTA; *3* Boronmodified phenol resin (18% B); *4* Poly-*p*-phenylene; — *5* Polycarbonate

an oxidative or inert atmosphere. The high oxygen content of phenolic resins is the reason for this process. The reaction steps described below have been proposed by Conley[5].

The methylene bridge is the thermodynamically most stable cross-link in PF-polymers and therefore the prevailing one in cured resins. The degradation process is dependent upon the stability and concentration of the dihydroxyphenylmethane units so that different resins show an almost similar behavior if the curing process is complete.

The first step of the thermo-oxidative degradation is the formation of hydroperoxide followed by decomposition to dihydroxybenzophenone and benzhydrol structures (8.1 and 8.2).

$$(8.1)$$

$$(8.2)$$

$$(8.3)$$

$$(8.4)$$

The sequence for this reaction is in accord with the observed increase in the IR-absorption at 3.0 μm (hydroxyl group) and 6.05 μm (carboxyl group) and the identification of volatile products. The decarboxylation and decarbonylation reactions occur at approximately 300 °C and 500 °C respectively.

A second fragmentation reaction and hydrogen abstraction leads to cresol and methane formation. The formation of methane occurs in increasing amounts above 400 °C.

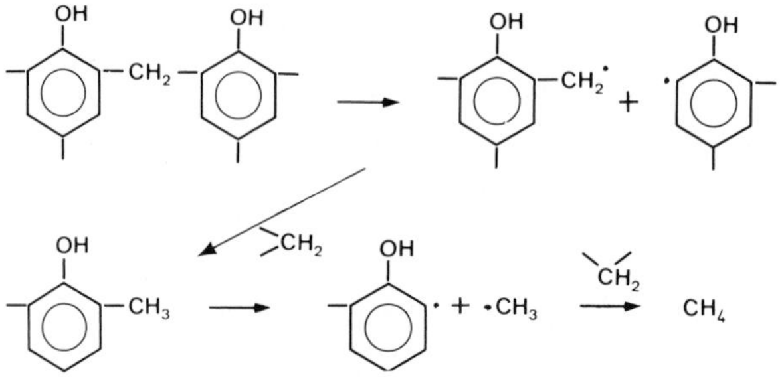

(8.5)

(8.6)

For the formation of benzene, toluene and benzaldehyde the scission of the hydroxyl group is a necessary prerequisite.

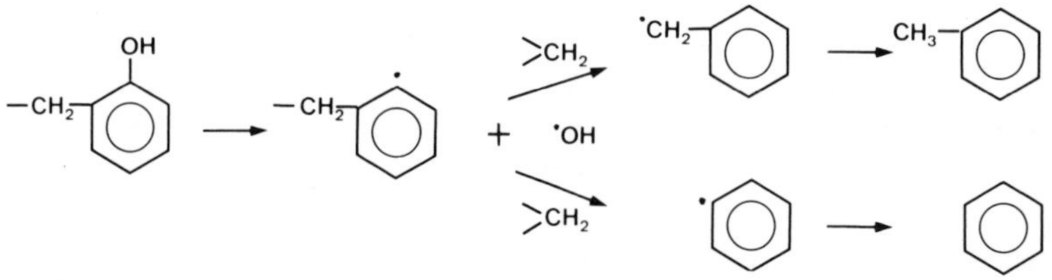

(8.7)

(8.8)

The unusual splitting of phenols to aromatic hydrocarbons is rather complex mechanistically and requires special reaction conditions. The formation of aromatic hydrocarbons is observed in metal casting with phenolic resin bonded sand molds.

Electrical conductivity changes associated with thermal decomposition between 25–700 °C have been correlated with TGA and mass spectroscopy of the decomposition products[7]. The thermal decomposition of cured resol resins has been recently investigated by solid state NMR spectroscopy (Chap. 7) and closely parallels the earlier mechanistic pathways except methylol groups are expelled or oxidized to aldehyde:

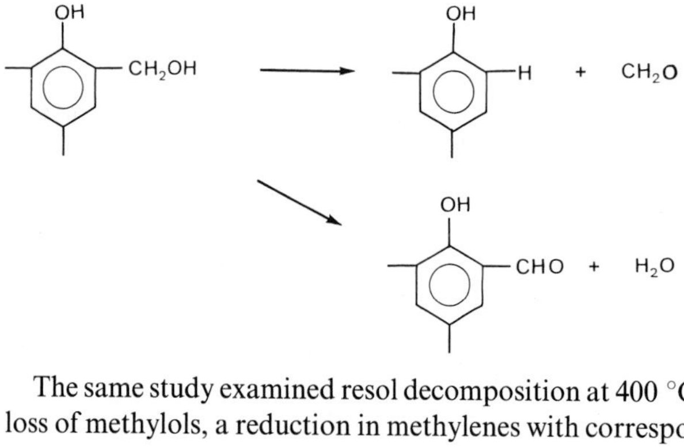

(8.9)

The same study examined resol decomposition at 400 °C under vacuum. A complete loss of methylols, a reduction in methylenes with corresponding appearance of methyl groups was noted. No evidence of oxidation was observed with structural and chemical integrity of the sample preserved[8]. In another study[9] the thermochemical characterization of some thermally stable thermoplastics and thermoset polymers including a modified phenolic resin was performed to assess their potential for use in aircraft composites.

8.2 Oxidation Reactions

Phenolic compounds can easily be oxidized to higher MW materials containing phenolic and quinone structural elements. The first step in this one electron abstraction reaction is the formation of relatively stable phenoxy radicals[10]. Their lifetime can vary from 10^{-3} s, in the case of the non-substituted aryloxy radical, up to hours or days by suitable substitution in 2, 4, 6 positions due to increased resonance stabilization and steric hindrance. Some of these aryloxy radicals are highly colored (and paramagnetic) as monomers.

Phenoxy radicals undergo a variety of reactions depending on substitution and on the presence of other compounds. Reacting with other radicals C–C or C–O coupling may occur; this is illustrated with p-cresol as an example (8.10), (8.11).

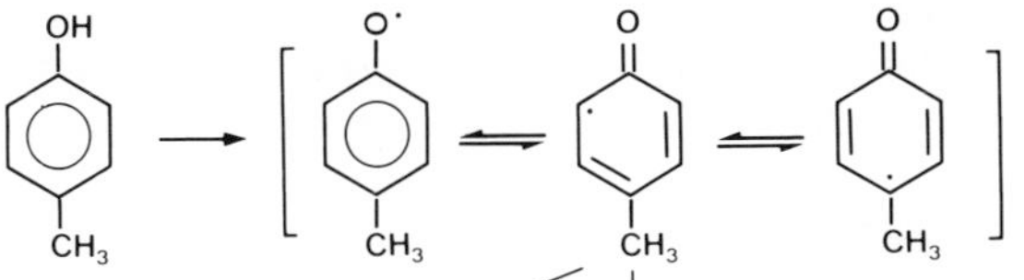

(8.10)

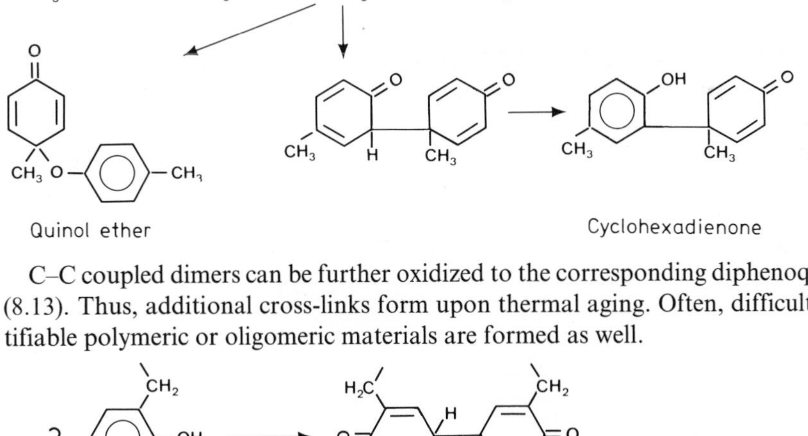

(8.11)

Quinol ether   Cyclohexadienone

C–C coupled dimers can be further oxidized to the corresponding diphenoquinones (8.13). Thus, additional cross-links form upon thermal aging. Often, difficultly identifiable polymeric or oligomeric materials are formed as well.

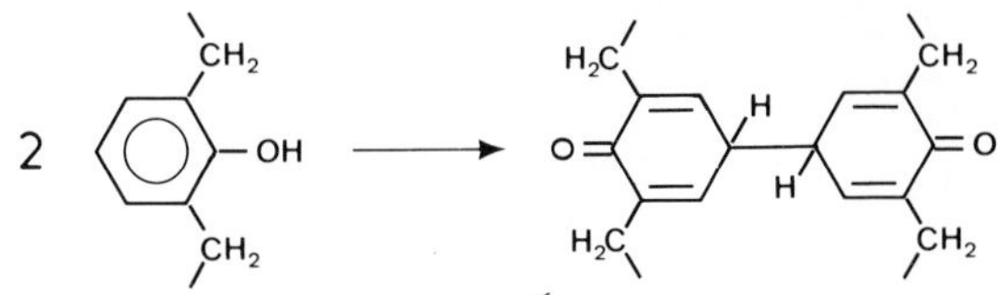

(8.12)

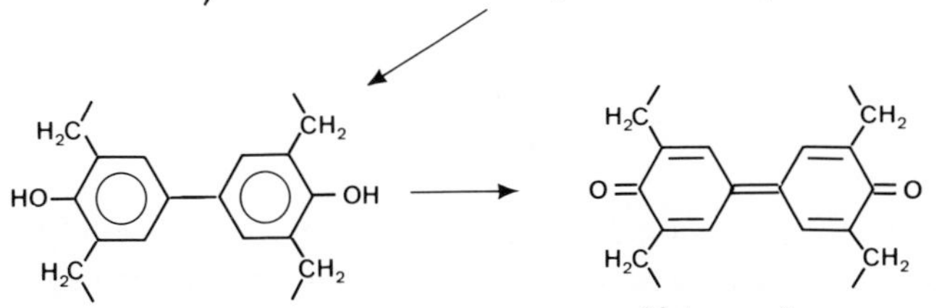

(8.13)

Diphenoquinone

Thus, there are two highly susceptible species in cured phenolic resins when exposed to thermo-oxidative stress: they are the methylene bridge and the phenolic hydroxyl group. Thermal resistance can be improved by crosslinking phenols with heteroatoms or transformation of the phenolic hydroxyl group (see Chap. 9).

8.3 Degradation by High-Energy Radiation

Phenolic resins and phenolic resin molding compounds are of considerable interest for use in electrical components in nuclear power equipment and high-voltage accelerators, equipment components for handling radioactive materials, electrical and structural elements of space vehicles and as protective coatings in nuclear power plants. In these applications high thermal resistance is required in conjunction with high-energy radiation.

When high-energy radiation [γ- and X-rays, neutrons, electrons, protons and deuterons], passes through matter, a strong interaction occurs within the nucleus or within the orbital electrons resulting in the dissipation of a large portion of the incidental energy[11, 12]. The final result of such interaction in polymeric materials is the formation of ions and radicals followed by rupture of chemical bonds. Concurrently, new bonds are formed, followed by crosslinking and degradation at different rates. The difference in the corresponding rate constants determines the resistance to high-energy radiation. The rates of degradation are much lower for polymers containing aromatic rings because of resonance stabilization of the transient species. Generally, rigid molecular structures, i.e. thermoset materials, are

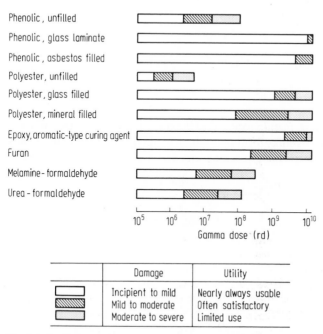

Fig. 8.2. Radiation resistance of thermosetting resins[13]

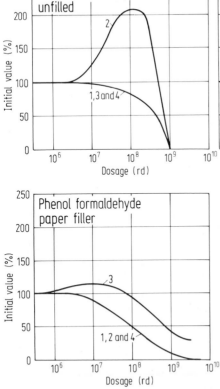

Fig. 8.3. Effect of radiation on mechanical properties of phenolic resins according to van de Voorde and Restat[14].

Curve No	Property
1	Tensile strength
2	Elongation
3	Elastic modulus
4	Shear strength

more resistant than flexible thermoplastic and elastomeric structures. TGA provides satisfactory, but only a qualitative indication of the radiation resistance of the material under study. Steric effects contribute to the probability of ion- or radical recombinations. Comparable to the effect of antioxidants, some substances added in minor quantities have a well defined stabilizing effect, but the mechanism of this stabilization is not well understood. Radiation resistance is usually improved by the addition of mineral fillers[16]. Conversely, there are known additives which accelerate the deterioration. They are called radio-sensitisers. An important example of a material inducing this effect in phenolic resins is cellulose.

A summary of the radiation resistance of different thermosetting materials is shown in Fig. 8.2.

According to the schematic, phenolic resins reinforced by glass- or asbestos fibers can be regarded as very radiation resistant synthetic materials. Non-filled phenolic resins have relatively low radiation resistance. The oxygen content of phenolic resins has a considerable adverse effect on radiation resistance.

The main causes of failure of electrical insulation in nuclear radiation environment are primarily loss of mechanical strength or the evolution of gaseous by-products, rather than changes in electrical properties[11]. Van de Voorde and Restat[14] have reported the relationship of high-energy radiation and the mechanical strength of filled and unfilled phenolic resins (Fig. 8.3).

Table 8.1. Amount of gaseous by-products released by phenolic molding compounds and other plastic materials during high energy radiation impact[15]

Polymer	Filler	Gas evolved cm^3/g Radiation dose Mrd				
		50	100	200	500	1000
Phenolic resin	Wood flour	0.58	1.07	2.09	4.91	8.85
Phenolic resin	Asbestos	0.11	0.18	0.39	1.07	1.97
Polyester	Mineral flour/					
	cellulose	1.19	2.22	4.00	10.52	19.40
Polyamide 6	–	1.32	2.39	4.68	10.75	–
Polycarbonate	–	1.12	2.00	3.60	8.09	13.69
Polyethylene terephthalate	–	0.22	0.40	0.72	1.58	–
Polystyrene	–	0.07	0.11	0.17	0.30	0.42

The evolution of gaseous by-products for a variety of polymeric materials by high-energy radiation[15] is summarized in Table 8.1.

Principally, gas evolution is dependent on the type of filler and proceeds parallel to the loss of mechanical strength, as shown in Fig. 8.2. A chemical identification of the gaseous degradation products was not included in the study. However, they probably consist mainly of hydrogen and minor quantities of methane and higher hydrocarbons, carbon monoxide and carbon dioxide.

8.4 References

1. Marks, B. S., Rubin, L.: ACS-Organic Coatings and Plastics Chemistry, 28/1, 94 (1968)
2. Goldstein, H. E.: ACS-Organic Coatings and Plastics Chemistry, 28/1, 131 (1968)
3. Jones, R. A., Jenkins, G. M.: Volume Change in Phenolic Resin During Carbonization, Carbon 76, Baden-Baden 27.6–2.7.1976
4. Johnson Jr., R. T., Biefeld, R. M.: Sandia National Laboratories Report SAND 81–0786, 1984
5. Conley, R. T.: Thermal Stability of Polymers, Chap. 11, New York: Marcel Dekker Inc. 1970
6. Gautherot, G.: Contribution a l'étude de la dégradation des résines phénoliques. Office national d'études et de recherches aérospatiales 1969
7. Johnson Jr., R. T., Biefeld, R. M.: Polymer Engineering and Science 22, 147 (1982)
8. Fyfe, C. A. et al.: Macromolecules 16 (7), 1216 (1983)
9. Kourtides, D. A., Gilwee, W. J., Parker, J. A.: Polymer Engineering and Science, 19 (1), 24 (1979) and 19 (3) 226 (1979)
10. Mihajlovič, M L., Cekovič, Z.: Oxidation and Reduction of Phenols. In: S. Patai (ed.): The Chemistry of the Hydroxyl group, New York: Interscience 1971
11. Van de Voorde, M. H.: Effects of Radiation on Materials and Components: CERN 70–5, European Organization for Nuclear Research, Geneva 1970
12. Parkinson, W. W.: Radiation Resistant Polymers. In: Encyclopedia of Polymer Science and Technology, Vol. 11, New York: Wiley 1969
13. Batelle Memorial Institute, Columbus: Radiation-effects, State of the Art, 1965–1966, REIC Report 42 (1966)
14. Van de Voorde, M. H., Restat, C.: Selection Guide to Organic Materials for Nuclear Engng. CERN 72–7, European Organization for Nuclear Research, Geneva 1972
15. Morgan, J. T., Stapelton, G. B.: Gas Evolution from Plastic Materials by High Energy Radiation, Sc. Res. Council RL-74-021, Rutherford Laboratory, 1974
16. Gilfrich, H. P., Wilski, H.: Chemie-Ingenieur Technik 42, 19 (1970)

9 Modified and Thermal-Resistant Resins

Polymers are considered to be moderately thermally stable, if they survive incremental increases in temperature in an inert atmosphere without a significant change in properties (Table 9.1).

Plastics which meet these requirements have the following structural characteristics:
- high proportion of heterocyclic or aromatic rings
- high bond energies between the atoms
- high cohesive strength between the polymer chains
- oxidation resistant bonds.

Phenolic resins possess these prerequisites to a considerable extent. Although non-modified, inorganically filled phenolics are already considered temperature resistant, their thermo-oxidative resistance can be further improved by chemical modification[3]. The weak point of phenolic resins is the oxidative susceptibility of the phenolic hydroxyl group and the methylene linking group. Poly-*p*-xylylene is known for its excellent high temperature stability.

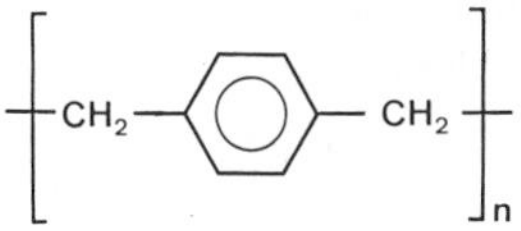

$$\left[CH_2 - \bigcirc - CH_2 \right]_n \tag{9.1}$$

The following methods are utilized to improve the thermo-oxidative resistance of phenolic resins:
1. Etherification or esterification of the phenolic hydroxyl group,
2. Complex formation with polyvalent elements (Ca, Mg, Zn, Cd, ...)
3. Replacement of the methylene linking group by heteroatoms (O, S, N, Si, ...)

Table 9.1. Guidelines for rating of plastics as thermally resistant[1]

°C	Hours
175	30,000
250	1,000
500	1
700	0.1

Table 9.2. Thermal resistance of phenol resols as related to temperature and time[2]

°C	Time
1,000–1,500	Seconds
500–1,000	Minutes
250– 500	Hours
< 200	Years

The most important general modification reactions in phenolic polymer chemistry are the etherification (O-alkylation) and C-alkylation (Friedel-Crafts) reactions[4]. Both reactions are commonly used to enhance flexibility and compatibility with polymers and solvents and to adjust reactivity and performance. Because of the strong nucleophilic activity of the phenolic group, mild catalysts and operating conditions are usually employed for alkylation with olefins. However, the easy formation of ethers, especially during mild conditions, and the tendency of the hydroxyl group to complex the catalyst must be considered. Diisobutylene, terpenes and tung oil are the olefins most frequently used (see 14.1.1, 18.1, 19.4).

9.1 Etherification Reactions

The hydroxymethyl group in phenols and phenol prepolymers can easily be etherified (9.2) with alcohols because of their tendency to form hydroxybenzylcarbonium ions. High hydroxymethylated phenols and an excess of alcohol are used to avoid the self-condensation reaction.

In general the reaction is performed at pH 5–7 and at temperatures between 100–120 °C with monoalcohols like methanol, butanol and isobutanol. Butanol is most frequently used; water is separated via azeotropic distillation with an excess of butanol[5]. Such etherified resols exhibit a higher solubility in aromatic solvents and improved flexibility. They are used mainly in coatings, impregnating resins for electrical laminates and adhesives. Their reactivity, however, is reduced. Also polyhydroxy compounds are recommended for flexibilization, e.g. glycol, glycerine, polypropylene glycols, hydroxypolyesters and polyvinylacetals.

On the other hand, the etherification of the phenolic hydroxyl group (O-alkylation) (9.3) leads to improved alkali resistance. Better flexibility, light fastness and with allyl compounds, enhanced air drying properties are also obtained. The reactivity of phenol ethers towards formaldehyde is drastically reduced in comparison to phenols. Therefore, the resol is first prepared and the phenolic hydroxyl group etherified (alkylated) with stronger electrophiles, e.g. allyl chloride, alkyl bromides, alkyl sulfates, epichlorohydrin and epoxide compounds in the presence of sodium hydroxide. In general, mild conditions are necessary to avoid polymer formation and C-alkylation.

Novolak resins can also be O-alkylated for higher performance. Resins with improved light fastness, which are soluble in alkali, are obtained by O-alkylation with monochloroacetic acid.

(9.2)   R = H, CH_2OH, CH_2X (9.3)

Allyl prepolymers, produced by O-alkylation with allyl chloride can be used as additives for can and drum coatings, and for electrodepositive paints based on epoxy-, polyvinylacetal- or polybutadiene resins because of their increased thermooxidative stability, excellent resistance to chemicals and relatively good flexibility[6].

9.2 Esterification Reactions

The esterification of phenol novolak resins with inorganic polybasic acids, such as phosphoric and boric acid, or the reaction with phosphorus oxyhalides are of particular importance in increasing the heat and flame resistance of phenolic resins. Because of the high OH-functionality of novolaks, the reaction with polyfunctional compounds often leads to gelation. Linear polymers have been obtained using bisphenols and bi-functional derivatives of phosphoric acid or phosgene[7]. Dannels and Shepard have shown[8] that "high-*ortho*" novolaks can be quite readily esterified with bi-functional compounds, while novolaks with random isomer distribution gel at a low level of esterification. Similar behaviour is observed during reaction with boric acid or diphenyl silyl dichloride. The intermolecular (crosslinking) reaction 9.4 takes place predominately with random novolaks[9], whereas with *ortho* novolaks the intramolecular esterification seems to be the preferred reaction yielding 8-membered rings (9.5).

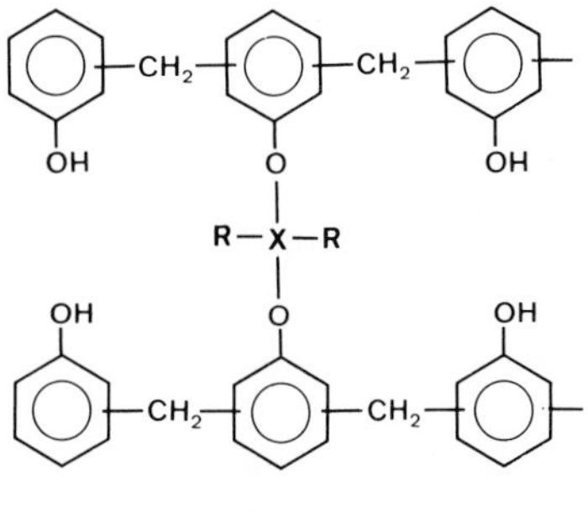

(9.4)

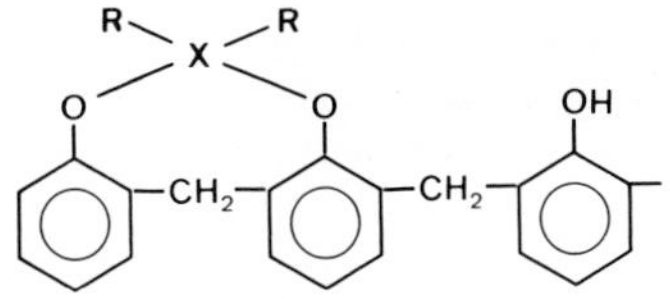

(9.5)

9.2.1 Boron-Modified Resins

The increased thermal resistance of phenolic resins modified with boron is attributed to structures (9.6) and (9.7)[10]. Structure (9.7) is based on a hexa catalyzed resin.

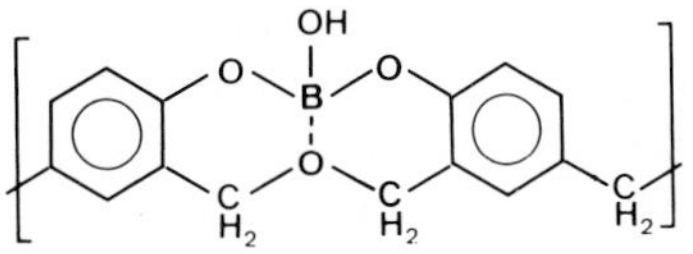

(9.6)

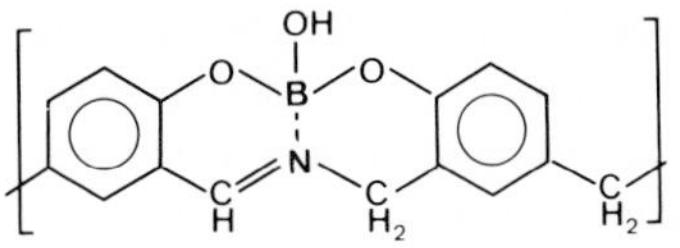

(9.7)

Phenyl borates are prepared initially according to Huster by reaction of phenol and boric acid or B_2O_3[11] with removal of water (15 h at 280 °C). By reacting aryl borates with paraformaldehyde or trioxane at temperatures between 80 and 120 °C, yellow colored, solid prepolymers are obtained via a moderately strong exothermic reaction, and then formulated in the usual manner to molding compounds with HMTA, fillers, etc. Relatively high temperatures of 200 °C are required for cure to obtain the desired high temperature resistance. The curing can be performed at lower temperatures, between 100–120 °C if epoxy compounds are used instead of HMTA[12].

Ablative materials with outstanding temperature resistance are obtained from boric acid modified p-aminophenol/formaldehyde resins[13]. These resins are prepared by heating 3 mol p-aminophenol with 1 mol boric acid in boiling xylene. Water is removed by azeotropic distillation. The tris-p-aminophenyl borate compound is soluble in water and colored blue. After further reaction with trioxymethylene or formaldehyde at 70 °C by addition of an acidic catalyst (3 h), a red solid resin is obtained. Cure occurs with HMTA. Very little weight loss occurs at very high temperatures. Above 2,500 °C boron nitride-like structures are obtained.

Although boron-modified phenolic resins are recommended for many different applications, they are presently used in the manufacture of brake linings and ablative materials.

9.2.2 Silicon-Modified Resins

The addition of silicon compounds for the improvement of the thermal resistance of phenol-formaldehyde resins was recommended as early as 1941 by E. G. Rochow[14]. Since then, a series of patents have been granted[15–18]. Although the thermal resistance is improved by this modification, silicon modified phenolic resins have as yet been of very little commercial importance. The reason may be attributable to the high price of suitable organic silicon compounds. A phenolic resin becomes two to three times as expensive, if only 10% of a silicon compound is added. The addition of smaller quantities does not bring significant improvement in thermal resistance. The modification can be performed by chemical reaction of silicones or siloxanes containing reactive groups with phenolic compounds or by a mixture of the components.

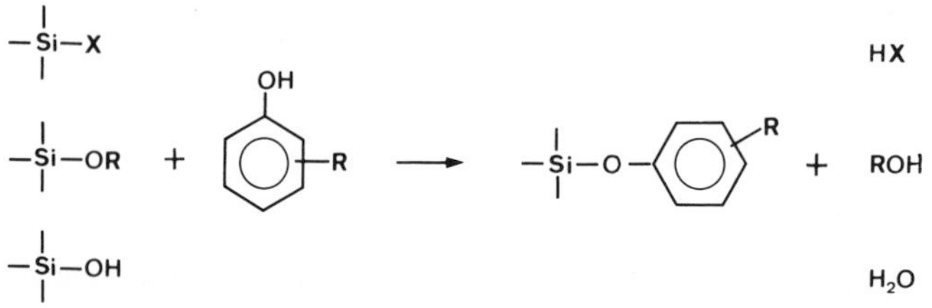

$$\text{(9.8)}$$

During the reaction of siloxanes [X = halogen, hydrogen) with phenolic compounds as O–H acidic components, a Si–O-bond is formed[19]. The reaction is not limited to monomeric compounds; prepolymers also provide many possibilities. The reaction conditions are selected in such a way that the co-condensation of both systems is preferred instead of the self-condensation reaction. The formation of block polymers cannot be completely excluded. The reaction between phenol or phenol novolaks and methoxyphenylpolysiloxanes (9.9) according to the general formula (9.8) by methanol

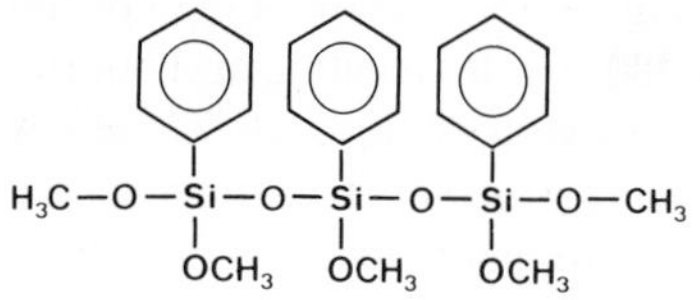

$$\text{(9.9)}$$

displacement is the simplest to achieve. The incorporation of silicon atoms in the polymer chain instead of the methylene linkage is obtained by reaction of *p*-silylphenols, which are available by hydrolysis of phenoxysilanes, with formaldehyde[15].

The addition of silanes as adhesion promoters is widespread, for instance in the manufacturing of mineral wool mats, foundry sands, silica microsphere composites[20, 20a]. The added quantities are, however, very small, mostly far below 1%, so that such resins cannot be designated as silicon-modified resins. Amino-functional silanes are preferred, for instance *γ*-aminopropyltrimethoxysilane, less frequently epoxy functional type like glycidoxypropyltriethoxysilane.

9.2.3 Phosphorus-Modified Resins

Phosphorus-modified novolaks are obtained by esterification of novolaks with phosphoric acids or by reaction with phosphorus oxychlorides. The reaction with difunctional phosphorus oxychlorides (9.10) is conducted at 20–60 °C in dioxane[13, 21, 22].

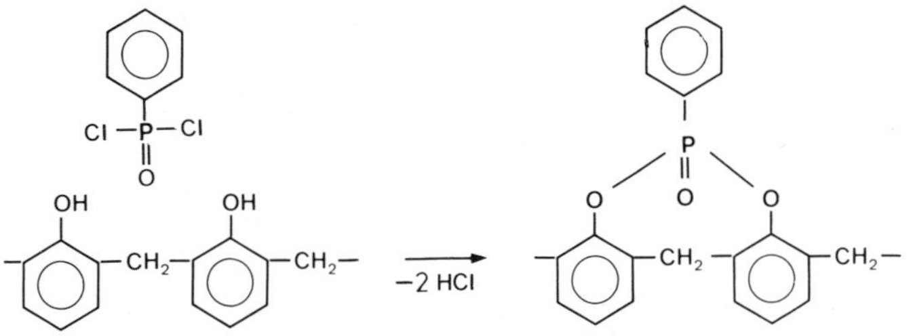

(9.10)

With *ortho*-linked novolaks, cyclic structures prevail (8-membered ring); with normal *ortho*, *para* isomer distribution, intermolecular esterification is preferred[8]. In this case, early gelation may occur. Although phosphorus-modified resins exhibit excellent heat resistance in oxidizing media and outstanding flame resistance, they are relatively rare on the market. During thermal stress phosphine is evolved in a reducing medium[13].

9.3 Heavy Metal-Modified Resins

Heat- and flame resistant resins are obtained by the reaction of phenols or phenolic resins with metal halides (molybdenum trichloride, titanium tetrachloride, zirconium oxychloride, tungsten hexachloride), metal alcoholates (aluminum trimethoxide, titanium tetramethoxide) or metal-organic compounds (acetylacetonates). If such metal containing phenolic resins are subjected to high temperatures, they decompose considerably slower than conventional resins. Further it is assumed that metal carbides are formed with the carbon atom of the resin. The resins are deeply colored and may contain up to 20% of ionic bound metal[13]. Red-colored titanium-modified resins for instance are obtained by heating phenol and titanium tetramethoxide and displacing methanol at 80 °C for 1 h. Prepolymer formation occurs by reaction with paraformaldehyde.

Otherwise[23], titanium tetrachloride is added to molten phenol, and the hydrogen chloride which is formed is removed by passing nitrogen through and raising the temperature up to the boiling point of phenol. The intensely red tetraphenyltitanate (MP 130 °C) is converted in the second stage by reaction with trioxymethylene into a solid prepolymer (MW ca. 500; soluble in dimethylformamide), which can be cured with 10% of HMTA. The reaction of phenolic resins with Ca-, Mg-, Zn-, and Cd-oxide and their value in the formulation of adhesives is described in Chap. 19.

9.4 Nitrogen-Modified Resins

Some polycondensation products of formaldehyde with aromatic amines are characterized by significant thermal stability. When heated to 330 °C 15–30% weight is lost, but when heated further to 900 °C, they form a nonvolatile residue constituting up to 65% of the weight of the original polymer. The most thermostable materials are polycondensation compounds of formaldehyde with p-aminophenol[24, 25].

By reacting aniline with formaldehyde in a strong acidic medium, p-aminobenzylalcohol is obtained and undergoes linear polycondensation by loss of water[26–29].

$$H_2N-\langle\bigcirc\rangle-CH_2OH \longrightarrow -NH-\langle\bigcirc\rangle-CH_2-NH-\langle\bigcirc\rangle-CH_2-NH-\langle\bigcirc\rangle- \qquad (9.11)$$

The chains are crosslinked with methylene groups to some extent. A co-condensation reaction with phenols is easier to perform[30, 31]. Methylenediphenyldiimide is formed from aniline and formaldehyde in alkaline solution. In a neutral or weak acidic pH environment resin-like products which contain azomethine groups, result. They are soluble, brittle and high-temperature resistant[27].

Aniline-formaldehyde resins or phenol-aniline mixed resins have been used in the past to produce laminates and molding compounds for electrical applications. These resins possess higher tracking resistance and favorable electrical properties[26, 32]. Aniline resins have little technical value today. The reasons for this are difficulties in processing, low flow and physiological hazards which are connected with the use of aniline. Aniline is very easily absorbed by the skin and even in minor concentration acts as a strong blood pigment poison which inhibits ability of haemoglobin to transfer oxygen because it interacts competitively in the redox process. Formerly, aniline was reported to cause cancer. However, cancer is attributed to higher aromatic amines which were present as contaminants[33].

The addition of urea or UF-resins is conducted on a large scale to modify wood and mineral fiber binder and foundry resins. Melamine, dicyandiamide and sulfonamides may also be used to produce nitrogen-modified resins[27]. Reaction conditions favouring copolymerization with amino compounds are indicated in Chap. 3.

9.5 Sulfur-Modified Resins

The direct reaction of phenols and sulfur occurs relatively easy in the presence of alkaline catalysts at temperatures between 130 and 230 °C. The resulting materials are liquid or solid resins with an unpleasant hydrogen sulfide odor. The softening range depends upon the type of phenolic component and molar ratio. The simplest compound which results from this reaction is the dihydroxydiphenylpolysulfide, which is either further reacted with formaldehyde or can be immediately crosslinked with resols[27, 34]. Similar structures can be obtained if phenols are first reacted with aldehydes and then with sulfur and alkali hydroxides. Phenolic resins modified with sulfur are not of any commercial importance. The sulfur containing phenolics possess high plasticity and relatively high water solubility.

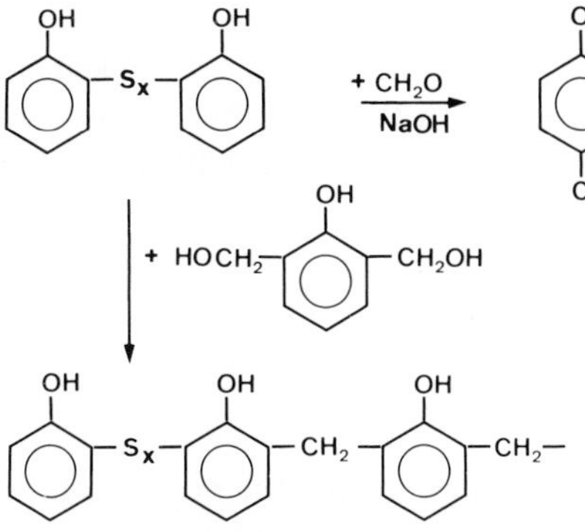

$$ (9.12) $$

$$ (9.13) $$

4,4-thio-*bis*-(3-methyl-6-tert-butyl) phenol is a known antioxidant (Chap. 20). Resins with improved temperatures resistance are obtained by the reaction of dihydroxydiphenylsulfone with formaldehyde[41].

9.6 Others

Condensation resins of an aralkylether and phenol (9.14) (Xylok, developed by Midland Silicones Ltd.) are intermediate between phenolic- and poly-*p*-xylylene resins. The increased temperature resistance compared to that of nonmodified phenolic resins[35, 36] is the result of the reduced oxidative susceptibility of the methylene linkage.

Polynuclear aromatic and heterocyclic compounds (naphthalene, carbazol) also react easily with *p*-xylylene-dimethylether. The prepolymers can be cured with HMTA (10–15%) or with (cycloaliphatic) epoxy resins. Xylok resins are recommended for the production of class H glass laminates or in combination with phenolic resins for

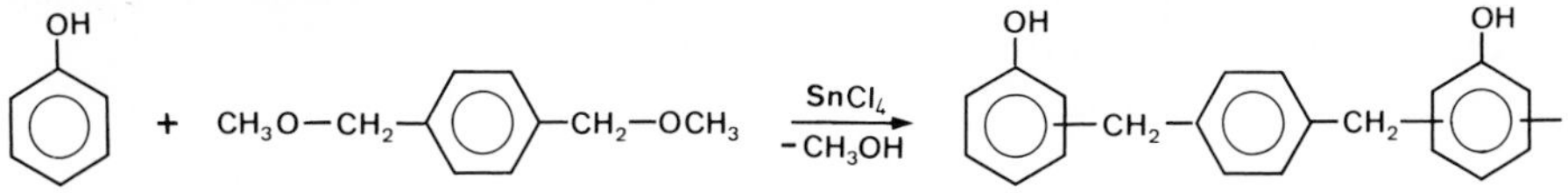

p-Xylylene dimethylether   Xylok prepolymer   (9.14)

molding compounds. A cost factor of four to six times higher than that of phenol resins has inhibited their commercial development. Minor quantities are used for the manufacture of friction linings.

Polydehydration of difunctional phenols in the presence of zinc chloride under pressure above 220 °C yields polyhydroxyphenylenes (9.15); benzyne and hydroxybenzyne being intermediates[37].

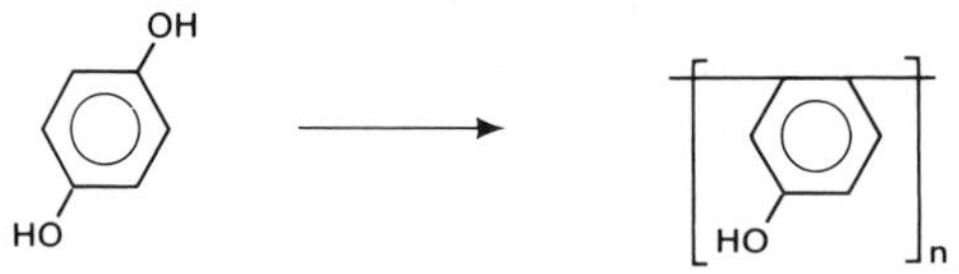

$$(9.15)$$

Linear, thermoplastic phenolic resins (polyhydroxystyrene or poly-4-vinylphenol) are obtained by polymerization of 4-hydroxystyrene (9.16) which can be crosslinked either with HMTA[38] or preferably with epoxy resins[39]. In addition, they are used as host resins for photo-reactive compounds in photo resists[40] (see Chap. 10.5).

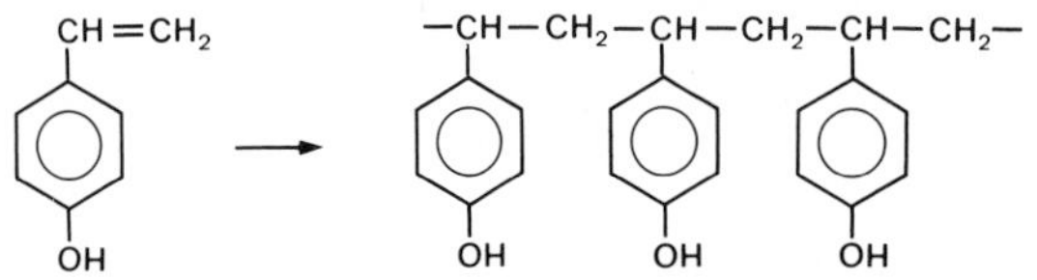

$$(9.16)$$

Polyhydroxystyrene/epoxide combinations are recommended for the production of multilayer circuit boards.

The corresponding α-methyl hydroxystyrene has been prepared[42, 43] by a base catalyzed cleavage of BPA (9.17)

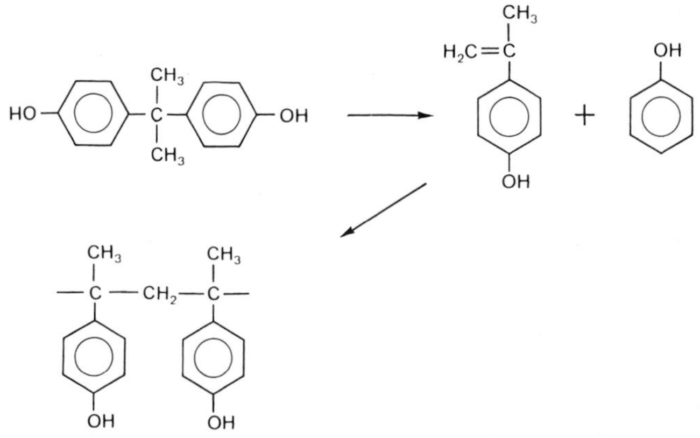

$$(9.17)$$

High MW polymer is obtained[42, 43] via special conditions, such as derivatizing the hydroxyl group, followed by BF_3/SO_2 polymerization conditions. The blocking group is removed thermally or by CF_3COOH treatment.

9.7 References

1. Levine, H. H.: Ind. Engn. Chem. 54, 22 (1962)
2. Techel, J.: Plaste u. Kautschuk 10, 137 (1963)
3. Bachmann, A., Müller, K.: Plaste u. Kautschuk 24, 158 (1977)

4. Olah, G. A.: Friedel Crafts Chemistry. New York: Wiley 1973
5. Lemmer, F., Greth, A.: Phenolharze. In: Ullmanns Encyclopädie d. techn. Chem. Vol. 13, 3. Ed. München: Urban und Schwarzenberg 1962
6. General Electric Co: Methylon Resins, Technical Bulletin, US-PS 257 330 (1951)
7. Wright, H. R., Zentfmann, H.: Chem. Ind. (London) 244, 1101 (1952)
8. Dannels, B. F., Shepard, A. F.: Inorganic Esters of Novolaks. J. Polym. Sci.: Part. A-1, Vol. 6, 2051 (1968)
9. Cass, U. E.: US-PS 2 616 873
10. Hoechst AG: DE-OS 24 36 358 (1974)
11. Dynamit Nobel AG: DE-PS 1 233 606 (1960)
12. Dynamit Nobel AG: DE-OS 2 214 821 (1972)
13. Nord-Aviation Société Nationale de Constructions Aéronautiques: DE-AS 1 816 241 (1968)
14. General Electric Co.: US-PS 2 258 218 (1941)
15. Dow Corning Corp.: DE-PS 937 555 (1956)
16. Stenbeck, G.: DE-OS 1 694 974
17. Westinghouse Electric Corp.: US-PS 2 836 740 (1958)
18. Dow Corning Corp.: US-PS 2 842 522 (1958)
19. Noll, W.: Chemie und Technologie der Silicone. Verlag Chemie, Weinheim: 1968
20. Price, H. L., Ku, J.: Polymer Preprints 24 (2) 198 (1983)
20a. Union Carbide Corp.: Technical Bulletin
21. Helferich, B., Schmidt, K. G.: Chem. Ber. 92, 2051 (1959)
22. General Electric Co.: GB-PS 1.031 908 and 1.031 909 (1966)
23. Hitachi Chem. Co. Ltd.: J-PS 76 58493 (1974)
24. Sergeev, V. A., Korshak, V. V., Kozlov, L. V.: Plasticheskie Massy 3, 57 (1966)
25. Korshak, V. V., Vinogradova, S. V.: Russian Chem. Rev. 37 (11) 885 (1968)
26. Frey, K.: Helv. chim. Acta 18, 491 (1935)
27. Scheiber, J.: Chemie und Technologie der künstlichen Harze. Stuttgart: Wissenschaftl. Verlagsges. 1943
28. Ellis, K.: The Chemistry of Synthetic Resins. New York: Reinhold 1935
29. Scheuermann, H.: Anilinharze. In: Ullmanns Encyclopädie d. techn. Chem. Vol. 3, 3. Ed. München: Urban und Schwarzenberg
30. Fibre Diamond: FR-PS 644 075
31. E. I. du Pont de Nemours & Co.: US-PS 2 098 869
32. Imhof, A.: Kunststoffe 27, 89 (1939)
33. Oettel, H.: Anilin, Toxikologie. In: Ullmanns Encyclopädie d. techn. Chem. Vol. 7. 575, 4. Ed. Weinheim: Verlag Chemie 1974
34. Cherubim, M.: Kunststoff Rundschau 13, 235 (1966)
35. Midland Silicones Ltd.: Xylok 210, Technical Bulletin
36. Harris, G. J.: Werkstoffe u. Korrosion 22, 227 (1971)
37. Jones, J. I.: J. Macromol. Sci.-Revs. Macromol. Chem. C 2 (2), 303 (1968)
38. Takahashi, A., Yamamoto, H.: Polymer J. 12, 79 (1980)
39. Maruzen Oil, Jap.: Technical Bulletin
40. Iwayanagi, T., et al.: Polymer Engineering and Science 23, 935 (1983)
41. Union Carbide Corp.: US-PS 3,185, 666; US-PS 3,230,198; US-PS 3,225,104
42. Ito, H., et al.: Macromolecules 16, 510 (1983)
43. Ito, H., et al.: ACS Polymer Preprints 25, 158 (1984)

10 High Technology and New Applications

In the 80's the wave of high technology has fostered the active participation of phenolic resins in "high tech" areas ranging from electronics, computers, communication, outer space, aerospace, biomaterials, biotechnology and advanced composites. From the sophistication of microchip technology in communication systems, or the delicate heart beat maintained by a pacemaker to the excessive temperature environment of outer space, phenolic resin chemistry plays an indispensable role in these burgeoning high technology areas.

10.1 Carbon and Graphite Materials

Elemental carbon is a unique engineering material which is quite temperature resistant (to 3,000 °C) in an inert atmosphere. Carbon exhibits excellent resistance to highly corrosive liquids[1] such as phosphoric acid, hydrochloric acid, sulfuric acid, organic acids as well as corrosive gases like hydrogen chloride and sulfur dioxide. It does not melt at atmospheric pressure. Carbon as graphite is corroded only by strongly oxidizing chemicals like nitric acid, chromic acid, or by fluorine and sulfur vapor at high temperatures. This explains its wide utility in engineering applications[2-4], especially in the fabrication of chemical equipment (heat exchangers, reactors, columns and pipes), as lining or cladding material for blast furnaces, melting furnaces, foundry molds and high temperature insulation material. Additional application areas are moderators, reflectors, cladding for fuel rods in nuclear power equipment and matrices for carbon fiber composites.

In 1878, Thomas A. Edison succeeded in thermally degrading bamboo fibers into carbon fibers; this was the break-through that was necessary for the development of the filament-lamp. However, the thermal degradation of natural polymeric materials leads to carbon-rich residues of porous structure and low strength. Carbon yield seldom exceeds 30% by this method. Phenolic resins with ordered macromolecular structures and high carbon content open new horizons for the production of high-strength carbon materials.

Today phenolic resins are systematically being incorporated into carbon and graphite technology. Phenolic resins are used as impregnating resins to increase density, i.e. impermeability gas, and strength of the molded part, or as carbon precursor for the production of glassy or polymeric carbon and carbon foams[2, 5].

For special applications phenolic resins are used as temporary binders for the production of shaped parts rather than coal tar pitches. In comparison to coal tar pitch, phenolic bonded structures possess higher strength and lower gas permeability.

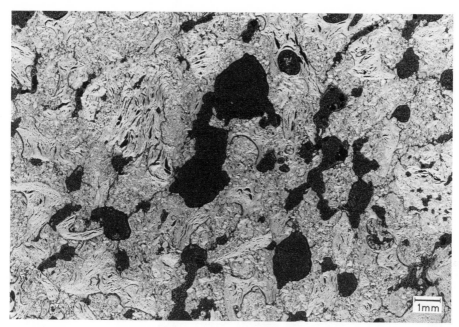

Fig. 10.1. Phenol resin impregnated graphite materials. (Photo: Sigri Elektrographit GmbH, D-8901 Meitingen); dark = phenol resin; light = graphite

Table 10.1. Physical properties of phenol resin treated "artificial" carbon and graphite[2]

	Unit	PF-resin treated graphite	PF-resin bonded carbon
Density	g/cm^3	1.75	1.60
Flexural strength	N/mm^2	50	70
Compressive strength	N/mm^2	100	180
Modulus of elasticity	N/mm^2	$2.7 \cdot 10^4$	$3 \cdot 10^4$
Coefficient of thermal expansion K^{-1}	K^{-1}	$2 \cdot 10^{-6}$	$3 \cdot 10^{-6}$
Permeability coefficient	cm^2/s	$1 \cdot 10^{-5}$	$5 \cdot 10^{-5}$
Thermal conductivity	W/m K	70	4.6

Uniform particulate petroleum coke or pitch coke is coated with coal tar pitch (at 100–170 °C) or phenolic resins and the "dough-like" material is extruded or molded. The preform is then protected from oxygen by a coke layer fired in a furnace. The bonding agent which varies from 15–25% is carbonized and binds the carbon particles to a microcrystalline solid material with high porosity. The latter is due to the emission of gases which occur during the pyrolysis of the binder. Homogeneity and impermeability are obtained by repeated impregnation with either coal tar pitch or with phenolic resins followed by a firing cycle (Table 10.1). The cycle requires about 20 days. Carbon materials which are heated to approximately 1,300 °C are called "artificial" carbons;

Fig. 10.2. Compressor impeller made of polymeric carbon. (Photo: Sigri Elektrographit GmbH, D-8901 Meitingen)

Fig. 10.3. Fuel elements in a nuclear power reactor THTR 300, made of polymeric carbon. (Photo: Hochtemperatur-Reaktorbau GmbH, D-6800 Mannheim)

those treated to 3,000 °C are "artificial" graphites. Graphitized structures possess considerably higher thermal and electrical conductivity.

Glassy carbon which is obtained by carbonization of phenolic resins displays excellent impermeability to gases and chemical resistance. Numerous new applications have been developed for glassy carbon in addition to conventional uses such as laboratory equipment (tubes, crucibles).

The outstanding biocompatibility of polymeric carbon has been effective in the development of vital biomaterials such as heart pacemaker electrodes, percutaneous leads, orological protheses, dental and joint implants. These are a few examples of recent medical applications.

An important application area (besides the aerospace industry) that offers commercial opportunity for carbon composites is in the automotive and related high performance engineering areas. An example of a highly sophisticated automotive component is the compressor impeller of a diesel vehicle turbo-charger. The impeller assures a lower fuel consumption and reduced emissions.

The fuel elements of the high-temperature thorium reactor THTR 300 shown in Fig. 10.3 are manufactured by unique technology using high-purity ion-free phenol

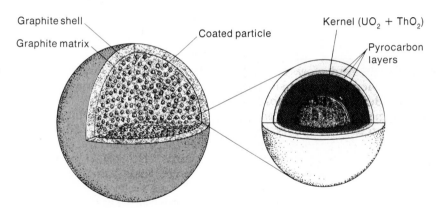

Graphite shell

Graphite matrix

Coated particle

Kernel (UO$_2$ + ThO$_2$)

Pyrocarbon layers

Fig. 10.4. Fuel element composition containing embedded fuel particles. (C & EN, March 5, 1984)

resol resin[6]. The spherical fuel elements consist of nuclear fuel particles individually coated with layers of pyrocarbon dispersed in a graphite matrix and an outer graphite shell. A molding compound is prepared from graphite, phenolic resin and coated fuel particles, cured in spherical molds and then carbonized via a defined heating schedule from 800 to 2,000 °C. The resulting hard graphite shell and matrix provide the fuel element with resistance to mechanical breakdown inside the reactor core and during bonding and unloading[2].

Carbon foams are obtained by carbonizing phenolic foams (Chap. 13). The resulting foams demonstrate thermal resistance to 3,000 °C in an inert atmosphere or vacuum and superior to the thermal resistance of refractory bricks. Important applications of carbon foam are high-temperature insulation, filters for corrosive agents and catalyst supports.

The propensity of phenolics to form high carbon and high strength foam is the rationale for the use of phenolics in ablative polymeric systems that require ultra-high thermal insulation[6]. The phenomenon of ablation consists of a variety of endothermal processes implemented by thermal decomposition of materials leading to heat absorption, dissipation and obstruction. Elastomer modified (e.g. nitrile rubber) phenolic resins reinforced with asbestos, carbon fibers or silica fibers are most widely used.

Figure 10.5 demonstrates the various changes of a charring ablator during high temperature impact. Such ablative polymeric materials are used in very high temperature aircraft, heat sinks in ballistic vehicles, propulsion[7], and ground base support applications. A highly technical and informative presentation of the early developmental efforts to provide ablative heat protection is reviewed by Sutton[8].

Carbon fibers obtained by the carbonization of phenolic fibers are generally amorphous. Even treatment to 2,000 °C does not result in the formation of ordered graphite structures. A phenol hexamine resin can be extruded from the melt to produce fibers and carbonized to form fine glassy carbon fibers (5 µm) with a tensile strength up to 2 GNm^{-2}, a specific modulus of 70 GNm^{-2} after 900 °C heat treatment[9]. Both strength and modulus increase rapidly with decrease in diameter. Carbon fibers from Novoloid phenolic fibers (American Kynol Inc.) have been described[10]. Pyrolysis to

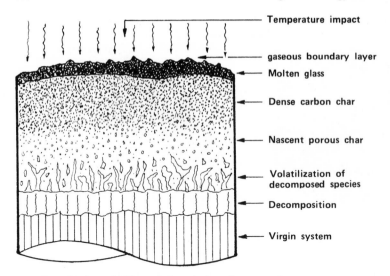

Fig. 10.5. Ablation of a fiber reinforced phenolic composite

about 800 °C yields carbon fibers with 95% carbon content while heat treatment to 2,000 °C increases carbon content to 99.8%. The tensile strength of these pyrolyzed fibers is reported in the range of 400 MNm^{-2} with a modulus of 14 GNm^{-2}.

Carbon-carbon composites[3], i.e. carbon fiber reinforced polymeric carbons, have been developed for use in aerospace technology. Such composites do not show any reduction in strength to 2,500 °C (in inert atmosphere). Besides their use for rocket propulsion and reentry systems, they are employed for airplane disc pads. Carbon brake linings show a very uniform braking performance and do not bind. Liquid resols with a 75–85% solids content are preferred for these previously mentioned applications; novolak-HMTA can also be used. The impregnation with resols is conducted with vacuum and pressure. The different structural relationships of phenolic resins – resol or novolak, formaldehyde ratio, and crosslink density – have only a moderate effect on carbon yield[11, 12]. Functional groups and other groups which are non-contributing to crosslinking are released as volatile products. Thus porosity, shrinkage[13] and even the carbonization process, itself, are the critical parameters in the development of optimum mechanical properties of polymeric carbon.

10.2 Phenolic Resin – Fiber Composites

The outstanding performance of phenolic resins under extremely hazardous and highly flammable conditions is discussed in Chap. 6.5. New, promising application areas for fiber reinforced phenolic resins are being rapidly developed as result of phenolic resin resistance to ignition, low smoke generation, high temperature resistance and strength. Application areas consist of the aircraft industry, transportation and construction. Fiber reinforced structures and composites[14] with phenolics can be produced by the conventional hand lay up, hot molding, vacuum injection, filament winding, pultrusion or prepreg technique.

Table 10.2. Typical properties of phenol resol resins for fiber reinforced composites[16]

		Laminating resin	Prepreg resin
Characterization	–	Resol	Resol
Solids content	%	85	65
Viscosity	mPas	2,000 (60 °C)	2,500 (20 °C)
Solvent		H_2O	MEK
pH		5.8	8.3
Free phenol	%	5	3
Free formaldehyde	%	0.5	1.5
Gel time 130 °C	min	8	15
Storage stability 20 °C	weeks	4	12

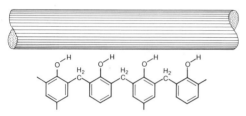

Fig. 10.6. Orientation of high-*ortho* structures towards reinforcing fibers

Phenolic resins designed for laminating and molding processes in the low temperature range are high solids aqueous resol resins cured by organic sulfonic acids, e.g. Tg. Important properties required[15] are high and constant reactivity towards acids, high solid content, low viscosity, and low content of free phenol and formaldehyde. The scope of flexibilizing additives compatible with aqueous resols is very limited. Flexibilizers reduce flame resistance and acid reactivity; therefore they are not utilized.

Typical properties of commercial phenolic resins for fiber composites are shown in Table 10.2. Catalysts during prepolymer formation particularly influence the adhesion of the cured resin to reinforcing fiber and other substrates. High-*ortho* structures offer significant advantages (Fig. 10.6).

Cure of aqueous laminating resols is performed preferably between 60–90 °C (10–15 min); a postcuring treatment in an oven at 80 °C/4 h is necessary to obtain the ultimate high Tg and flexural strength and to remove entrapped volatiles.

Molds for acid cured phenolics are made from epoxies; metal molds require an acid resistant coat. Shrinkage of glass fiber reinforced phenolics is very low, to relieve outmolding internal or external release agents may be used. Binders applied in glass fiber mats or filament sizing agents may influence the acid reactivity of resols; it is therefore advisable to check the utility of the reinforcing material. Epoxy based binders and silanes are preferred. Cured phenolics are dark in colour, for bright appearance gel coats are applied.

The outstanding thermal performance of fiber reinforced phenolic materials are indicated[15] in Figs. 10.7 and 10.8. Torsional braid analysis of a non-reinforced and

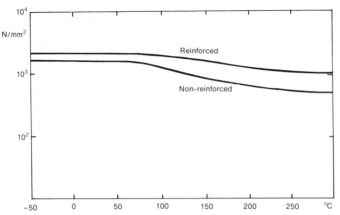

Fig. 10.7. Torsional modulus of phenolic resins in relation to temperature[15]; glass fiber reinforced and non-reinforced

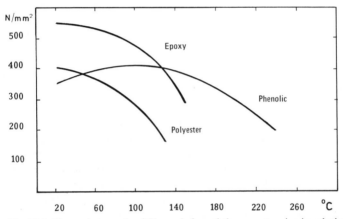

Fig. 10.8. Flexural strength of fiber reinforced thermoset resins in relation to temperature[15]

Table 10.3. Thermal resistance and Tg of thermoset materials[15]

		Polyester, unsaturated	Epoxy	Phenolic
Thermal resistance °C, Martens	DIN 53458	115	170	180
Thermal resistance °C, Iso/R 75	DIN 53461	145	180	210
Glass transition temperature °C	DIN 53445	170	200	> 300

glass fiber reinforced acid cured phenolic shows (Fig. 10.7) that modulus is practically unchanged up to 300 °C. Figure 10.8 compares epoxy, polyester and phenolic thermoset systems regarding their flexural strength in different temperature regions. While phenolics exhibit a lower flexural strength at RT, they maintain this level at elevated temperatures and exceed polyesters and epoxies in the ≥ 150 °C temperature range. Maximum glass transition temperatures and thermal resistance data reported for various thermosetting resins are shown in Table 10.3.

Mechanical property characterization and comparisons of these newly developed phenolic composites has been sparse (see Chap. 10.3). Attempts to compare interlaminar shear strength of glass reinforced phenolics with polyester made by hand lay-up method are reported to be 20% lower than the polyester composite[17]. The comparison is doubtful because of differences in the fracture mechanism. Yet the values of interlaminar shear strengths of both acid cured phenolic and polyester resins are relatively low compared to the values for advanced composites such as carbon fiber reinforced epoxies. Yet in another study[18] carbon fiber reinforced phenolics have, indeed, exhibited equivalent mechanical properties and much superior thermal stability to epoxy-carbon fiber analogs. An amine catalyzed resole resin was utilized in the study.

A recent pultruded reinforced phenolic product is reported[19] to have substantially uniform cross-section throughout its length. The product consist of a filament impregnated with glycol modified phenolic resin catalyzed by p-toluenesulfonic acid. Flexural strength and modulus values are comparable to typical polyester products but the phenolic pultruded products possess superior heat distortion and flammability properties.

The low smoke generation and excellent fire retardancy identify fiber reinforced phenolics as candidates for building applications and transportation. Currently one of the most ambitious uses in transportation is the construction of subway cars for the Caracas Metro[20]. Seat shells and interior parts are made by hand lay up with 40% glass reinforced phenolic resin.

The prepreg technology is used for production of flat and contoured shapes with a wide variety of fibers, fabric styles, and core materials for structural and non-structural parts in the aircraft and aerospace industry. Phenolic woven prepregs are prepared by impregnation on conventional vertical impregnating machines via solution method. Prepregs with a defined tack are stored at −18 °C for longer periods. Exact specifications, careful control of all raw materials and process parameters involved and extensive physico-chemical characterization including HPLC, GPC, TGA, and DSC of resin batches are indispensable. Laminates and shaped structures with a defined fiber volume and low void content are fabricated in an autoclave using a vacuum bag. Cure temperature is minimally 130 °C for 60 min.

Table 10.4. Minimum strength values of phenolic resin laminates according to MIL-R-9299 C and Normenstelle Luftfahrt[21]. Glass fabric 181 style, 37% resin content

		Phenolic MIL-R-9299 C		Epoxy MIL-9300 B Structural
		Grade A	Grade B	
Flexural strength	N/mm^2	350	510	525
Tensile strength	N/mm^2	280	320	330
Compression strength	N/mm^2	245	400	350
Modulus of elasticity	N/mm^2	21,000	24,500	22,500
Interlaminar shear strength[21]	N/mm^2	20[a]	20[a]	20[b]

[a] WL 8.4361.6
[b] WL 8.4321.6

Table 10.5. Properties of phenolic and epoxy adhesive prepregs (normal temperature range type, resin content 50%)

Property		Unit	Test method	Phenolic adhesive prepreg Bakelite PL 2485	Epoxy adhesive prepreg Bakelite 0146N/LS 82
Flexural strength	23 °C	N/mm²	DIN 53452	500	560
	82 °C	N/mm²	DIN 53452	400	385
ILSS	23 °C	N/mm²	DIN 29971	25	46
	82 °C	N/mm²	DIN 29971	16	30
Peel strength	23 °C	N.mm/mm	MIL-A-25463	130	150
	82 °C	N.mm/mm	MIL-A-25463	82	98
Smoke density		D_s	NBS-chamber	< 30	>280
		D_s	ATS 1000.001	< 50	>350

The US-specification MIL-R-9299 C encompasses two grades of phenolic resins (Table 10.4) which are supplied in liquid form or as prepreg. Grade A is specified as having normal properties and grade B as having improved properties. Similar specifications have been issued in West Germany by Normenstelle Luftfahrt[21] (WL 5.3600, I and II for resins; WL 8.4360 and WL 8.4361 for prepregs). In addition to mechanical properties, a defined tack of the prepreg and adhesion to aluminum and Nomex honeycombs is required. These properties and improved toughness are achieved by modification with thermoplastic or elastomeric additives, however, the additives cause a decrease in flame resistance.

In modern civil aircraft programs, phenolic prepreg materials are increasing in use for interior structural parts instead of epoxies because of their very low smoke density and low toxicity of gaseous products in case of fire. Table 10.5 shows selected properties of phenol and epoxy prepregs used in the Airbus program indicating the significantly lower smoke density but also somewhat lower mechanical and adhesive properties of the modified phenolic material.

10.3 Reaction Injection Molding

Reaction injection molding (RIM) is a fast molding technique whereby two or more components (monomers or low viscosity oligomers) are rapidly transformed in a low pressure mold to a fabricated part. The RIM process is carried out by the rapid injection of the highly-reactive material streams under impingement mixing into a hot mold at relatively low pressures. Polymerization takes place in seconds to minutes to produce a finished plastic part. There are many advantages to the RIM process, i.e. low mold pressure, lower capital investment, lower energy requirements (part is produced directly from monomers or low MW oligomers), greater mold design flexibility, and high fiber loading. Materials processed by RIM are polyurethanes, polyamides, and epoxies[22]. Very fast reactions by impingement mixing occur within 30–60 s for urethane components. Longer reaction times of 2–3 min are required for ε-caprolactame and epoxy RIM systems. Phenolic RIM systems have been recently reported by

Brode, Chow, and Michno[23]. The key features of phenolic RIM involve the utilization of a high solids, high-*ortho* (Chap. 5.3.2) liquid resol resin with selective latent catalysts such as phenyl hydrogen maleate, phenyl trifluoroacetate[24] or butadiene sulfone[25].

The resol-latent catalyst system possesses satisfactory ambient room temperature stability. The resin/catalyst combination is injected into preformed fiberglass mats. Preferred cycle time is 1–2 min. Mechanical properties of phenolic RIM are comparable[26] to vinyl ester or polyester materials.

Phenolic RIM composites are characterized as materials which possess balanced static mechanical properties, retain modulus to 300 °C, exhibit good room temperature and −35 °C damage tolerance, very good fatigue resistance and UL-94 VO flammability rating[26].

Some high-*ortho* liquid resol resins or phenol hemiformal solutions (Chap. 3) can be combined with other polymeric materials such as solid resols, novolaks, or polyesters[27]. The resulting polymer-liquid resol/hemiformal solutions are stable, moderately low viscosity compositions and are added to fiber glass mat and cured with phenol sulfonic acid at a temperature range of 120–150 °C. Non flammable glass reinforced composites with excellent mechanical properties, unusually high glass content and high heat resistance are obtained. However, it is not known to what extent volatile emissions are released during rapid RIM cycle conditions.

10.4 Phenolic Fibers

Phenolic resin fibers were developed by Economy at Carborundum (USA) in the early 1970's[28, 29]. At present Nippon Kynol, Inc. (Japan) and American Kynol Inc. (licensees of Carborundum) manufacture fibers under the trademark Kynol[10].

Phenolic fibers are produced by melt spinning a novolak resin with an average molecular weight of 800–1,000 and very low free phenol content (0.1%). The melt-spun fibers are then cured by immersion in an acidic aqueous solution of formaldehyde with gradual heating. A skin is formed on the fiber surface and reaction proceeds in the fiber interior as the acid and formaldehyde diffuse into the substrate. Partial cure and extraction of the uncured core would lead to hollow fibers. A special laboratory still as a quick processing method was described for the preparation of fibers from novolak HMTA compositions[30]. Short and fine phenolic fibers with random diameters can be produced by an air blowing process.

The most important property of phenolic fibers is their high flame resistance; the rigid fiber does not burn, but is carbonized in the flame while retaining its form and a certain degree of strength. The fibers cannot be designated as high-temperature fibers, since the temperature limit for long-term application is 150 °C in air and 200–250 °C in the absence of oxygen.

Phenolic fibers are light gold in colour and darken gradually to deeper shades on exposure to light and temperature. Acetylation of the hydroxyl groups leads to white and light stable fibers. The limitation in dyeability of phenolic fibers is overcome and punking tendency diminished, but tensile strength and flame resistance are reduced. To prevent excessive fiber attrition, phenolic fibers must be processed by a modification of normal textile spinning operations because of moderate tensile strength and

Table 10.6. Flame resistant fibers compared to wool

Property	Wool	Nomex (polyarylamide)	Kynol
Oxygen Index (% O_2)	0.28	0.28	0.38
Vertical flame (char length, cm)	–	3.2	0.1
Smoke (3 min ignition, l/m)	1.99	0.86	0.002
(15 min ignition, l/m)	3.4	1.95	0.20

lack of inherent crimp. The combination of infusibility, excellent flame and chemical resistance, low evolution of smoke or toxic gases of phenolic fibers (Table 10.6) identifies Kynol as a replacement of asbestos in many uses. Examples are fire protection suits, insulating gloves, clothing for foundry workers, curtains, flame resistant papers and felts. They are recommended as carbon fiber precursor and polypropylene reinforcing agent[31].

10.5 Photo Resists

The recent dramatic growth of the electronics industry has been paralleled by advances in photo resist technology among other high technology developments. Photo resist technology was originally developed for the printing industry[32] about 50 years ago, but has undergone extensive refinements to meet the continuing needs of the semiconductor industry to increase the density of microelectronic devices.

Photo resists are light sensitive resins that are used for either positive or negative imaging. From the electronics industry and the conventional printed circuit boards to the highly sophisticated integrated circuits of the rapidly developing microelectronics industry, various resins, sensitizers and developers are being transformed into extremely detailed patterns with high resolution and submicron geometries[33]. Radiation energy of near UV (350–410 nm), mid UV (280–350 nm), deep UV (230–280 nm), X-rays (10–0.1 nm), and electron beam (0.2–0.01 nm) is utilized.

Commercial positive photo resists use a phenolic host resin such as cresol novolak[34] in combination with a photoreactive compound, predominately 5-substituted diazonaphthoquinone[35–37]. The unexposed resist is not soluble in buffered aqueous alkali solution. When exposed to UV radiation, the o-diazonaphthoquinone is converted to indene carboxylic acid via a ketene intermediate[35, 36].

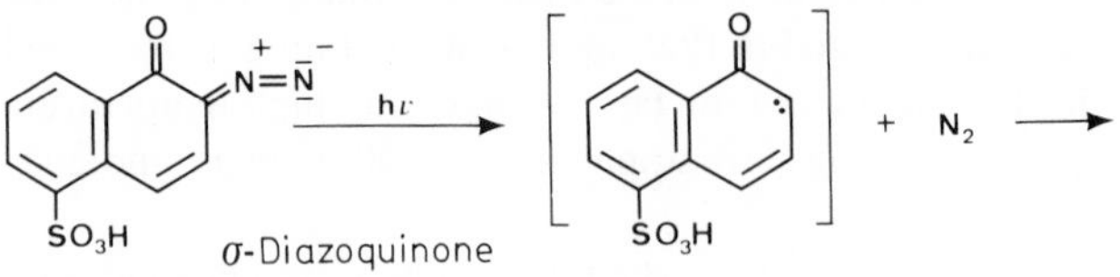

o-Diazoquinone   (10.1)

Ketene   Indene-1-carboxylic acid   (10.2)

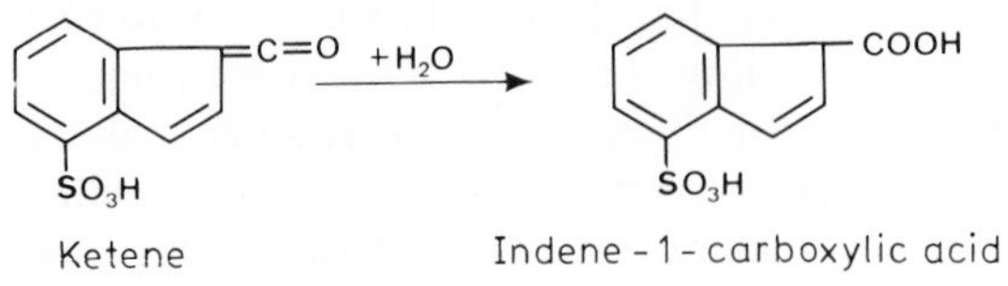

The insolubility of the unexposed coating is attributable to a proposed azo coupling between the phenolic resin and the sensitizer or possibly hydrogen bonding between phenolic hydroxyl and the carbonyl group of the diazo compound. Narrow MWD as well as high MW and high softening point p-cresol novolak resins are effective as deep UV resists[38].

More recently[39] Hiraoka and coworkers evaluated ketene reactions of the diazo sensitizer with cresol novolak by distinguishing the hydration of ketene to carboxylic acid from the ketene reactions with novolak. Suppression of the resin crosslinking reaction increased lithographic sensitivity and was accomplished by the introduction of an *ortho* halogen substituent (10.3), or use of benzaldehyde (10.4) instead of formaldehyde or a naphthol p-hydroxy benzaldehyde resin (10.5).

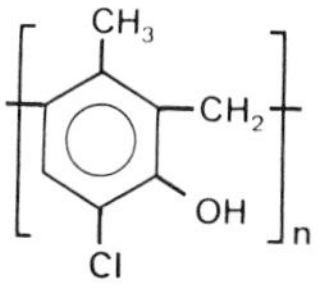

(10.3) 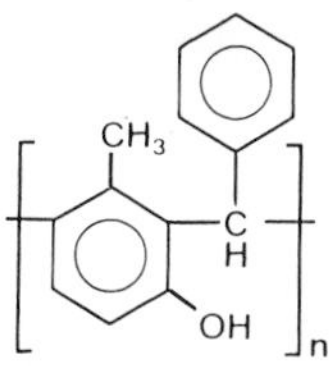   (10.4) 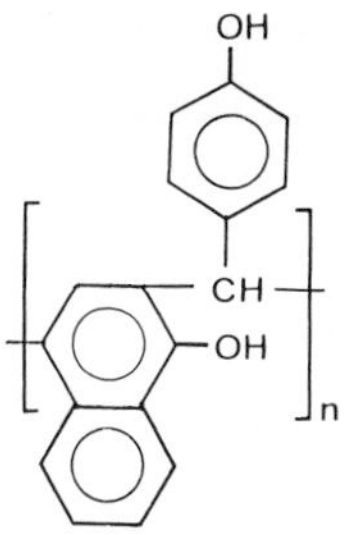   (10.5)

Other sensitizers besides diazonaphthoquinones have been utilized with selective novolaks as positive resists. Positive resist patterns are obtained from o-cresol novolak with selected peroxides as sensitizer[40]. One particularly important sensitizer is the olefin-sulfur dioxide copolymer such as poly 2 methyl-1-pentene sulfone.

A negative UV resist which consists of 3,3-diazidodiphenylsulfone (10.6) as photosensitive compound and poly-4-vinylphenol (9.16) as a phenolic resin matrix was recently described[41]. The exposed patterns become insoluble in organic solvents as well as in aqueous bases. This is explained by the expected decomposition of the bisazide to a strongly basic secondary amine that interacts with the phenolic hydroxyl group and leads to insolubilization of the resist. Trilevel resist patterns are obtained when epoxy novolak is added to the previous resist system[42].

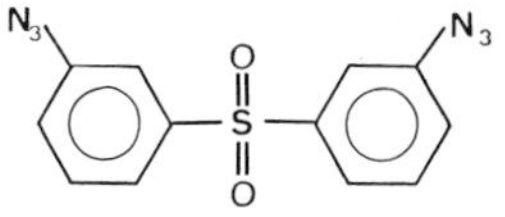

(10.6)

Further improvements in sensitivity, contrast, resolution, and etch resistance are required. Resolution improvement can be realized by use of higher energy, shorter wavelength radiation than is currently used in the popular mid-UV lithograph range (300–400 nm), i.e. dry-etch techniques in which etching is effected by the charged particles present in a plasma.

For these techniques, a dry-etch resistant positive electron resist was formulated[43] by crosslinking a linear phenol-formaldehyde resin with methyl methacrylate/methacryloyl chloride resin (10.7). The reaction between the acid chloride and phenolic hydroxyl group leads to high-energy radiation-degradable phenolic ester adduct. *Para-* or *ortho*-substituted phenols are used to ensure a linear condensation rather than a

Fig. 10.9. Resist patterns in the 1 μm range. (Photo: Shell Polymers 7, 28, 1983)

completely crosslinked structure which is quite difficult to degrade by radiation.

$$(10.7)$$

The electronics industry is increasing the use of positive resists at the expense of negative ones. Furthermore resolution of positive resists is not affected by polymer swelling as negative resists during solvent development. Positive resists can be imaged by projection rather than contact printing and are less oxygen sensitive. For finer resolution the trend is toward dry processing techniques rather than the customary solvent development.

10.6 Carbonless Paper

Carbonless paper is a specially treated paper for use in business forms and allows copies to be made without the use of carbon paper. It is an industry wide, fully accepted modern method of written communication. Carbonless paper volume is estimated[44, 45] in the Western world including Japan at 500,000 t/year (1980) including up to 10% chemicals.

Briefly it is the reaction between a colorless electron donating organic compound (color former) and an electron accepting color developing material (color developer). A dye intermediate such as crystal violet lactone, benzoyl leucoauramine dyes or spiropyran dyes is micro-encapsulated to form particles less than 10 μm and coated on the underside of the top sheet of paper. The bottom sheet or copy is coated with the color developing material such as inorganic materials like acid clay, activated clays, zeolite, bentonite, metallic salts of salicylic acid or selected phenol formaldehyde resins. Under normal handling and storage conditions, the dye capsules resist break-age but are easily ruptured under fixed pressure of a writing device. Capsule rupture allows the two colorless materials to react and produce a dye copy of the original top sheet.

Phenolic novolaks based on *p*-phenyl phenol were initially identified as excellent color developers in carbonless multiple paper sheets due to good color developing per-formance, rheological coating properties and water resistance. The use of *p*-phenyl novolaks in single carbonless paper sheets is unsatisfactory because of "background-ing" or the spontaneous color development during manufacture and/or handling. *p*-Alkylphenol (C_1–C_{12})/phenol novolak copolymers are improved color devel-opers; however, papers become yellow by air oxidation. Yellowing can be minimized by using a mixture of alkyl phenol formaldehyde copolymers with hindered phenols[46] some of which are recognized as polyolefin antioxidants (Chap. 20).

New formulations are based on zinc or aluminum salts of *p*-phenyl, *T*-butyl or *T*-oc-tyl phenol novolaks that are combined with higher functional phenols as zinc salts. The latter are either bisphenol-A or 4,4 dihydroxy diphenylsulfone. These phenolic color developers have provided improved color developing ability, weather resistance, diminished yellowing and better oil resistance[47].

10.7 Ion Exchange Resins

The earliest ion exchange resin (IE) was a sulfonated PF resin developed by Adams and Holmes[48]. Relatively stable synthetic cation and anion-exchange resins were uti-lized to soften or demineralize water and quickly replaced the unstable aluminum sili-cates. Different active groups such as $-SO_3H$, $-COOH$, and $-NH_2$ were incorporated within the phenolic resin[49, 50]. The chelating ability of unmodified phenolic resins was already discussed in Chap. 3.4.3. The high resistance to the usual regeneration agents (dilute acids and bases), solvents and chemicals is the advantage of cured phenolic resins as the matrix material for functional groups with exchange character-istics. The desired mesh size is obtained by pulverization and screening or by a suspen-sion condensation process in a reversed-phase[50]. Other features of PF ion exchange resins are resistance to breakage, hydrophilic behaviour, and difunctional character-istics due to phenolic hydroxyl and sulfonyl acid or amine group.

Depending on whether the IE resin functions as a cationic, anionic, non-ionic or chelating species, separate categories exist for the use of these distinct IE resins[51].

Strong Acid Cation Exchangers

These materials are prepared by reaction of formaldehyde with phenol sulfonic acid or by phenol and formaldehyde with sodium sulfite. An important application of this strong acid IE resin is the selective removal of Cesium 137 from low level radioactive wastes[52, 53].

Weak Base Anion Exchangers

The derivatization of a resol resin with an oligomeric amine[51] results in the amine IE resin:

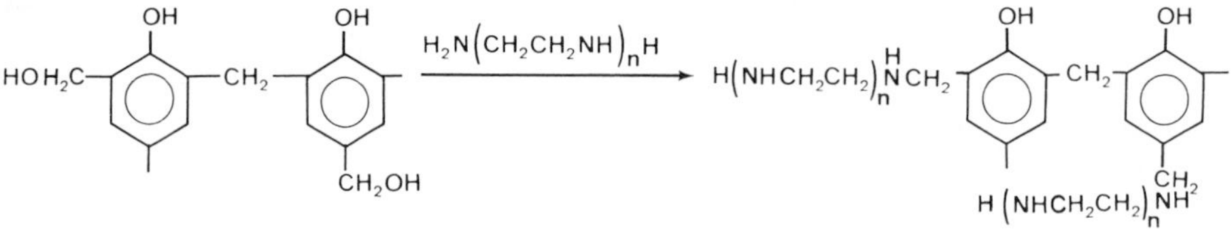

$$(10.8)$$

The majority of uses of these weak base anion exchangers are for food related processess[54] such as deacidification of fruit juices, de-ashing or demineralizing of gelatin, carbohydrates (sorbitol, sugar, lactose in whey), removal of undesirable tannins and brown coloration from white and red wines, and in the production of mono sodium glutamate. PF anionic exchangers also supplement decolorization processes by activated carbon. A major application is the deionization and decolorization of both glucose and fructose derived from corn starch hydrolyzate, a rapidly growing industry[55]. A particularly useful non-food application is the color removal from a paper pulp mill bleach plant waste in Sweden[56].

Non-Ionic Exchangers and Adsorbents

PF resols can be prepared under selective conditions which give optimum porosity and adsorption characteristics; both attributable to phenolic hydroxyl and methylol components. These resins are primarily used in the decolorization of agricultural and fermentation products[57].

Chelation Ion-Exchangers

PF resins are uniquely designed for a particular mode of metal chelation. In most cases these special resins with the proper architecture achieve remarkable metal removal or separation capability.

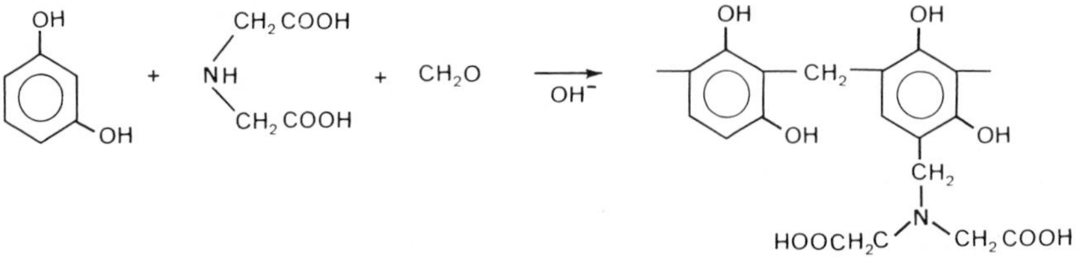

$$(10.9)$$

Erba and Wallace[58] have developed an improved phenolic cation exchange resin with highly selective chelating characteristics by reacting resorcinol with iminodiacetic acid and formaldehyde (10.9). The resulting resin is highly selective for simultaneous recovery of cesium and strontium from aqueous alkaline nuclear waste solutions. By combining the removal of both species in the same bed, the newly developed resin allows significant reduction in size and complexity of facilities for processing nuclear waste. The incorporation of tyrosine, a phenolic amino acid, into PF resin (10.10) has been shown to be effective in the recovery of uranium from various sources like sea water, phosphoric acid/fertilizer solutions, low grade uranium ore, etc.[59].

$$
\text{(phenol)} \quad + \quad \text{(p-substituted phenol with } CH_2\text{-CH(}NH_2\text{)COOH)} \quad + \quad CH_2O \quad \longrightarrow \quad \text{(polymer, repeat unit } n\text{)} \qquad (10.10)
$$

Newer applications yet to achieve commercialization are enzyme immobilization[60], especially for lactase (hydrolysis of lactose in acid whey) and glucose isomerase (isomerization of glucose to fructose); gold recovery by adsorption of cyanide complex, and use as a cigarette filter removing acetaldehyde, acrolein, methanol and MEK from cigarette smoke but not aroma or flavor components[61].

10.8 Interpenetrating Polymer Networks and Polymer Blends

Interpenetrating polymer networks (IPN) are polymeric structures that penetrate one or more other polymers to immobilize macromolecular species together. The technique combines thermoplastic or elastomeric and thermosetting resins to obtain toughness of the thermoplastic and elastomeric material as well as the high temperature performance of the thermoset component. The matrix of the IPN can be thermoplastic, elastomeric or thermoset depending on properties desired for ultimate use.

The phenolic resins provide the crosslinking characteristics to selective carrier resins which are predominately thermoplastic such as PMMA/PS copolymers[62] and polyesterimides[63]. The resulting IPN's exhibit better heat stability due to the thermosetting unit acting as a rigid reinforcing matrix in which the thermoplastic units are embedded within the interpenetrating network. Some modest amount of polymer/polymer compatibility is believed to occur.

IPN's based on amorphous thermoplastic resins such as polysulfone, polycarbonate or polyester-carbonate together with a modified phenolic for carbon fiber prepreg exhibit unlimited shelf life, short cure cycles and minimal shrinkage on curing. HDT of cured composites approach 240 °C. Excellent carbon fiber adhesion and properties equal or superior to epoxy thermoset systems identify these IPN systems for a variety of high performance composite applications[64].

Besides thermoplastic-thermoset IPN systems mentioned previously, wholly compatible thermoset systems which may involve IPN structure have been reported. Compatible blends of resols with polyimides have been prepared[65]. Intermediate polyimide precursors, polyamic acids, have been shown to be fully compatible with resols by examining different resol/polyamic acid compositions in water/dimethyl acetamide. Thermal stability of the cured blends by TGA paralleled resol/polyimide composition. It has been proposed that compatibility may be due to an interpenetrating structure and/or a resol extended polyimide network. Examples of elastomer/phenolic IPN's are the various pressure sensitive and contact adhesives and reinforced rubber systems described in Chap. 19.

Polymer blends or multi-component polymer systems are of great commercial importance especially those thermoplastic or alloy systems identified as engineering plastics. These include blends, grafts and block copolymers which exhibit varying amounts of compatibility depending on temperature, and composition. Multicomponent systems for electronic resists (Chap. 10.5) are based on novolaks with selected addition polymers or copolymers. The morphology of some of these systems have been examined (novolak with poly 2-methyl pentene-1-sulfone) by a variety of microscopy techniques such as interference contrast and SEM[66].

Solvents such as methyl cellosolve acetate, xylene, chlorobenzene and butyl acetate provide variable contrast as a result of differing solubility of the two polymeric phases present. Further novolak resin compatibility studies with other thermoplastic materials[67, 68] have been carried out by examining hydrogen bonding characteristics of the novolak/addition polymer blend as well as glass transition temperatures.

Although the concept of IPN has been utilized indirectly in many phenolic systems (high-impact molding materials, rubber reinforcement, coatings, adhesives, laminates, friction, others), it is anticipated that better characterization of IPN may provide better understanding of the morphological and chemical bonding of the matrix in the network.

10.9 Enhanced Oil Recovery

Ethoxylated nonylphenol novolaks are attractive additives in enhanced oil recovery[69] and function as emulsion breakers. Type and origin of oil requires special molecular weight and level of ethoxylation in each case for optimum performance.

10.10 References

1. Fitzer, E., Heym, M.: "High-temperature mechanical properties of carbon and graphite". High Temperatures-High Pressures 19, 29 (1978)
2. Vohler, O., Reiser, P. L., Overhoff, D., Martina, R.: Angew. Chem. Int. Ed. 9, 414 (1970); Sigri Elektrographit GmbH: Diabon/Durabon R, Technical Bulletin
3. Fitzer, E., Hüttner, W.: "Structure and strength of carbon/carbon composites". J. Phys., D: Appl. Phys. 14, 347 (1981)
4. Fitzer, E., Heym, M.: Chem. Ind. 29, 527 (1977)
5. Fitzer, E.: Pure & Appl. Chem. 52, 1865 (1980)
6. NUKEM (Nuklear Chemie und Metallurgie), D-6450 Wolfgang b. Hanau: Private communication
7. Powers, L. B.: J. Spacecraft 19, 104 (1982)
8. Sutton, G. W.: J. Spacecraft 19 (1) 3 (1982)
9. Kawamura, K., Jenkins, G. M.: J. Materials Sci. 5, 262 (1970)
10. Hayes, S. J.: Novoloid Fibers. In: Kirk-Othmer, Encyclopedia of Chemical Technology, Vol. 16, P. 125 (1981)
11. Conley, R. T.: Thermal Stability of Polymers, Chap. 11, New York: Marcel Dekker Inc. 1970
12. Fitzer, E., Schäfer, W.: Carbon 8, 353 (1970)
13. Jones, R. A., Jenkins, G. M.: "Volume Changes in Phenolic Resin During Carbonization"; Carbon 76, Baden-Baden, 27.6.–2.7.1976

14. Lubin, G. (Ed.): Handbook of Fiberglass and Advanced Composite Materials. New York: Van Nostrand Reinhold Co. 1969
15. Schik, J.-P., Schönrogge, B., Perrier, A.: Kunstharz Nachrichten (Hoechst), P. 26 (1983)
16. Bakelite GmbH, D-5860 Iserlohn-Letmathe. Company Brochure
17. Kaminski, A.: J. Appl. Polym. Sci. 27, 2511 (1982)
18. Chang, E. P., Kirsten, R. O., Harrington, H. J., Slagowski, G. L.: J. Appl. Polym. Sci. 27, 4759 (1982)
19. Occidental Chemical Corp. (1983), US-PS 4,419,400
20. N. N.: Modern Plastics Int., P. 34, Jan. 1982
21. Werkstoffleistungsblatt WL 8.4361.6 (Jan. 1978), Köln: Beuth-Verlag
22. Macosko, C. W.: Plast. Eng., P. 21–25, April 1983
23. Brode, G. L., Chow, S. W., Michno, M.: ACS Polymer Preprints, 24 (2), 192 194 (1983)
24. Union Carbide Corp., US-PS 4,395,521 (1983)
25. Union Carbide Corp., US-PS 4,395,520 (1983)
26. Union Carbide Corp., US-PS 4,403,066 (1983)
27. Union Carbide Corp., US-PS 4,430,473 (1984)
28. Carborundum Co., US-PS 3,650,102 (1972): US-PS 3,723,588 (1973)
29. Economy, J.: Chem. Tech. P. 240 (1980)
30. American Kynol Inc., US-PS 4,076,692 (1978)
31. Broutman, L. J.: Polym. Eng. Sci. 23, 776 (1983)
32. Kosar, J.: Light Sensitive Systems, New York: Wiley 1965
33. Thompson, L. F., Willson, C. G., Bowden, M. J.: "Introduction to Microlithography: Theory, Materials and Processing", ACS-Symposium Series No. 219 (1983)
34. Howson-Algraphy Ltd., DE-OS 1809248
35. Steppan, H., Buhr, G., Vollmann, H.: Angew. Chem. 94, 471 (1982)
36. Rosshaupter, E., Hundt, D.: Chemie unserer Zeit, 1971, P. 147
37. Willson, G., et al.: Polym. Eng. Sci. 23, 1004 (1983)
38. Gipstein, E., Ouano, A. C., Tompkins, T.: J. Electrochem. Soc. 129, 201 (1982)
39. Hiraoka, H., Pacansky, J., Schumaker, R., Harada, A.: ACS Polymer Preprints 25 (1) 322 (1984)
40. Fahrenholtz, S. R., Kelley, L. C.: ACS Polymer Preprints, 24 (2), 191 (1983)
41. Iwayanagi, T., et al.: Polym. Eng. Sci. 23, 935 (1983)
42. Shiraishi, H., Ueno, T., Suga, O., Nonogaki, S.: ACS Polymer Preprints, 25 (1), 293 (1984)
43. Roberts, E. D.: Polym. Eng. Sci. 23, 968 (1983)
44. Kondo, A.: "Microcapsule Processing and Technology", J. W. Van Valkenberg Ed. Marcel Dekker N.Y. 1979, Chapter 2
45. Sparks, R. E.: Carbonless Copy Paper, In: Kirk-Othmer Encyclopedia of Chem. Techn. Vol. 15, P. 470 (1981)
46. Mitsui Toatsu Chemicals, Inc., US-PS 4,271,059 (1981)
47. Sumitomo Durez Co., Ltd., US-PS 4,409, 374 (1983)
48. Adams, B. A., Holmes, E. L.: J. Soc. Chem. Ind. 54 1-GT (1935)
49. Wegler, R., Herlinger, H.: Polyaddition u. -kondensation v. Carbonyl- u. Thiocarbonylverbindungen. In: Houben-Weyl: Methoden d. Org. Chem. Vol. XIV/2, Thieme: Stuttgart 1963
50. Unitaka Ltd., DE-OS 2403158 (1974)
51. Abrams, I. M., Benezra, L.: "Encyclopedia of Polymer Science and Technology" Vol. 7, 692, Mark and Gaylord, Ed. John Wiley, N.Y., 1967
52. Dickinson, B. N., Higgins, I. R.: Nucl. Sci. and Eng. 27, 131 (1967)
53. Wiley, J. R.: I & EC Process Design and Development, 17, 67 (1978)
54. Cristal, M. J.: Chem. and Ind. 814, Nov. 7, 1983
55. Abrams, I. M.: Private Communication (1984)
56. Lindberg, S., Lund, L. B.; TAPPI, 63 (3), 65 (1980)
57. Abrams, I. M.: I & EC Product R & D, 14, 108 (1975)
58. US-Dept. of Energy, US-PS 4,423,159 (1983)
59. Unitaka Ltd., US-PS 4,414,183 (1983)
60. Abrams, I. M.: Duolite Paper, 121 (1983)
61. Daicel Chemical Industries, Ltd., Japan-PS 81 35,435 (C.A. 96 31861t)

62. Jayabalan, M., Balakrishnan, T.: Angew. Makromol. Chem. 118, 65 (1983)
63. Maiti, S., Das, S.: Polym. Sci. Technol., 21, 257 (1983)
64. N. N.: Plastics Technology, P. 9, June, 1983; Plastics Engineering P. 31, April, 1984
65. Adduci, J., Dondge, D. K., Kops, J.: ACS Polymer Preprints 22 (2) 109 (1981)
66. Bowden, M. J.: J. Appl. Polym. Sci., 26, 1421 (1981)
67. Fahrenholtz, S. R., Kwei, T. K.: Macromolecules, 14, 1076 (1981)
68. Cangelosi, F., Shaw, M. T.: Polym. Plast. Technol. Eng. 21 (1), 13 (1983)
69. Poettmann, F. H.: Enhanced Oil Recovery: Encyclopedia of Chemical Processing and Design. McKetta J. J. Ed. Vol. 19, P. 111, Marcel Dekker Inc., N.Y. 1983

11 Composite Wood Materials

Phenol resin bonded wood materials – particle boards (PB), plywood, fiber boards (FB) and glued wood construction products – are used for outdoor construction and in high humidity regions because of the high moisture and weathering resistance of the phenolic adhesive bond and its high specific strength.

The rationale for the use of these materials[1–6] is
– to improve the anisotropic strength of natural wood,
– to better utilize inferior wood and wood waste from the lumber industry,
– to produce better and more economical shaped wood products.

The competitiveness and development of the lumber industry are of utmost importance for the development of thermosetting resins. This industry is the largest consumer of urea-, melamine- and phenol resins. Approximately 85% of UF- and more than one third of PF-resin production is used for wood materials (see Chap. 1.2).

11.1 Wood

The main chemical components of wood[2, 4] are cellulose (40–45%), hemicellulose (20–30%) and lignin (20–30%). Further, varying portions of water, different natural resins, fats (mono-, di-, and triglycerides), waxes, tannins and sterols are found. Conifers in particular contain low molecular compounds like mono- and diterpenes which are extractable by organic solvents.

Lignin is a polymeric substance[7] with a phenolic structure. This is the reason for its reactivity with formaldehyde. Lignin is formed from coniferyl alcohol (11.1) and the related sinapin- and cumaryl alcohols by a dehydrogenating, oxidative coupling mechanism (11.2) via free radicals which ultimately lead to a wholly random composition.

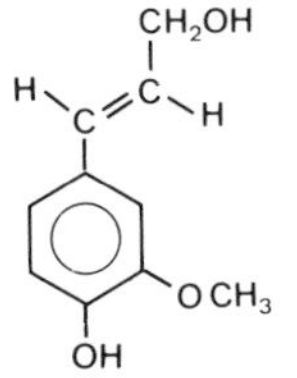

Coniferyl alcohol (11.1)	Lignin prepolymer (11.2)	Abietic acid (11.3)

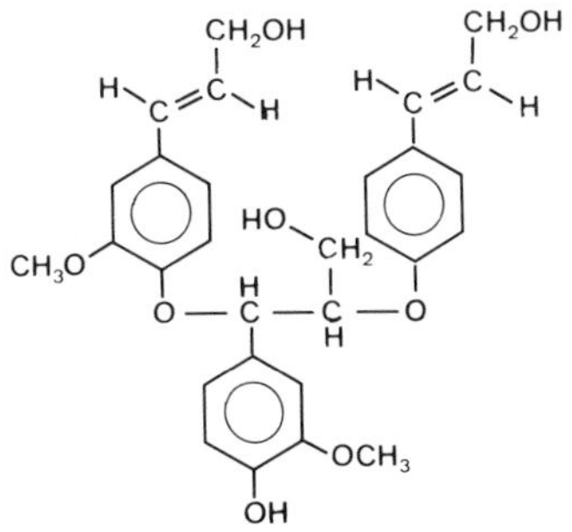

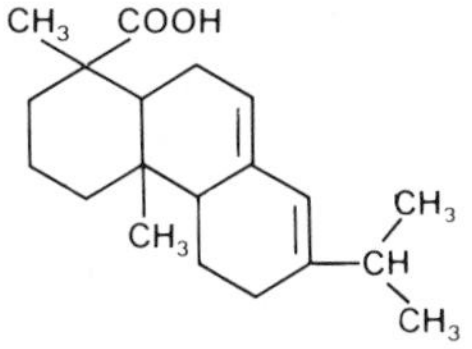

The resins found in conifers, called rosin, can easily be extracted with organic solvents. Rosin consists of approximately 90% free rosin acids, the remainder being neutral substances like oxidized terpenes and minor quantities of esters and anhydrides. Among the rosin acids there exists a structural equilibrium with respect to the position of the double bonds, leading to laevo-pimaric, neo-abietic, palistric and abietic acids. Abietic acid (11.3) is the main component of the different commercially available rosins.

Wood contains a series of reactive compounds and functional groups, so that aside from hydrogen bond formation and acid-base interaction between rosin acids and sodium phenoxide, a number of chemical reactions with PF-resins can occur.

11.2 Adhesives and Wood Gluing

All large volume wood gluing processes such as plywood and particle board manufacturing are influenced by adhesives costs[8]. Higher costs of the latter must be compensated by processing advantages or significant product improvements. The applied thermosetting adhesives are listed in decreasing order of consumption:
- urea formaldehyde and melamine formaldehyde resins
- phenol resins (including resorcinol resins)
- diisocyanates
- lignin sulfonates

Further, adhesives based on natural resources containing phenolic structures, i.e. tannin extracts[9] (pine) and lignins are finding industrial interest.

An important characterization of composite wood materials follows in accordance with their use and are designated as exterior and interior grades. Comparative water resistance of various wood adhesives under carefully controlled laboratory conditions, that reflects outdoor performance, was determined by Kreibich and Freeman[10]. The results are shown in Fig. 11.1.

The strength of the adhesive bond formed by thermosetting resins can be attributed to several factors[2]. Due to the high polarity and low viscosity the adhesive penetrates

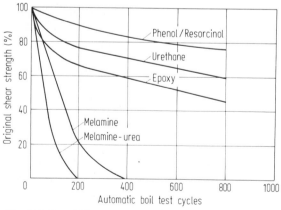

Fig. 11.1. Durability of various adhesive types by boil-dry cycling[10]

deeply into the micro-pores of the wood resulting in an initial mechanical fastening. Because of high polarity, thermosetting resins are known to form very strong hydrogen bonds with hydroxyl groups. Hence strong dipole-dipole interactions and van der Waals forces are developed. Chemical reactions further occur between wood components and PF- and UF-resins or urethanes as a result of the chemical composition and temperature. The cohesion force of the cured resin exceeds the tensile strength of the wood. Roughness and cleanliness of wood surface have considerable influence on bond strength.

11.3 Physical Properties of Composite Wood Materials

A compilation of the physical properties of several composite wood materials is shown in Table 11.1. Fireproofing of composite wood materials was reviewed by Satonaka[11]. Wood as a natural polymeric organic material is an ideal construction material. It has a high modulus of elasticity (MOE) in fiber direction with a corresponding low specific gravity (SG). Strength, additionally, is temperature independent over a wide range and unusual for an organic material. In this respect, wood is far superior to some synthetic organic polymers. Its thermal conductivity is low, making wood an excellent insulating material. The unusual anisotropic strength, high water absorption and swelling are obvious disadvantages.

11.4 Particle Boards

Wood waste of the lumber industry provided the stimulus of particle board (PB). Industrial production started relatively late because parallel to a reasonable theoretic and practical concept (Fahrni, Kollmann, Klauditz), highly efficient machines and process automation methods had to be developed (Himmelheber-Steiner/Behr/Novopan AG)[1, 3, 6].

Today 90% of PB's are used for indoor application, mainly furniture, and are bonded with urea-formaldehyde resins. Construction panels for outdoor use or high humidity areas require a water resistant bond which is obtained with phenol-formaldehyde resins[12]. In recent years the growing concern about formaldehyde emissions[13] has introduced a multitude of new regulations and developments in the particle board industry. Besides continuing competition between modified amino resins[14, 15, 18], phenolics[12] or diisocyanates[16-18], the latter two emit little or no formaldehyde, a wide range of resins has been developed with a diversity of modifications to reduce or eliminate formaldehyde emissions as well as decrease binder cost.

In West Germany formaldehyde emission standards[19, 20] for urea particle boards were established in 1980. Nevertheless, formaldehyde emissions of urea bonded particle boards recently enhanced a controverse emission/toxicity dismission. Table 11.2 shows the formaldehyde emission of various formaldehyde resin bonded materials according to the widely accepted Perforator test. It is important to note that even non coated wood particles exhibit aldehyde liberation at test conditions applied.

Table 11.1. Mechanical properties of composite wood materials according to Deppe[22]

Type of board Properties		Plywood		Fiber board		Particle board	
		Veneer plywood	Core plywood	HDF	MDF	Furniture V 20	Construction V 100
Thickness	mm	12–15	16–19	4	6–19	19	19
Specific gravity (SG)	kg/m³	550–700	450–650	900	680–750	620	700
Flexural strength ∥	N/mm²	60–100	10–15	40–60	20–40	19	22
Flexural strength ⊥	N/mm²	20–40	–	43	18–35	18	20
Modulus of elasticity (MOE) ∥	N/mm²·10³	7.0–12.0	1.0–1.5	–	2.0–2.2	3.0	4.5
Modulus of elasticity ⊥	N/mm²·10³	1.5–3.0	–	–	2.0–2.1	2.8	3.3
Internal bond (IB)	N/mm²	–	–	0.45	0.18–0.70	0.38	0.45
Internal bond, wet	N/mm²	–	–	0.10	0.02–0.20	–	0.20
Internal bond, dry again	N/mm²	–	–	0.35	0.15–0.65	–	0.32
Thickness swelling, 2 h	%	–	–	12	3–12	4	8
Thickness swelling, 24 h	%	–	–	18	6–17	11	14
Thickness swelling, 120 h	%	–	–	22	22–80	16	20
Water absorption, 24 h	%	–	–	45	15–35	–	–
Water absorption 240 h	%	–	–	70	–	–	–
Swelling (lengthwise)	%	0.1–0.2	0.2–0.3	0.2–0.4	0.2–0.3	0.2	0.3
Shrinkage	%	0.2	0.2	0.3	0.3	0.2	0.2

Table 11.2. Formaldehyde emission of untreated wood particles, UF and PF bonded particle boards and mineral/glass fiber materials according to the FESYP Perforator Test

	mg Formaldehyde/ 100 g material
Urea PB, E 1 class	≤ 10
Urea PB, E 2 class	10 –30
Urea PB, E 3 class	30 –60
Wood particles, non coated	4 – 5
Isocyanate PB	3 4
Phenolic PB	4 6
PF mineral fibrous insulation	0.1 – 0.5
PF/UF mineral fibrous insulation	0.2 – 3

Recent results indicate a comparable level of aldehyde emission (the Perforator test only identifies the carbonyl functional group) of phenolic and isocyanate bonded particle board compared to non coated wood particles.

Thus, phenol resin bonded particle boards according to DIN V 100 exhibit, especially if corrected for wood emissions, significantly lower formaldehyde emission compared to the lowest E 1 class level allowed for UF-particle boards. E 2 and E 3 class UF-boards are acceptable for use with facings only to maintain a low formaldehyde emission within the acceptable ranges. The emission of different organic compounds from PB and formaldehyde determination methods have been reviewed by Deppe[13] and Jellinek and Müller[12].

11.4.1 Wood Chips, Resins and Additives

Apart from wood chips and resin a number of additives like water repellants, extenders, flame retardant agents[11] and fungicides are used to improve other properties.

Wood Chips

A PB consists of about 90% wood chips or equivalent wood particles. While the use of a certain kind of wood is a matter of economy, the form of the chips and the struc-

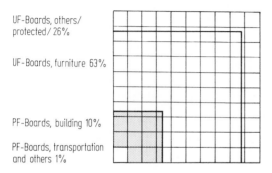

UF-Boards, others/ protected/ 26%

UF-Boards, furniture 63%

PF-Boards, building 10%

PF-Boards, transportation and others 1%

Fig. 11.2. Application of particle boards in West Germany in 1983

ture of the board are critical factors with regard to quality. Thus, all kinds of wood grown in Europe are employed – for instance, fir, Scotch pine, red beech, poplar, birch, etc., either solely or in combination. The use of larger portions of bark result in considerable reduction in board strength. Saw dust or wood shavings are also utilized in PB. Optimum properties are obtained from chips which have parallel and smooth boundary surfaces leading to high adhesive strength at low binder addition. This desirable chip form is obtained by cutting machines, while the chopping and tearing action in mills yields irregular splintered particles.

Resins

The use of PF-resins for the production of PB's leads to a weather resistant bond. This statement is supported by about 35 years of successful experience in the field of various plywood glues. Diisocyanates are also satisfactory for this type of application[16, 23]. Urea-melamine-phenol resins have also been recommended for outdoor application[14, 15, 18] but they have shown some definite deficiencies with regard to long term weather resistance.

Particle board resins are obtained by condensation of phenol and formaldehyde in a molar ratio of 1:(1.8–3.0) in an aqueous solution with NaOH as catalyst. Apart from its catalytic action sodium hydroxide has other important features[24] – such as providing a low viscosity for a high average molecular weight resin and excellent water solubility. In contact with the acidic wood components the resin is probably neutralized and exhibits a strong increase in viscosity and develops bonding properties. Therefore, sodium hydroxide is often used in an equimolar ratio based on phenol.

However, the high caustic content of conventional phenol resins results in high water absorption, thickness swelling, discoloration, and staining of light color facings because of the high hygroscopic behavior of sodium hydroxide and potassium carbonate. The latter is often added to enhance the cure rate. Further, incineration of the resin containing dust which results from finishing operations causes problems in the power plant. The pH of water extract of these boards is in the range of 9–10. Therefore, limits for the NaOH-content have been issued[21]. Fast curing resins with a relatively low caustic content (Table 11.3) are currently available and minimize these aforementioned problems.

The molecular weight distribution and the very low content of unreacted phenol and formaldehyde and mononuclear phenol derivatives is evident by examining Fig. 11.3.

Table 11.3. Characterization of a phenol resin for the production of particle boards V 100[25]

Characterization	phenol resol, water solution
Dry solids	$45 \pm 1\%$
Viscosity, 20 °C	130 ± 20 mPa·s
Specific gravity, 20 °C	1.17 g/cm^3
Alkali content (solution)	5%
Dilutability with water	∞ ratio
Storing ability, 20 °C	3–4 weeks
B-time, 100 °C	20 min
Content of free formaldehyde	$< 0.1\%$
Content of free phenol	$< 0.1\%$

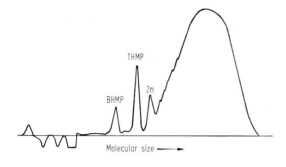

Fig. 11.3. Molecular weight distribution (GPC) of a phenol resin for particle boards
BHMP = bis-(hydroxymethyl)phenol
THMP = tris-(hydroxymethyl)phenol
2 n   = dihydroxydiphenylmethane derivative

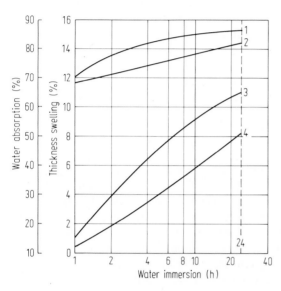

Fig. 11.4. Thickness swelling and water absorption of PF-boards during immersion. 16 mm single layer laboratory PB made of Scotch pine chips, 8% of solid resin based on dry chip weight. (Drawing: Mobil Oil AG, D-2000 Hamburg). *1* Thickness swelling without additives; *2* Water absorption; *3* Thickness swelling with 1,2% paraffin; *4* Water absorption

Often special catalysts are used to accelerate the curing rate. Organic or inorganic carbonates (FR-PS 15 5087), formamide (DE-AS 21 10 264), lactones (DE-PS 10 65 605), resorcinol, and resorcinol resins and isocyanates (DE-OS 27 16 971) have been proposed. These accelerators are added in the mid-layer only and reduce curing times to 12 s/mm board thickness at a press plate temperature of 220 °C.

As the binder represents the major cost of PB, phenol resins are sometimes modified by addition of urea, lignosulfonates, or other cost reducing additives (Chap. 3.8).

Hydrophobic Agents

The swelling of PB's mainly occurs vertically to the board surface due to chip orientation during the molding process. Hydrophobic or water repellant agents are added to reduce the wettability and water absorption. The amount of bonding agent is, by far, insufficient to coat the chip completely.

Aqueous dispersions of paraffins with a MP between 50–60 °C and 30–65% solids content are almost exclusively used as hydrophobic agents[26]. The addition of paraffin is very effective in preventing the absorption of water (wetting problem) but is far less effective against dampness. The water absorption and deformation processes are only temporarily delayed.

Generally, the paraffin dispersion is mixed homogeneously with the resin. Separate addition of the two components is preferred because of more effective distribution, but is more expensive because of the equipment needed. The amount of paraffin used in PF-boards is higher than that in UF-boards and is, in general, approximately 1–1.5% based on dried wood. A higher percentage of paraffin reduces binder wetting and adhesive strength[27].

Fungicides and Insecticides

PF-boards provide better resistance to wood harming insects and fungi than wood itself. Conversely, UF-resins promote the growth of fungi. Furthermore, the kind of wood, binder content and board influence fungi resistance[28]. Several cases of damage by fungi in the construction industry have dictated that additional protective measures are necessary. In West Germany, PB's resistant to fungi are designated as V 100 G (DIN 68763). Similar boards are made in France according to CTB-G standard. The increased resistance is achieved by the addition of fungicides, preferably tributyl tin oxide or fluorine compounds[29] to 1.5%.

Flame Retardants

In West Germany, standard quality PB's used as construction material are classified as B_2 "normal flammability", according to DIN 4102. The requirements of class B_1 can be achieved by adding flame retardant materials[11]. Ammonium hydrogen phosphate or ammonium polyphosphate are preferred. They may be used in combination with halogen compounds. Boron compounds are not very effective because of poor compatibility with resols. The material costs are dramatically increased by flame retardants so that the PB economics are effected. The addition of inorganic substances like vermiculite or perlite has been recommended, but board strength is reduced by these materials. Inorganic bonding agents (cement) and fillers[30] also increase thermal conductivity so that their use would impede other requirements.

11.4.2 Production of Particle Boards

A description of PB-technology is given in the literature[1, 3, 6, 31–33]. Only the gluing and pressing processes are mentioned briefly as far as they decisively influence the effectiveness of the gluing operation.

The chips are dried to a relative humidity (RH) of 1–5% prior to blending. The chips of the middle layer should have 2% RH. The resin content in three-layer boards amounts to 10–12% in the surface layers and 7–9% in the middle layer.

After blending, the chips have a RH of 8–18%.

The formulation and chip size for the middle and surface layers are different, so that separate storage tanks and gluing machines are required for a continuous operation, as shown in Fig. 11.5.

In modern high speed chip gluing machines the glue is uniformly distributed by centrifugal force through a hollow shaft out of several distribution tubes. The uniform binder spread is achieved by a rolling and wiping effect facilitated by the high rotational

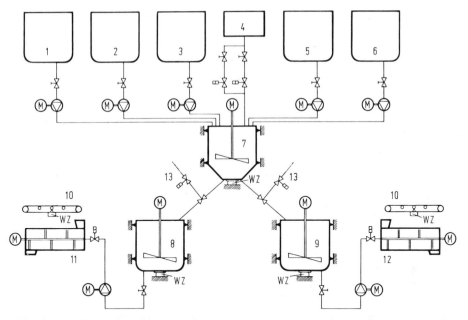

Fig. 11.5. Flow sheet of a PB-glue formulation plant. (Drawing: C. Schenck AG, D-6100 Darm-stadt.)
1 PF-resin; *2* UF-resin; *3* Paraffin dispersion; *4* Water; *5* Ammonia; *6* Hardener (for UF-boards) or additives; *7* Agitated vessel and scale; *8* Glue formulation, middle layer; *9* Glue formulation, surface layer; *10* Continuous scale; *11* Chip blender, middle layer; *12* Chip blender, surface layer; *13* Compressed air; *M* Motor/Pump; *WZ* Scale

speed. The glue is not captured by an individual chip but rolls from chip to chip leaving a thin glue track. The glue is distributed in that way so that only discrete chip areas are covered and glued together.

Gluing with resin powder has been recommended[34, 35]. Although this method would result in considerable savings of material and energy, the technical problems have not been satisfactorily resolved (see also Chap. 11.5).

Pressing is performed either in an one-daylight press (temperature up to 220 °C) or in a multi-daylight press at 160–190 °C. The pressure applied is, generally, between 1.5–3.5 N/mm² according to desired board density. The molding time is between 12–30 s/mm thickness depending on press plate temperature. Water content of the chips critically influences the heating rate. Due to the low thermal conductivity of the system, heat is mainly transferred by the steam formed internally.

11.4.3 Properties of Particle Boards

The requirements of PB's to be used for furniture are different from those for the construction industry[22, 36, 37]. High surface quality is required in the manufacture of furniture UF-boards because bonding to various top layers must be flawless and optically perfect. PB's for construction must have high strength (MOE, flexural and tensile

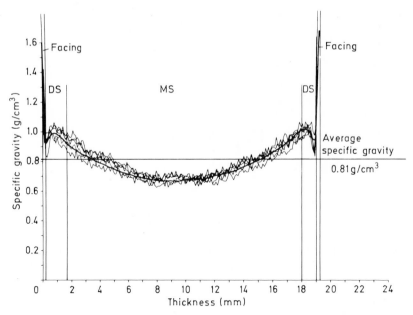

Fig. 11.6. Plotter-drawing of the density profile of a three-layer particle board clad with decorative facings, measured by gamma ray transmission[37]. (Kossatz, May, Wilhelm-Klauditz-Institut für Holzforschung, D-3300 Braunschweig) MS = middle layer, DS = top layer

Table 11.4. Properties of single-layer particle boards with varying resin content. Chips 80% conifer, d = 20 mm, molding conditions 0.35 min/mm at 160 °C, 2 N/mm^2 pressure

Properties		% Resin content			
		5	7.5	10	12.5
Specific gravity	kg/m^3	650	645	640	660
Internal bond, dry	N/mm^2	0.48	0.57	0.69	0.76
Internal bond, 2hH_2O/100 °C	N/mm^2	0.04	0.12	0.22	0.27
Thickness swelling	%	28.2	17.7	14.2	11.9

strength) and high weather resistance (swelling, IB, etc). The SG of a PB (Fig. 11.6) may be affected by production parameters (chip form, RH, press diagram and reactivity of the resin). Even if there are no specifications for the roughness of PB surface – PF-boards are not covered with decorative facings – a high density of the surface layer is desirable (sandwich principle, water resistance).

Board strength depends upon chip form, board structure and binder content. Mechanical properties are indicated in Table 11.1. Table 11.4 shows the strength of single-layer boards depending upon resin content. These boards were made without paraffin addition, this explains the relatively high thickness swelling. The influence of paraffin is shown in Fig. 11.4.

The high water resistance of PF-boards[12, 38, 39] is shown in Fig. 11.7. Comparable values are observed if the boards are boiled in water. After 20 h at 100 °C the thickness

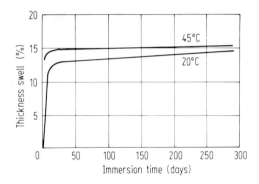

Fig. 11.7. Thickness swelling of a PF-bonded three-layer PB dependent upon time and temperature during water immersion. resin content – top layer 11%; resin content – middle layer 9%; paraffin content 1.2%; pressing time 0.30 min/mm; pressing temperature 170 °C; chips: fir; spec. gravity 665 kg/m³

swelling amounts to about 15%. Phenolic resin bonded plywood shows a similar behavior.

Formaldehyde emission during board production[40] and use[41] is a very important factor for further market acceptance of different PB-resins. The exceptionally low formaldehyde emission of PF-boards in comparison to UF-bonded boards will favour further market development of PF-boards. Even standard types FPY and FPO will be subject to E-classification (Table 11.2). Methods for the evaluation of the formaldehyde separation of PB's have been published[42, 43].

Future developments within the particle board industry considering raw material supply, new board developments and competitive materials like MDF-board, waferboard, or strandboard have been discussed by Deppe[44].

11.5 Wafer Board and Oriented Strand Board

An entirely new wood composite industry has emerged during the past 15 years in North America and is attributable to its traditionally strong plywood position. The composite is known as wafer board. Large diameter trees suitable for veneer production are getting scarce and more expensive as well. Low grade and fast growing softwood with diameters as small as 10 mm is used in the wafer board process. The use of these smaller diameter trees allows proper forest management by controlled "thinning". Thus far, there are 14 wafer board plants in operation in the US and Canada; a series of further plants is planned. In West Europe a plant is being built in Scotland.

Wafer boards are construction panels competitive with plywood and made of large area wood particles, called wafers, bonded with spray dried phenolic resin to provide a satisfactory weather resistant bond. They are used as sheeting in home and industrial applications such as roofing, siding, crating, panelling, and fencing.

Single layer boards with large square wafers, developed in 1954 by J. Clark, did not have sufficient flexural strength. By the early 70's DHYM Co. succeeded in the development of three-layer boards with a mid-layer of square wafers, approx. 35 mm long, 0.7 mm thick, as well as cover plies with approx. 75 mm long and 0.5 mm thick wafers (DHYM system). For wafer boards aspen is preferred in relation to birch and maple[45–47]. Inside the plant facility the logs are debarked and cut to convenient

Table 11.5. Property comparison of wafer boards (aspen), CDX-plywood, and ANSI-208.1-1979 requirements

		Wafer board (aspen)	CDX-plywood	ANSI A 208.1-1979
Gravity	kg/m³	650	500	
Flexural strength ∥	N/mm²	25	50	23.6
Flexural strength ⊥	N/mm²		15	
MOE ∥	N/mm²	3 400	8 000	3 200
MOE ⊥	N/mm²		1 200	
Internal bond	N/mm²	0.42	0.85	0.42

length wafers with specially designed drum-flakers. The wafers are dried (5–6% RH) and are then sprayed with molten wax to enhance water repellent behavior and adhesion of the powdered resin. The powdered resin is blown into a drum by compressed air and forms a fine mist which is deposited uniformly on the wafers. Approximately 0.3–0.8% of wax and 2.5–3.5% solid resin based on the dry wafers are added to the wafers. In spite of the relatively low resin content, remarkable weather resistance is achieved because of large area wafers and favourable resin distribution. Compared with weather resistant particle boards with an average resin content of about 10%, higher pressures are necessary during molding; 40–50 bar are common. The molding time is also longer with 24–30 s/mm board thickness.

Table 11.5 shows the properties of wafer boards in comparison to 3-layer CDX-plywood used in the USA and Canada. Wafer boards meet the requirements of the US-specification ANSI A 208.1 for construction boards; however, their flexural strength is lower as compared to plywood. Because of the low resin content the internal bond (IB) is relatively low and is insufficient for the German V-100-specification. Yet the strength is satisfactory for many building applications where plywood has been used until now. In the US and Canada wafer boards have achieved considerable importance as a low cost alternative to plywood[46, 47].

On the other hand, for many applications where the high strength like plywood is necessary, a clear increase in strength is achieved by scatter orientation of the wafers. By orienting the wafers in the surface layer the flexural strength increases parallel to the wafer orientation, strength is naturally decreased in vertical direction.

Another approach[45] is scatter orientation in the mid layer and the use of finer chips in the surface layer allowing a higher compression of the surface yielding an additional increase in strength. During all orientation processes either electrostatic, scattering grid or oscillating conducting surface, a higher length to width ratio of wood particles favours orientation. For these reasons less expensive wood can be used, such as pine and jaw among aspen. These materials are known as oriented strand board (OSB). By the use of liquid resin and higher dry resin content in the board of up to 7%, the tensile strength of OSB's approaches that of plywood.

11.6 Plywood

In general, plywood is a material which consists of at least three layers of wood, mostly veneers, with application of adhesives, pressure and heat. The fiber direction of the

Table 11.6. Types of plywood according to DIN 68 705 with respect to their humidity resistance and type of adhesive

Class acc. to DIN 68 705	Type of adhesive	Property	Pre-treatment of samples
IF 20	Urea resins	Resistant to low humidity, usual in closed rooms	24 h in H_2O at 20 ± 2 °C
IW 67	Urea and urea/melamine resins	Resistant to higher humidity, resistant to water up to 67 °C for a short time	3 h in H_2O at $67 \pm 0,5$ °C and 2 h in H_2O at 20 ± 3 °C
A 100	Urea/melamine resins melamine resins	Outdoors limited resistance, resistant to cold and hot water	6 h in H_2O at 100 °C and 2 h in H_2O at 20 ± 5 °C
AW 100	Phenol resins Phenol/resorcinol resins Resorcinol resins	Resistant to all climatic influences	4 h in H_2O at 100 °C, 16–20 h in air at 60 ± 2 °C and 4 h in H_2O at 100 °C, 2–3 h in H_2O at 20 ± 5 °C

individual layers is mostly vertical to each other, but can also be parallel (veneer plywood, wood core plywood, block board, laminar board and composite plywood) or diagonal (star plywood). Plywood is generally distinguished as interior and exterior grade[1, 5]. Exterior type plywood will retain its glue bond when repeatedly moistened and dried or otherwise subjected to vigorous weather conditions for permanent exposure.

In the USA the most commonly used specifications for plywood are the product standards established by the industry with the assistance of the U.S. Department of Commerce, NBS. Softwood plywood is covered by U.S. Product Standard PS 1–74[48], hardwood and decorative plywood is covered by PS 51–71[49]. These specifications include requirements for species, grade, thickness of veneer and panels, glue bonds, moisture content, etc. They also provide standardized definitions of terminology, suitability, and performance test specifications, expected minimum test results, and explanations of the codes which normally are stamped on each panel. The classification in West Germany according to DIN 68 705 is indicated in Table 11.6. Typical uses for plywood include containers, marine-grade materials, concrete form plywood, exterior sidings, kitchen and bathroom cabinets, door skins, interior parts in transportation vehicles, and foundry patterns.

11.6.1 Resins, Additives and Formulations

Normal resols catalyzed by sodium hydroxide are used for weather resistant plywood. The MWD of those resins is shown in Fig. 5.5, Chapter 5. The viscosity of the resins varies between 700 to 4,000 mPa·s according to the gluing process – dry or wet. Typical resin properties are compiled in Table 11.7.

As with PB-resins the content of free phenol and free formaldehyde is extremely low and practically non detectable. To adjust wetting and avoid excessive penetration for

Table 11.7. Phenol resol resins for AW 100 plywood manufacture[50]

Resin type		Dry process	Wet process
Dry solids content	%	45 ± 1	44 ± 1
Viscosity at 20 °C	mPa·s	4,000 ± 500	700 ± 200
Alkali hydroxide content	%	3	7
Gel time at 100 °C	min	20	33
Water solubility		∞	∞
Free phenol content	%	0.1	0.1
Free formaldehyde content	%	0.1	0.1
Storage stability at 20 °C		3 weeks	12 months

Table 11.8. Glue formulation for the production of weather-resistant plywood[50]

Formulation and processing		Dry process	Wet process		
Phenol resin	pbw	100	100	100	100
Chalk	pbw	–	10	10	10
Coconut flour (300 mesh)	pbw	5–10	8	8	8
Water	pbw	10–20	–	–	–
Hardener (paraformaldehyde)	pbw	–	–	2	2
Accelerator (resorcinol resin)	pbw	–	–	–	6
Glue service life		1 d	1 d	1 d	4 h
Press temperature	°C	130–150	130–140	120–125	100–110

a uniform joint thickness, diluents and/or fillers are almost always used. While rye or wheat flour is used in combination with UF- and MF-resins, only inert (chalk) or non-swelling fillers like coconut shell flour can be used for exterior grades. Apart from cost reduction, the brittleness of the adhesive joint is reduced by these additives. An excessive quantity of fillers, however, may lead to strength reduction. Examples of glue formulation are listed in Table 11.8. Flame retardants may be added[11].

11.6.2 Production of Plywood

Both softwood and hardwood are used in the manufacture of plywood (these terms make no reference to the actual hardness of the wood itself). In general, softwood plywood is intended for construction and industrial use and hardwood plywood is used where appearance is important. Softwood plywood is made of Douglas fir, western hemlock, larch, white fir, ponderosa pine, redwood, southern pine, and other species. For hardwood plywood, red beech is prominent, followed by birch, pine, and fir; of tropical woods limba and okoume are mainly used.

Before gluing, the veneers are carefully dried. The relative humidity should be between 4 and 8% for aqueous glues. Excessive wet or very dry veneers can cause faulty gluing. Too high humidity of veneer favours excessive resin diffusion of the glue joint or the formation of blisters. Water also reduces the curing rate. Veneers which are too dry compete for water in the resin and reduce resin flow[50–53].

In general, the adhesive (Table 11.8) is spread on two sides of one ply which is placed between two plies of unspread veneer in the panel assembly. The adhesive is applied e.g. by roll spreaders in an amount (spread rate) of approximately 160 g/m². A rate of 130–140 g/m² is often enough for low porous veneers. A thin glue line reduces costs and may offer technical advantages. However, the risk of gluing defects is increased. After gluing, the veneers are dried in a veneer dryer to a RH between 8–12% (dry process). The drying temperature should not exceed 70–75 °C. Veneers are assembled prior to plywood pressing. By far the largest production uses a hot hydraulic multidaylight press.

The molding temperature is – according to the glue formulation – 100 to 140 °C, the pressure is 0.8–1.5 N/mm² for softwood (SG up to 0.55 g/cm³) and 1.5–2.5 N/mm² for hardwood (SG above 0.55 g/cm³).

The curing temperature can be reduced to about 100 °C by the addition of accelerators. The practice of hot stacking is common. Gluing with resorcinol resins can be accomplished at room temperature. After being press-cured, the plywood is trimmed to size and sanded to the desired thickness. The glue bond quality (exterior type) is tested by a vacuum-pressure test, boiling test, scarf and finger joint tests, and heat durability test[48, 49]. Mechanical properties are listed in Table 11.1.

11.6.3 Compressed Laminated Wood

Compressed laminated wood is closely related to plywood. It is made by coating and impregnating wood veneers with a higher proportion of resin[54]. It is then pressed under high pressure to laminates with a SG between 1.0–1.4 g/cm³.

Compressed laminated wood has excellent mechanical strength, water resistance and very good machinability. In the field of mechanical engineering, screws, bolts, drives, cog wheels and parts for weaving machines are made from it. Further fields of application are chair seats, trays, dash boards, knife handles, bearing shells, sliding ledges for conveyer plants etc. Molded parts with excellent electrical properties are made by using phenol resins which are free of inorganic ions. They are used for the construction of transformers and control devices because of their good insulating properties, high specific strength and resistance to transformer oil.

Red beech veneers 0.2 to 1.0 mm thick are commonly used. They are predried to a RH of 8% maximum and the surface is then either coated with the resin or impregnated uniformly depending on future use. A resin content of approximately 20% can be applied by coating. In the dipping (impregnation) process the veneers are flooded with aqueous resin for several hours. Resin addition of about 30% is obtained. Uniform homogeneity can be achieved by the vacuum-pressure impregnation in an autoclave. A maximum resin addition of 60% can be obtained by this process.

After gluing or impregnating, the veneers are dried to 2–10% RH depending on product type in a veneer dryer. A tolerance of ±0.5% is not exceeded. This can be achieved by post conditioning. The pressing temperature is generally 140–160 °C. The pressure is dependent upon the wood type, the desired degree of compression and the part shape. In general, it is between 5 and 40 N/mm². For high pressures, highly polished chromium-nickel steel molds are required, while at low pressure aluminum molds with a modified surface are adequate[55].

Table 11.9. Classification of fiber boards according to specific gravity[1]

Fiber board		Specific gravity g/cm^3
Non-compressed:	Semi-rigid insulation board	0.02–0.15
	Rigid insulation board	0.15–0.40
Compressed:	Medium density fiber board, MDF	0.60–0.80
	Hardboard (high density FB), HDF	0.90–1.20
	Special density hardboard	1.20–1.45

11.7 Fiber Boards

As early as 1872 Clay obtained a patent which suggested the use of thick paper of mediocre quality – "papier mache" – for the construction field and furniture manufacture. Later, the strength of the paper was improved by additives.

A large number of fiber board classifications can be identified according to appearance, method of production, kind of application and SG[1, 4, 5]. According to SG, five types can be differentiated (Table 11.9).

The market for insulation boards is relatively small and is further diminishing in importance, acoustical boards are readily being replaced by non-flammable, inorganic materials. Medium density fiber boards (MDF), developed in the USA and Scandinavia especially for use in the construction industry, may also be used for the manufacture of furniture because of their remarkable homogeneity and smooth surface. MDF's of comparable thickness attain the same strength as particle boards.

The most important application areas for hardboards are the furniture industry (about 40%), followed by the manufacture of doors (about 25%), and the automotive and house trailer industry (about 20%). In the furniture industry, they are mainly used for rear cabinet enclosures and drawer bottoms. For this purpose, the boards are coated; lamination with decorative papers is unnecessary.

Weyerhaeuser Co, USA, developed a special form of a fiber mat which is used for the production of shaped parts. These mats for the "pres-tock" process contain a higher amount of non-cured PF-resin in combination with a thermoplastic resin.

After steam conditioning the mats are compression molded. In the automotive industry, for instance, these mats are used for switch boards, ceilings or coverings – parts which are mostly clad with plastic foils or textile.

11.7.1 Wood Fibers, Resins and Additives

Only low grade wood and industrial wood waste is used for the production of fibers. Conifer wood is more suitable for the wet fiber process, because it is easier to dehydrate than wood from deciduous trees. The latter is, however, better suited for the dry process and is preferred in Europe because of its availability. More and more hard- and softwood mixtures are used for both processes for economic reasons. Bark is not completely removed because as much as 15% does not affect the properties of the fiber board. Sawdust can also be added with amounts up to 30%. In addition, non-wood fibrous materials such as residues from perennial plants can be used. The fiber ef-

Table 11.10. Characteristic properties of a phenol resin for fiber boards[58]

Solid resin content	%	41 ± 1
Content of precipitating resin	% min.	35
Alkali content	% max.	6
Viscosity at 20 °C	mPa·s	500 ± 100
B-time in precipitated condition (pH 4) at 100 °C	min.	1.5 –3
Dilutability with water	–	∞
Storage life at 20 °C	weeks	4

ficiency of this latter material is rather low and the overall properties are inferior to those of wood fibers.

In general, fiber boards can be produced without bonding agents, using only the bonding potential of the fiber. To improve the mechanical properties and to reduce the water absorption and swelling, phenol resols, sometimes combined with natural resins like colophonium, are added as bonding agents. Urea resins are used in the dry process for boards which are not subjected to high humidity, such as boards used in the furniture industry[56]. The portion of synthetic resins in fiber boards is very small – resin specifications are described in Table 11.10. In general, the dry resin content is between 1 and 3% and calculated on dry fiber weight. This low resin coat can only be achieved by precipitation of a diluted aqueous resin solution by dilute sulfuric acid or aluminum sulfate at pH 4.

MDF's produced in the USA for building purposes contain a considerably higher PF-resin amount of between 8–10%. The main problem during MDF production is the uniform gluing of the voluminous, non-free-flowing fiber mass[57]. The process is performed with centrifugal gluing machines.

Hydrophobic agents like wax and paraffin reduce water absorption and swelling. The amount of wax is normally within the range of 1% based on the weight of the dry fibers.

Furthermore, flame retardants, fungicides and insecticides, release agents and other improving materials like drying oils can be added.

11.7.2 Production of Fiber Boards

The technology of fiber board production is described in detail in the literature[1, 4, 5]. The preparation of the fiber material can be performed thermo-mechanically or chemo-mechanically. It is obvious that fiber structure retention and the relation of length to cross section are the factors responsible for the quality and strength of the fiber board. The wet and dry process are mainly differentiated by the medium for fiber sheet formation. In the wet process the sheet is made with Fourdrinier or cylinder machines from an aqueous fiber suspension similar to the paper manufacturing process.

A dewatering screen is placed under the sheet while molding in a multi-daylight press. Because of this, screen marks develop on one side of the fiber board. Press time is approximately 2.0–3.5 min/mm of board thickness at temperatures of about 180–200 °C. Press diagram includes several pressure stages. In the dry process an air suspension of fibers is used for the sheet formation. The products are somewhat inferior in quality to those obtained by wet felting, but the dry process offers significant advantages with regard to environmental problems and reduced water consumption. Bonding agents and additives can be introduced at different stages, for instance prior to or during the defibrillation or after defibrillation as in special gluing machines like Drais or Lödige types. Pressures and temperatures for semi-dry and dry process are higher than those for the wet process; 7 N/mm^2 and 200–250 °C, respectively. Mechanical properties of fiber boards are listed in Table 11.1.

11.8 Structural Wood Gluing

The gluing of wooden construction elements[59] with cold setting adhesives is gaining considerable importance compared to the conventional techniques using nails and screws. Wood faults can be eliminated by lamination of thin layers; swelling and shrinkage of the wood can be reduced; and a wide range of design variations is obtainable. The performance of wooden structures in fire is surprisingly good because of the low thermal conductivity and insulating char formation. The type of glue does not influence flammability.

Softwood is preferred for structural elements. Wood quality must be thoroughly evaluated. Prior to its use, it is dried at 100 °C and conditioned. Depending on exposure, cold setting glues based on urea, resorcinol or phenol/resorcinol combinations are used. For interior parts UF-resins are satisfactory.

Structural elements subjected to weather or other climatic conditions must be glued with resorcinol/phenol- or resorcinol resins. In the marine field, glued parts are also subjected to strong dynamic stresses; the frequent change from wet to dry as well as the attack of salt water must be considered[60].

11.8.1 Resorcinol Adhesives

The reactivity with aldehydes is greatly increased by the introduction of a second hydroxyl group into the phenol nucleus[61–63]. Resorcinol reacts with formaldehyde at room temperature even without the addition of catalysts. The rate of reaction is at a minimum at pH 3.5 (phenol 4–4.5) as shown in Fig. 11.8. DSC, HPLC, and ^{13}C-NMR study of the reaction between resorcinol and formaldehyde was performed by Šebenik et al.[64].

However, alkaline catalysts are often used to produce resorcinol-formaldehyde resins. Prepolymers are first made with a low formaldehyde molar ratio (<1). Those prepolymers, which are completely stable, are cured by the addition of paraformaldehyde or formaldehyde solutions (now called hardeners). A total amount of 1.1 mol of formaldehyde is sufficient for curing, which is performed at ambient temperature and neutral conditions. The application of acids is not required so that the wood is not damaged. The bonds are gap filling and resistant to boiling water (Fig. 11.1), acids, mild alkali and common solvents. The temperature resistance of resorcinol adhesives is approximately 200 °C. The adhesion to various materials is very good so that they can also be used to bond leather, rubber, plastics, ceramics, etc. For glue preparation, a prepolymer solution (55–60% solids) is mixed in appropriate mixers with the har-

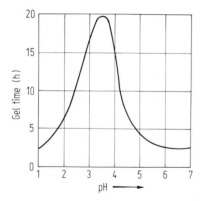

Fig. 11.8. Reactivity of a resorcinol-formaldehyde prepolymer with formaldehyde. Relationship between time of gelation and pH[62]

dener (paraformaldehyde) shortly before application. If chalk is used as filler, it is recommended that the paraformaldehyde is first mixed with the chalk. The service life of the mixtures is approximately 1.5 h at 20 °C; if chalk is added, up to 2.5 h, according to the quantity of the hardener. The use of a solid powdered hardener (paraformaldehyde) involves several shortcomings, i.e. difficult mixing, short pot life, dust contamination of the working place and irritation to the lungs and mucous membranes. Liquid hardeners have been developed[65, 66] to avoid these disadvantages.

Parts to be glued must have a moisture content between 6 and 15%. If the glue is applied on both sides, the rate amounts to 200–400 g/m² for soft wood, and 150–300 g/m² for hard wood. The glue spread applied with modern glue dispensing machines on one side only is approximately 400 g/m². The open assembly time is 30 to 45 min maximum; the closed assembly time can be extended up to 1.5 h. The pressure can be applied by clamps or a press. Depending upon the construction and the application, approximately 0.6 to 1.5 N/mm² are necessary. Press time is about 8–14 h at 20 °C; cure however, is completed within 2 to 3 days. Higher temperature reduces the curing time considerably.

A new glueing process[67] was developed by Casco AB. The resorcinol resin and a liquid hardener are applied to the wood surface in separate relative thick stripes. Mixing and cure commence during assembly and application of pressure. A more effective utilization of glue component is achieved.

11.9 References

1. Kollmann, F. P., Kuenzi, E. W., Stamm, A. J.: Principles of Wood Science and Technology, Vol. II, Berlin, Heidelberg, New York: Springer 1975
2. Bosshard, H. H.: Holzkunde, Vol. 3, Basel: Birkhäuser 1975
3. Deppe, H. J., Ernst, K.: Fortschr. d. Spanplattentechnik, Stuttgart: DRW Verlag 1973
4. Lampert, H.: Faserplatten, Leipzig: VEB Fachbuchverlag 1967
5. Kehr, E. et al.: Werkstoffe aus Holz, Leipzig: VEB Fachbuchverlag 1975
6. Froede, O., Witt, H.: Holzwerkstoffe. In: Ullmanns Enzyklopädie d. techn. Chem. Vol. 12, 4. Ed. Weinheim: Verlag Chemie 1976
7. Sarkanen, K. V., Ludwig, C. W.: Lignins. New York: Wiley-Interscience 1971
8. Bird, F., van den Straten, E.: J. Inst. Wood Sci. (London) 29, 41 (1971)
9. Pizzi, A.: Ind. Eng. Chem. Prod. Res. Dev. 21, 359 (1982)

10. Kreibich, R. E., Freeman, H. G.: Forest Products J. No. 12, 24 (1968)
11. Satonaka, S.: Advan. Fire Retardants 3/2, 74 (1973)
12. Jellinek, K., Müller, R.: Holz-Zentralblatt 108, No. 52, 727 (1982)
13. Deppe, H.-J.: „Emissionen organischer Substanzen aus Holzwerkstoffen". Holz-Zentralblatt 108, No. 10, 123 (1982)
14. Clad, W., Schmidt-Hellerau, Ch.: Holz-Zentralblatt 102, No. 24 (1976)
15. Deppe, H.-J.: Holz-Zentralblatt 109, No. 49, 677 (1983); Adhesion 1983, No. 10, P 16
16. Sachs, H. J.: Holz-Zentralblatt 102, No. 96 (1976). – Deppe, H. J.: Technical Progress in Using Isocyanate as an Adhesive in Particleboard Manufacture, 11th Washington State University Symposium on Particleboard, Pullman, March (1977)
17. Tinkelenberg, A., Vaessen, H. W., Suen, K. W.: J. Adhesion 14, 219 (1982)
18. Wittmann, O.: Holz als Roh- und Werkstoff 41, 431 (1983)
19. Ausschuß für Einheitliche Technische Baubestimmungen (ETB, April 1980): „Richtlinie über die Verwendung von Spanplatten hinsichtlich der Vermeidung unzumutbarer Formaldehydkonzentrationen in der Raumluft", Beuth-Verlag, Berlin
20. Verband der Deutschen Holzwerkstoffindustrie, Gießen: Spanplatten und Formaldehyd-Erläuterungen zu den Formaldehyd-Richtlinien (July 1980)
21. Normenausschuß Holz im DIN: DIN 68763, July 1980, Beuth-Verlag, Berlin
22. Deppe, H. J.: Möglichkeiten und Grenzen der Weiterentwicklung von Holzwerkstoffen. In: Verbund von Holzwerkstoff und Kunststoff in der Möbelindustrie. Düsseldorf: VDI-Verlag 1977
23. Johns, E. W.: J. Adhesion 15, 59 (1982)
24. Wittmann, O.: Holz als Roh- und Werkstoff 31, 419 (1973)
25. Bakelite GmbH, HW-2504, Technical Bulletin
26. Amthor, J.: Holz als Roh- und Werkstoff, 30, 422 (1972)
27. Ranta, L.: Holz als Roh- und Werkstoff 36, 37 (1978)
28. Gersonde, M., Deppe, H.-J.: Holz als Roh- und Werkstoff 41, 323 (1983)
29. Metzner W., Bollmann, H.: Holzschutz. In: Ullmanns-Enzyklopädie d. techn. Chem. Vol. 12, 4. Ed. Weinheim: Verlag Chemie 1976
30. Deppe, H. J.: Holz-Zentralblatt No. 49/50, 737 (1973)
31. Hutschnecker, K.: Holz als Roh- und Werkstoff 33, 357 (1975)
32. Himmelheber, M., Kull, W.: Holz als Roh- u. Werkstoff 22, 28 (1969)
33. Hickler, H. H.: Holztechnologie 18, 84 (1977)
34. Luthardt, H.: Holztechnologie 13, 135 (1972)
35. Deutsche Texaco AG, DE-AS 23 64 251
36. Neußer, H.: Holztechnologie 18, 88 (1977)
37. Kossatz, G.: Holzwerkstoff und Kunststoff – ihre wirtschaftliche Bedeutung für den Möbelbau. In: Verbund von Holzwerkstoff und Kunststoff in der Möbelindustrie. Düsseldorf: VDI-Verlag 1977
38. Gressel, P.: Holz als Roh- u. Werkstoff 26, 140 (1968) and 30, 347 (1972)
39. Meierhofer, U. A., Sell, J.: Holz als Roh- u. Werkstoff 33, 443 (1975)
40. Bernett, J.: Wasser, Luft u. Betrieb 20, 75 (1976)
41. Frank, M., Thiemann, A.: Kunststoffe im Bau 11, 36 (1976)
42. Roffael, E.: Holz-Zentralblatt 111, 1403 (1975)
43. Roffael, E., Melhorn, L.: Holz-Zentralblatt 154, 2202 (1976); L. Melhorn, E. Roffael, H. Miertzsch: Holz-Zentralblatt 20, 345 (1978)
44. Deppe, H.-J.: Holz als Roh- und Werkstoff 41, 403 (1983)
45. Walter, K.: Holz-Zentralblatt 107, No. 110, 1665 (1981)
46. Moelntuer, H. G.: Holz als Roh- und Werkstoff 34, 353 (1976)
47. Trutter, G.: Holz als Roh- und Werkstoff 40, 361 (1982)
48. Construction and Industrial Plywood, Nat. Bur. Stand. Voluntary Product Standard PS 1–74, U.S., Department of Commerce, Washington, D.C., 1974
49. Hardwood and Decorative Plywood Nat. Bur. Stand. Voluntary Product Standard PS 51–71, U.S., Department of Commerce, Washington, D.C.
50. Bakelite GmbH: Bakelite-Harze für wetterfestes Sperrholz AW 100, Technical Bulletin
51. Kreibich, R. E.: Adhesives Age, Jan. 1974, P. 26
52. Reiter, L. L.: Holz- u. Holzverarbeitung 6, 440 (1975)

53. Neusser, H., Schall, W.: Holzforschung u. Holzverarbeitung 24, 108 (1972)
54. Wichers, H.: Holzschichtstoffe. In: R. Vieweg, E. Becker (ed.): Kunststoff-Handbuch, Vol. 10 – Duroplaste, München: Carl Hanser 1968
55. Bakelite GmbH: Bakelite-Harze für Preßlagenholz, Technical Bulletin
56. Deppe, H.-J.: Holz-Zentralblatt 104, No. 19 (1978); Wehle, H. D.: Holztechnologie 6, 37 (1965)
57. Kehr, E.: Holztechnologie 18, 67 (1977)
58. Bakelite GmbH: Phenolharze für Hartfaserplatten, Technical Bulletin
59. Kollmann, F.: Adhäsion 5, 134 (1974)
60. Noack, D., Frühwald, A.: Holz als Roh- und Werkstoff 34, 83 (1976)
61. Rhodes, P. H.: Modern Plastics, Aug. 1947, P. 145
62. Glauert, R. A.: British Plastics, Aug. 1947, P. 233
63. Koppers Co. Inc.: Resorcinol, Technical Bulletin
64. Šebenik, A., Osredkar, U., Vizovišek, I.: Polymer 22, 804 (1981)
65. Bakelite GmbH: Resorcinharzleime für den Holzbau, Technical Bulletin
66. Andersson, S.-E. (Casco AB): DE-AS 28 20 907 (1978)
67. Perciwall, E. W. (Casco AB): DE-AS 24 16 032 (1979)

12 Molding Compounds

Phenol molding compounds are best characterized as the first true engineering plastics[1] and provide the following key properties:
- high-temperature resistance;
- modulus retention at elevated temperatures;
- flame and arc resistance;
- resistance to chemicals and detergents;
- high surface hardness;
- good electrical properties;
- relatively low unit cost.

Based on these key features, they are used within a wide spectrum of applications in housewares and other appliances, electrical components and the automotive industry[2-4]. Typical examples of appliances are dishwashers, air conditioners, coffee makers, toasters, refrigerators, and flat- or steam iron handles. Light sockets, switch and transformer components, blower wheels, relays, connectors, coil forms, and wiring devices represent examples of electrical applications. In the automotive industry, phenolics are mainly used for under-the-hood components such as distributor caps, coil towers, commutators, fuse blocks, bulkheads, connectors, and brake components.

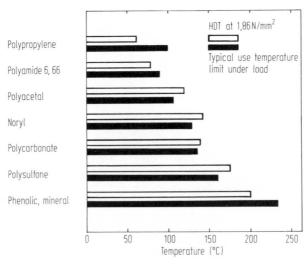

Fig. 12.1. Heat deflection and use temperature of phenolics and other engineering plastics including polypropylene[5]

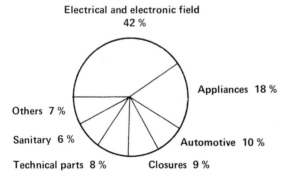

Electrical and electronic field
42 %

Appliances 18 %

Others 7 %

Sanitary 6 %

Automotive 10 %

Technical parts 8 %

Closures 9 %

Fig. 12.2. Market segments of phenol molding compounds[4] in Western Europe

Table 12.1. Production of phenol molding compounds in USA and West Germany

Production 1,000 to	1975	1976	1977	1978	1979	1980	1981	1982	1983
USA	102	165	148	136	141	110	116	87	100
West Germany	37	46	42	39	39	38	35	30	33
GB	–	19	19	20	22	17	13	13	14

Fig. 12.3. Phenolic brake piston.
(Photo: Bakelite GmbH, D-5860
Iserlohn-Letmathe)

Figure 12.1 provides a high-temperature resistance comparison of phenolic molding compounds with other engineering plastics. It is important to note that the heat deflection data according to ASTM D-648 are not adequate to predict performance at high temperatures.

Molding compound production in the USA[6], Europe[7], and Japan[8] has been declining in the past decade (Table 12.1). A variety of resons are attributable to the decline such as sluggish worldwide economy, replacement by thermoplastic materials, and a shift in the production of molded parts to other countries due to lower labour cost.

However, future growth will be in more critical and exact applications requiring heat resistance, dimensional stability under load and structural performance. The

198

12 Molding Compounds

Table 12.2. Properties[10] of a temperature resistant phenolic compound with a high dimensional stability, i.e. for brake pistons (Bakelite 100)

Gravity	g/cm³	2.1
Shrinkage	%	0.1
Post shrinkage	%	0.1
Thermal expansion coefficient	10^{-6}/K	13–15
Compressive strength	N/mm²	280
Flexural strength	N/mm²	90
Impact strength, notched	kJ/mm²	2.5
Thermal resistance/Martens	°C	170

Fig. 12.4. House power distributor box made of granular, high-impact phenolic molding compound (Type 84 acc. DIN 7708). (Photo: Bakelite GmbH, D-5860 Iserlohn-Letmathe)

world market for phenolic compounds is expected to increase considerably in the next five years.

In fact, there are several growing or new applications whereby phenolic compounds are replacing metal parts or thermoplastics where their unique combination of properties offers better performance and economics. For example[9], molded commutators in electric motors are made of copper-adhesive reinforced phenolics. Very high centrifugal forces are developed, speed can reach 40,000 rpm at temperatures up to 200 °C which can pull the commutator apart. If the copper components move out of line as little as 0.01 mm, failure occurs through arcing or wear of the carbon brushes.

New is the use of reinforced phenolic pistons in automotive disc brakes. Traditional steel pistons corrode or rust which causes them to seize up. They transmit heat, causing the brake fluid to vaporize. The steel pistons are also heavy and more expensive. New phenolic formulations solve these problems (Table 12.2).

Another automotive example is the transmission torque converter which is a large, complex, turbine shaped part. In this first application of plastic material for a major automotive transmission component, the phenolic part must withstand very high torque loadings at high speeds and high temperatures[9].

Table 12.3. Properties[10] of high-impact and temperature resistant granular phenolic molding compound (Bakelite VLP 7470)

Gravity	g/cm^3	1.55	
Impact strength, notched	kJ/mm^2	3.5 –	5.0
Compressive strength	N/mm^2	110	–120
Flexural strength 20 °C	N/mm^2	75	– 85
Flexural strength 180 °C	N/mm^2	60	– 65
Thermal resistance/Martens	°C	170	–180
Service temperature UL 746B	°C	180	
Water absorption	%	0.1	

Significant refinements have been accomplished in compounding. Asbestos has been substantially replaced as a filler because of the hazards during handling and production (see Chap. 18.1). Asbestos-free molding compounds have a majority of similar properties as products containing asbestos. Only the thermal resistance is not as high in all cases, but sufficient for most applications. The non-asbestos formulations contain mostly mixtures of glass-, mineral-, and organic fibers. Even commutator compounds that are asbestos-free have been developed with other fiber reinforcing materials.

Sophisticated filler combinations are applied in high-temperature resistant applications which are used for pot handles and molded objects that are dish washer resistant. Fiber reinforced, high-impact, and temperature resistant phenolics – which are now available in a free flowing pelletized form – command special attention. Their main application is the automotive industry. Rubber modified phenolics having a UL continuous use index rating of 180 °C and higher are expected to find also widespread use in the appliance industry. Impact modification has also been successful for asbestos-free compounds. The impact modifier additive boosts the material's impact strength without sacrificing heat resistance, hot rigidity, and appearance.

With carbon and graphite as fillers, molding compounds with exceptionally low thermal expansion have been developed and compare favourably to steel and other metals in very low thermal expansion.

To overcome color limitation, melamine-phenol compounds (MP) have been developed. Using light colored fillers like cellulose fibers, relatively bright colored molded parts are obtained. A further benefit is the improved tracking resistance making them particularly attractive for use in the electrical industry and for manufacture of household appliances. However, the demand for MP-compounds has not met volume projections. The phenolic resin content of these MP-compounds is generally about 10%.

12.1 Standardization and Minimum Properties

Due to the magnitude of common fillers and reinforcing agents, standards have been established to facilitate the comparison of properties and selection of materials. In general, the following differentiation is common: general purpose, improved impact, improved electrical and heat resistant grades.

In West Germany, standardization according to DIN 7708 is far more specific than any other designation. Further differentiations have been developed according to the

Table 12.4. Standardization of phenolic molding compounds according to DIN 7708

Type DIN 7708	Fillers/Reinforcing fibers	Flexural strength	Impact strength	Impact strength, notched	Heat resistance (Martens)	Water absorption
		N/mm^2	N mm/ mm^2	N mm/ mm^2	°C	mg
11	Mineral flour	50	3.5	1.3	150	45
12	Asbestos	50	3.5	2.0	150	60
13	Mica	50	3.0	2.0	150	20
15	Asbestos fibers, long	50	5.0	5.0	150	130
16	Asbestos cord	70	15.0	15.0	150	90
31	Wood flour	70	6.0	1.5	125	150
51	Cellulose fibers	60	5.0	3.5	125	300
71	Textile fibers	60	6.0	6.0	125	250
74	Textile chips	60	12.0	12.0	125	300
83	Textile fibers and wood flour	60	5.0	3.5	125	180
84	Textile chips and cellulose fibers	60	6.0	6.0	125	150
85	Cellulose fibers and wood flour	70	5.0	2.5	125	200

type of filler, quantity of resin and colour. Minimum physical properties requirements according to DIN 7708 are compiled in Table 12.4.

A further distinction is provided by "point" types, for example Type 31.5. The point designation is as follows:

0.5 improved electrical resistance
0.8 low organic acid content (<0.18%)
0.9 free of ammonia compounds.

Four more figures are added to this designation in order to indicate the resin content and coloration, for example: Type 31–1449 acc. to DIN 7708.

12.2. Composition of Molding Compounds

Phenol molding compounds consist of phenol resin, HMTA and selected catalysts, fillers or reinforcing fibers, colorants and pigments as well as various lubricants. The composition of a general purpose, wood flour filled molding compound, Type 31, is indicated in Table 12.5.

Compounds for compression molding have a lower resin content in comparison to injection grades which have a resin level approaching 50%. In addition, it is customary that each material is available in at least three types of flow: low, medium and high flow.

12.2.1 Resins

Phenol is the predominate resin raw material; technical cresol grades are seldom used. The latter delays the curing rate due to *o*- and *p*-cresol content. The flexibility of the

Table 12.5. Typical composition of a general purpose molding compound (Type 31), compression grade

40% Phenol novolak resin
 6% Hexamethylenetetramine
 1% Magnesium oxide
50% Wood flour
 1% Lubricants and release agents
 2% Colorants and pigments

Table 12.6. Novolak resin specification for the production of molding compounds

Melting range	°C	82–95
Viscosity, 50% sol. in acetone	Ford 4 mm, Sec	30–45
B-time with HMTA at 130 °C	min	12–16
Free phenol	%	≤ 2

molded parts can be improved by the addition of cresols to a limited extent. Novolaks for molding compounds (Table 12.6) are produced mainly by use of oxalic acid as catalyst; hydrochloric, sulfuric or phosphoric acid are seldom used.

The PF molar ratio is within the range of 1 : (0.75–0.85). The MWD of an appropriate resin is shown in Fig. 3.6, Chap. 3. Novolak resins with a low free phenol content ($\leq 2\%$) and relatively high flow are used in newer formulations.

High-*ortho* novolaks are added in some formulations to enhance the curing rate. Resols are used for special electrical applications when high hydrolytic resistance is required. Under the influence of moisture and temperature, ammonia is evolved from HMTA-cured novolaks and results in corrosion of copper or brass parts. The thermal shock resistance of resol based moldings is also better because less volatiles are released during cure. Resol/novolak mixtures, containing smaller quantities of HMTA, are intermediate in behavior.

Epoxide resins rather than HMTA are used to crosslink novolaks if the release of gaseous materials on curing are to be avoided and strong adhesion to copper is desired. Such compounds may be used for commutators and similar electrical applications (see Sect. 3.6). Also melamine-formaldehyde resins are used to crosslink phenolic resins for special applications. Epoxide-phenolic low-pressure molding compounds with selected, silane treated inorganic fillers have been developed for the microelectronic field. They are superior to pure epoxies [that are cured with diaminodiphenylmethane or hexahydroendomethylene tetrachloro anhydride, when dimensional stability, humidity resistance, storage stability and physiological behaviour are considered.

12.2.2 Fillers, Reinforcements and Additives

Wood Flour and Cellulose Fibers

Cellulose fillers, wood flour, nutshell flour or cellulose fibers are used in moldings to reduce shrinkage during cure, to improve impact strength and to provide flow control.

202

12 Molding Compounds

Table 12.7. Properties of wood flour for molding compounds

Fiber length	20–140 μm
Length/diameter ratio	~2.5 : 1
Apparent density	0.20–0.40 g/cm^3
Ash	<0.4%
pH	4.5 ± 0.5

Table 12.8. Properties of cellulose fibers for molding compound

α-Cellulose	90	% min.
Lignin	0.1	% max.
Calcium	0.03%	max.
Ash	0.15%	max.
pH	5.5	

Wood flour is by far the most attractive general purpose filler providing reasonably good all-around performance at relatively low cost. Soft wood species like pine, spruce and fir are preferred; hardwood flour may be used either solely or in combination. A somewhat lower water absorption is obtained with hardwood flour.

Properties of wood flour, prepared by wet grinding in stone mills or hammer mills, are indicated in Table 12.7. The retention of the wood fiber structure during the grinding operation is necessary for satisfactory mechanical properties, especially impact strength. Smaller granulation yields lower strength; impact strength is reduced significantly. Reduction in stiffness, however, is not significant.

Substitution of wood flour by nutshell flour results in distinct improvements whereby flow is enhanced. Walnut or coconut shell flour or apricot or olive pits, contain a considerable amount of lignin, resins, oils, and waxes as well as cellulose (~60%) and pentosans (~8%). These oils and waxes act as internal and external lubricants and, therefore, improve flow and surface quality. Similarly water absorption is reduced, i.e. the compounds behave as if they had a higher resin content. However, these fillers are not fibrous in structure and thus strength is reduced correspondingly. Normally, they are used to approximately 10%.

Cellulose fibers (Table 12.8) manufactured from carefully debarked softwood logs in the paper making industry, are used mainly for light colored melamine-phenol compounds or for general purpose compounds to improve impact strength.

Mineral Flour

Generally, mineral flour fillers in thermosetting plastics provide the following improvements:
– reduced shrinkage on cure and reduced exothermic heat during cure
– higher compressive strength and stiffness
– higher thermal resistance
– improved flame retardancy
– improved electrical properties
– adjusted flow
– machinability and surface quality
– lower costs

The physical properties of some important mineral fillers are shown in Table 12.9.

The abrasive action of inorganic fillers can cause serious problems. Not only the hardness of the mineral has to be carefully considered, but also the accompanying impurities. The addition of only 1% of an abrasive filler, e.g. quartz, can increase mold wear tenfold. Furthermore, the different basicity of the fillers may influence the curing

Table 12.9. Properties of mineral fillers[11] (* = in CO_2-free water)

	Chemical composition	Density g/cm^3	Mohs-hardness	pH*
Calcium carbonate	$CaCO_3$	2.71	3.0	9–10
China clay	$Al_4(OH)_8[Si_4O_8]$	2.69	2.0–2.5	5–6
Mica (Muscovite)	$KAl_2(OH, F)_2$ $[Al\,Si_3O_{10}]$	2.7–2.9	2–3	7–8
Silica flour	SiO_2	2.65–2.7	7.0	7–8
Talcum	$Mg_3(OH)_2$ $[Si_4O_{10}]$	2.65–2.8	1–2	9–10
Wollastonite	$CaSiO_3$	2.85	5.0–5.5	9–10

rate. Magnesium oxide, for instance, is frequently added to accelerate the curing reaction.

Mica, a sheet forming mineral, requires special mention since it provides unique characteristics to molding compounds. Muscovite, a calcium-aluminum silicate, is almost exclusively used as the mineral species. The mica sheets are flexible and possess outstanding dielectric properties as well as high thermal resistance. Mica filled compounds are used in electrical engineering areas such as commutators etc. Besides dielectric strength and thermal resistance, these compounds have low thermal conductivity, low water absorption and very good resistance to chemicals since diffusion processes are impeded considerably due to the sheet form of the mica filler.

Other Fillers and Fibers

Cotton fibers or chopped cloth are used as reinforcing materials in the production of large, flat parts with high impact and tensile strength. It is difficult to impregnate these materials uniformly with dry resin and therefore these molding materials are produced by the wet impregnation method using alcoholic novolak solutions or aqueous resols.

In addition to glass fibers which are used to a great extent, a series of organic fibers is used to increase impact strength. Organic fibers reduce the thermal resistance considerably; therefore, they are added in minor amounts. Polyester-, polyamide- or polyvinyl alcohol fibers have been used with some success. Carbon fibers and aromatic polyamide fibers (Kevlar, Arenka) have also been suggested as reinforcing fibers. Some organic fibers are either dissolved or decomposed by residual phenol at elevated temperatures.

The sliding properties of moldings are considerably improved by the addition of graphite or molybdenum sulfide. Such compounds are used to manufacture sliding rings, slide bearings and gaskets. Formerly, graphite was used as an additive to compounds for metal plated parts. Presently, better plating methods and compounds, free of conductive fillers, are available.

Colorants

The limited colorability of phenolic moldings is caused by the yellow coloration of the cured resin. Many important criteria must be considered in the choice of the colorant for molding compounds. Not only is cost a major factor but other critical parameters such as heat resistance, light fastness, weather resistance, migration resistance, physiological safety and ease of dispersibility are evaluated. Coloring is performed in the first phase of production. Attempts to color molding compounds with a weigh feeder in an injection molding unit have not been as yet successful. Also coating of naturally colored parts like flat- or steam iron handles, side parts of toasters, etc., is not feasible since the surface layer is not scratch resistant and marginal in adhesion. Also the technique of electrostatic powder coating is equally unsatisfactory.

Lubricants and Release Agents

In most cases, a combination of several lubricants is necessary to obtain optimal moldings. Formulations contain up to 1% of lubricants. Friction processes are alleviated by external and internal lubricants. External lubricants are used to reduce metal adhesion and facilitate the barrel feeding by lowering frictional heat development. They also act as mold release agents. Internal lubricants affect the melt flow (lower viscosity and injection pressure) and improve the homogeneity of the melt. Internal lubricants possess a highly polar chemical structure and must be readily soluble in the phenolic resin melt. Examples are fatty alcohols, acid esters or acid amides. Fatty acid salts like calcium or magnesium stearate are intermediate in performance. External lubricants should be non-polar in structure and virtually insoluble in the phenolic resin phase. This group includes paraffins and waxes.

12.3 Production of Molding Compounds

Molding compounds are produced by a combination of batch and continuous process steps. A phenol molding compound, for instance Type 31 or Type 12, consists of 6–10 components (Table 12.5). The mixtures are batch produced in appropriate mixing equipment and stored in an intermediate bunker. Plasticizing and homogenizing of the mixture, as well as the adjustment of the appropriate condensation grade, can be performed on a two-roll mill or in an extruder. The rolling process can be performed by a batch or continuous operation. The rolls are operated at different speeds to effect frictional heat at temperatures ranging between 90 and 130 °C. A rough sheet develops on the colder front roll which is homogenized by removing and turning several times, adjusted, and removed as a whole. Otherwise, the mixture is continuously fed to the center of the rolls and displaced to both sides where narrow tapes are cut off by stationary blades. After cooling, the compound is fed to a series of crushers, grinders and sieves where it is reduced and screened to the desired particle size and flowability. The rolling mill process has certain obvious disadvantages, such as dustiness, higher labour cost and higher energy consumption. In spite of these disadvantages, this process is successful for formulations containing abrasive fillers or small production quantities. The maximum throughput for continuous, high speed roll mills is approximately 800 kg/h. The production of molding compounds in extruders, on the other hand, offers economical, dust- and odor-free processing. Efficient homogenizing, which is necessary, can only be accomplished in special extruders like single screw extruder model "Buss-kneader" or in twin screw extruders with appropriate mixing sections.

The throughput of these extruders can be as much as 2,000 kg/h. The use of extruders is limited and unacceptable for compounds containing abrasive fillers. Final finishing and treating of the compounds is the same as for the mixing roll process. Production of highly reinforced or high-impact materials is unsatisfactory on differential rolls because of the grinding action of the rolls and attrition of the fibrous reinforcing material (glass fibers, cord or chopped cotton fabrics). This is resolved by the use of sigma blade mixers and impregnation with phenol resin solution and subsequent drying. Satisfactory product is obtained. With screw type extruders compounds of medium strength can be made. By use of suitable attachments, the molding compound can be produced in a pelletized, free flowing form.

12.4 Thermoset Flow

When heat and pressure are applied to a thermosetting molding material three processes occur: melting, flow and gelation. The viscosity change during these phases is a complicated function which depends upon the following variables:

$$n = f(\vartheta, t, \tau)$$

the temperature ϑ, the time or curing rate t and shear rate τ. The theoretical treatment of the rheology of thermosets[12, 13] is considerably more complex than thermoplastic flow because time is an additional and essential variable due to the curing reaction which occurs simultaneously. At first, the resin softens under the influence of temperature with accompanying viscosity reduction. Concurrently the crosslinking reaction causes a viscosity increase. The superposition of these two exponential functions yields an unsymmetrical parabolic curve. Viscosity also depends upon the shear rate, i.e. thermosetting molding compounds show non-Newtonian flow. Constant flow parameters are essential criteria for trouble-free and economical large scale production. The flow depends upon factors such as the resin content and reactivity, type and amount

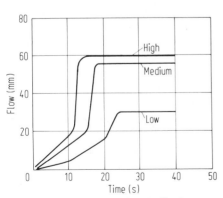

Fig. 12.5. Flow/time relationship[1] of phenol molding compounds according to DIN 53478

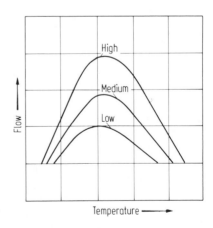

Fig. 12.6. Flow/temperature relationship[10] of phenol molding compounds according to DIN 53478

of fillers, lubricants, resin free phenol content, and especially on humidity content. The latter pertains to storage conditions and conditioning of the compound. Certain fluctuations from batch to batch occur. A multitude of tests have been introduced in order to predict the processability of a molding compound[1]. A number of methods describe the processing behavior by means of a molded part, e.g. a plate, cup, stick or spiral. A weighed quantity is molded in a plate configuration and the diameter measured (plate method) or a standard cup is molded and the closing time measured (DIN 53 465). According to ASTM-D-731, the minimum pressure is determined whereby the cup mold is filled. Therefore, several moldings are necessary. In the bar flow test a pellet is pressed into a flow channel from a heated chamber at a definite pressure. The length of the formed part is the measure of the flow duration. Further, there are two methods – according to Rossi-Peakes (ASTM-D-569) and DIN 53478 – which differ from each other by the dimensions of the flow mold.

The EMMI-spiral flow test developed by the Epoxy Molding Materials Institute is similar in principle, but is only suitable for low-pressure molding compounds[13], like epoxy-phenolic molding compounds. The flow distance is enlarged to 262 cm by an Archimedes winding.

All testing processes described have some apparent disadvantages. Test information as it relates to the total molding process is limited. Furthermore, the results are very dependent upon the testing devices, even the roughness of the mold must be considered. Test results of different molding compound producers are comparable to a certain degree. However, the information is satisfactory as long as the materials are processed by compression molding. For injection molding, it is necessary to know the melt viscosity and duration of the melt phase. These can be evaluated empirically with a torque rheometer[15] using sigma-type measuring heads (Brabender plastograph).

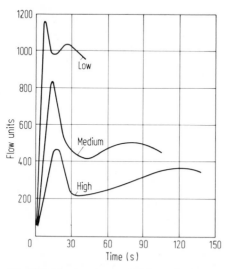

Fig. 12.7. Torsional moment/time relationship of several phenol molding compounds determined with the Brabender plastograph[1]

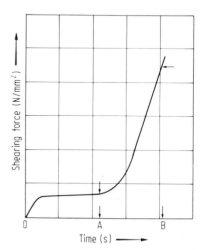

Fig. 12.8. Shear force/time relationship of a phenol molding compound determined with a rotational viscosimeter according to Kanavec[14]

Torsional moment (viscosity)/time recorder traces of different flow adjustments (soft, medium, hard) are shown in Fig. 12.7.

The final status of the molding compound cannot be determined completely since material becomes crumbly very early during the curing process. A related viscosimetric method has been quite successful in East Europe (Kanavec plastograph). A typical shearing force/time relationship is shown in Fig. 12.8. The empirical value B of the shearing force (viscosity) corresponds to the minimum curing time required[14]. The melt viscosity and the duration of the plasticized phase are very accurately determined by either methods. Satisfactory rheological data were obtained by a recently developed capillary rheometer[43].

12.5 Manufacturing of Molded Parts

The two most important parameters for the economical manufacture of molded parts[16-19] are the degree of automation and the cycle time. Molding time consists of mold temperature heating time and the time required for crosslinking. The latter usually occurs at 160 °C within 5–10 s. Since thermal conductivity of molding compounds is relatively low, heating time predominates. It is therefore reasonable to reduce the molding cycle by preheating the compound outside of the mold to a temperature slightly below that of the mold.

Compression Molding

The compression molding process yields stronger moldings than either transfer or injection molding processes[1]. Since material flow is relatively low, there is no fiber orientation or fiber attrition. The main disadvantage is the low level of automation of the conventional compression molding process. The material is difficult to feed, must be premeasured or preweighed and placed manually into the mold. Dust contamination in the plant is high. Process economics can be improved significantly by use of a screw-preplasticizing unit[20]. Pelletizing, handling and preheating of the pellets, and manual feeding of the press would be omitted, thereby minimizing dust contamination. A somewhat higher preheating temperature is possible with screw-preplasticization in comparison to HF-preheating.

Newly developed control systems promise to advance compression molding technology[21].

Transfer Molding

A preformed, portioned, preheated molding compound is pressed (~ 100 N/mm^2 pressure) from a heated transfer chamber by a plunger through a gate into the closed mold. The compound is preheated in the transfer chamber and, in addition, frictional heat is transferred during the flow through the narrow gate. Metal inserts or pins are unaffected due to the low melt viscosity. This process is especially effective for the manufacture of relatively thick-walled parts with different wall thickness. Since the material is subjected to a high pressure drop at the jet, an effective gas release occurs so that less shrinkage is observed. Since the material is fed into the closed mold, the metering accuracy is more precise with a corresponding reduction in flash. Some fiber orientation may occur but not as much as in the injection molding process[1, 22].

Injection Molding

Injection molding is by far the most important method for processing phenolic compounds. Progress in machinery, process control, mold design, and formulation in recent years is very impressive[4, 18, 23-25]. The operating sequence of a thermoset injection molding machine is similar to thermoplastic injection molding operation. Granular or pellet material is fed from a hopper, heated and plasticized by the shearing action of the screw. Heat is mainly generated by friction and the temperature is adjusted by hot water. As the preheated material builds up in the nozzle section, the screw is forced back into the barrel. Plasticized and homogenized material is then forced under high pressure into the hot mold by the forward movement of the screw. Further heat is generated by its flow through the narrow nozzle at high speed (Table 12.10).

Another molding technique currently gaining acceptance is a combined injection/compression molding process. A full shot of material is injected into a partially closed mold. The mold is then closed at full clamping pressure forcing the material to fill the mold cavity. This technique combines the advantages of compression molding and injection molding processes. Runnerless injection compression molding[26, 27] was developed in the USA. The orientation and anisotropic strength of the filler are reduced, but flash is increased. The open mold permits effective release of volatiles.

A typical injection compound differs from compression compound by a higher resin content (up to 50%) and higher internal lubricant content and therefore a longer flow. Ideal curing behavior would include a very low reactivity at temperatures up to about 110 °C, i.e. long "barrel" life, and a fast curing rate at temperatures above 140 °C. The use of latent catalysts is mentioned in the patent literature.

Molding time may be further reduced by the use of highly advanced or "harder" resins that are also predried thoroughly. The heat input required for the injection process is provided by barrel heating, transforming frictional energy into heat, adiabatic compression and partly by the reaction enthalpy of the curing reaction. In addition, cycle times can be reduced with hot cone technology and hyperthermal runner systems which permit the material to be super-heated and transferred quickly[28, 29]. The material loss with conventional injection molding averages about 15–20% due to sprues and runners which are not recycled. When multicavity molds are used for small moldings, such as in the electrical engineering and electronics fields, the loss may exceed 50%. The material cost represent a major portion of the production cost of molded parts. The amount of scrap is reduced considerably by application of cold-runner

Table 12.10. Typical operating conditions for injection molding of phenol molding compounds (Type 31)

Barrel temperature	
Middle	65– 85 °C
Front	85–110 °C
Nozzle	110–130 °C
Mold temperature	165–200 °C
Screw speed	35–100 rev/min.
Injection pressure	1,200–2,500 N/mm^2
Molding time	15–60 s

molding which was developed concurrently to hot runner molds for thermoplastic resins. Sprue and runner scrap is reduced or eliminated in multicavity molding. By maintaining the material non-cured in the sprue and runner, it does not cure and can be injected into the mold cavities on the ensuing shot. A third, heat insulated partition is inserted in the mold where the temperature is maintained between 80 °C and 110 °C by a liquid tempering technique. In this section the material is fed and evenly distributed.

A closed loop recovery system for reuse of thermoset scrap has been developed in Japan by Meiki & Co, Nagoya. Flash, runners and other discards are pulverized in special mills to 50–100 μm grain size, mixed uniformly with the virgin raw material in a predetermined ratio and then conveyed to the hopper of an injection molding machine[30]. It has been shown that up to 15% powdered phenolic scrap can be incorporated without deterioration of part quality[31, 32].

12.6 Selected Properties

Phenolic moldings offer remarkable advantages compared to other engineering plastics in those applications where higher temperature resistance under load is required. Resistance data to high-energy radiation are mentioned in Sect. 8.3. Figures for properties and strength under physical and environmental stresses are indicated in Tables 12.2, 12.3, 12.11, 12.12 and Fig. 12.9–12.11 in addition to literature[33–37]. Minimum requirements according to DIN 7708, measured on standard specimens, are included in Table 12.4. Overall chemical resistance is considerably higher as compared to thermoplastic materials.

Thermal Resistance

The following Figs. 12.9–12.11 demonstrate the influence of the filler and material pretreatment on strength and modulus at elevated temperatures for recently developed high-temperature resistant PF-compounds. Electrical conductivity changes during thermal decomposition of glass and polyamide reinforced phenolics between 25 and 700 °C have been compared to silicone compounds[38].

Table 12.11. Typical values for the shrinkage and postshrinkage of phenol molding compounds[1]

Type DIN 7735	Filler	Shrinkage %	Postshrinkage %
11	Mineral flour	0.3–0.5	0.2 –0.3
13	Mica	<0.2	0.05–0.15
16	Asbestos	<0.2	0.1 –0.2
31	Wood flour	0.4–0.8	0.15–0.3
51	Cellulose	0.2–0.5	0.2 –0.3
71	Textile	0.2–0.5	0.2 –0.3

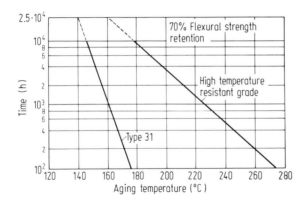

Fig. 12.9. Temperature/time limit relationship according to DIN 53446, VDE 0304 for general purpose and high-temperature resistant grade phenolics; 70% flexural strength retention

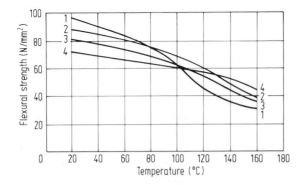

Fig. 12.10. Flexural strength/temperature relationship with variation of filler and pretreatment of the molding material[37]. *1* Type 31, wood flour; *2* as *1*, preheated for 10 min at 110 °C; *3* Type 12, asbestos; *4* HT-resistant type, carbon filled

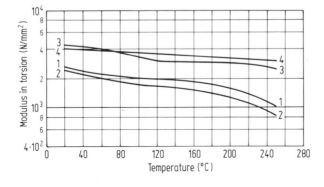

Fig. 12.11. Torsional modulus/temperature relationship with variation of filler and pretreatment of the molding material[37]. The designation of the curves is indicated in Fig. 12.10

Shrinkage and Post-Mold Shrinkage

For producing parts precisely to size, a number of factors that relate to thermoset part shrinkage[39] must be considered prior to mold design. Otherwise, a considerable amount of reworking or trial and error mold designing will be necessary. Shrinkage of a molded part is effected by two interrelated phenomena.

Shrinkage is typically the difference in dimensions of the cold mold and the cold molded part (after 24 h). The cold part can be subjected to further dimensional changes, such as heat treatment causing additional shrinkage. The following factors influence shrinkage:

Table 12.12. Linear expansion coefficients between 20–120 °C of some phenol molding compounds[37] compared to some metals ([x] preheated for 10 min at 110 °C)

Type DIN 7735	Filler	Resin content %	Linear expansion coefficient 10^{-6} K $^{-1}$
31	Wood flour	40–45	35–45
31[x]	Wood flour	40–45	30–35
71	Textile fibers	40–50	25–30
12	Asbestos	35–40	20–26
	Carbon	35–38	18–22
	Graphite	30–35	10–13
Phenolic resin, non-filled			60–70
Steel			11–12
Copper			16–17
Aluminum			20–23

- different coefficients of linear thermal expansion of the molded part and the mold;
- reduction in volume due to chemical reaction, crosslinking and release of volatile components;
- elastic deformation of the molded part upon removal from the mold due to energy storage;
- cavity pressure during processing.

Postmold shrinkage (Table 12.11) is due to the diffusion of volatile components through the surface layer, postcuring reaction, or reorientation of fillers.

Total shrinkage depends mainly on the volatile content of the molding material. Volatile content of wood flour phenol molding compounds is usually between 3 and 4%. For precision parts molding compounds with volatiles less than 1% can be used.

Thermal Expansion

Due to lower production costs compared to those of shaped metal parts, thermosetting plastic parts are receiving more attention in precision mechanics[40] and automotive industries[41, 42]. Since they are mainly used in conjunction with metal parts, comparable thermal expansion is an essential prerequisite. Expansion of polymers is slightly temperature dependent, however, the gradual increase with rising temperatures is so small that in most cases it is negligible. The volume-temperature relationship shows (Table 12.12) a sudden rise at the resin glass transition point; expansion is higher above the glass transition temperature due to probable structural motion and rearrangement of macromolecular segments.

12.7 References

1. Schönthaler, W.: Verarbeiten härtbarer Kunststoffe, Düsseldorf: VDI-Verlag 1973
2. Houston, A. M.: Materials Engng. 6, 1975
3. Knop, A., Müller, R., Schönthaler, W.: Kunststoffe 66, 633 (1976)

4. Jellinek, K., Schönthaler, W., Niemann, K.: Chem.-Ing.-Tech. 55, 30 (1983)
5. General Electric Co: Genal Phenolic Molding Compounds, Technical Bulletin
6. N. N.: Modern Plastics Int. Jan. 1984, P. 21
7. Verband Kunststofferzeugende Industrie e. V., Frankfurt: Geschäftsbericht 1984
8. N. N.: Japan Chemical Week, May 31, 1984, P. 7
9. Hayes, W. A.: Plastics Design & Processing, Oct/Nov. 1983, P. 8
10. Bakelite GmbH: Technical Bulletin
11. Katz, H. S., Milewski, J. V.: Handbook of Fillers and Reinforcements for Plastics. Van Nostrand Reinhold Co., N.Y. 1978
12. Bassow, N. I., et al.: Plaste u. Kautschuk 19, 507 (1972)
13. Pall, G.: Plaste u. Kautschuk 18, 665 (1971)
14. Ehrentraut, P.: Kunststoffe 56, 10 (1966)
15. Rothenpieler, A., Hess, R.: Kunststoffe 62, 215 (1972)
16. Rhyner, H. G., Bläuer, B., Leukens, U.: Prakt. Erfahrungen in d. Duromerverarbeitung. In: Wirtschaftl. Herst. v. Duromerformteilen. Düsseldorf: VDI-Verlag 1978
17. Müller, K.: Plaste u. Kautschuk 22, 225 (1975)
18. Bauer, W., Rörick, W.: Kunststoffe 71, 730 (1981)
19. Bauer, W.: Kunststoffe 74, 14 (1984)
20. Bovensmann, W.: Plastifizieren und Homogenisieren von Duromeren mit Schnecken-Plastifizier-Geräten. In: Wirtschaftl. Herst. v. Duromerformteilen. Düsseldorf: VDI-Verlag 1978
21. Todd, W. H.: Modern Plastics Internat., August 1976, P. 46
22. Schönthaler, W., Niemann, K.: Maschinenmarkt 82, 1638 (1976)
23. Schönthaler, W.: Plastverarbeiter 33, 113 (1982)
24. Niemann, K., Danne, W., Tüscher, Th.: Plastverarbeiter 32, 1200 (1982)
25. Schönthaler, W., Niemann, K.: Kunststoffe 71, 6 (1981)
26. Dannels, W. A., Bainbridge, R. W. (Occidental Chemical Corporation): US-PS 972182 (1978); US-PS 972189 (1978)
27. Bakelite GmbH: Angußloses Kaltkanal-Spritzprägen. Technical Bulletin
28. N. N.: Modern Plastics Internat., June 1973, P. 21
29. Asahi Organic Chemicals Ind.: Japan Plastics, June–July 1976, P. 6
30. Meiki & Co.: Recycling Machine for Thermosetting Resin Scrap, Technical Bulletin
31. Weißler, E. P.: Plastverarbeiter 34, 875 (1983)
32. Fischer, V.: Technische Akademie Wuppertal, Symposium Nr. 118033 Preprint, Jan. 1983
33. Bachmann, A., Müller, K.: Phenoplaste, Leipzig: Verlag 1973
34. von Meysenbug, C. M.: Kunststoffkunde für Ingenieure. München: Carl Hanser 1968
35. Gilfrich, H. P., Wallhäuser, H.: Kunststoffe 62, 519 (1972)
36. Verb. Dtsch. Ing./Verein Dtsch. Elektrotechniker: VDI/VDE Richtlinie 2478, Phenoplastformstoffe mit Holzmehl
37. Bakelite GmbH: Härtbare Formmassen, Technical Bulletin
38. Johnson, R. T., Biefeld, R. M.: Polymer Engineering and Science, 22, 147 (1982)
39. Sokolov, A.: Plaste und Kautschuk 29, 88 (1982)
40. Anemat, A.: Der Zuliefermarkt, September 1983, P. 105
41. Gardziella, A., Schönthaler, W.: Kunststoffe 71, P. 163–170 (1981)
42. Wendt, G., Decker, K.-H.: Plastverarbeiter 33, 666 (1982)
43. Malguarnera, S. C., et al.: Ind. Eng. Chem. Prod. Res. Dev. 23, 103 (1984)

13 Heat and Sound Insulation Materials

13.1 Inorganic Fiber Insulating Materials

Of the three heat transfer processes: heat conduction, convection and radiation, convectional heat transfer is reduced by fiber and foam insulation materials[1, 2]. Air circulation is prevented by compartmentalizing the air space into small sections which reduce the temperature gradients dramatically (Grasshoff-value). A heat conductivity close to that of motionless air is the final endeavor.

Thermal and acoustical insulation is one of the largest and fastest growing markets for phenolic resins. Thermal insulation markets (Fig. 13.1) include the use of phenolic resins as binder for mineral- and glass-fibers as well as smaller volumes in the manufacture of phenolic foam. Fibrous or porous absorbents are used in industrial construction to control sound and alleviate noise and sound, which are not only disturbing, but also harmful to human health (above 90 dBA).

Increasing energy cost and the opportunity to replace polystyrene foam contribute to further growth of phenolic resin bonded inorganic fibrous insulation. The flammability of most organic expanded foams is generally regarded as a prime factor for their replacement in the building construction area. In general, desirable material properties for minimizing fire hazards are:
– high thermal resistance and low heat conduction,
– high decomposition temperature and low combustibility of gases generated;
– low heat of combustion;
– low rate and amount of smoke generation;
– low toxicity of the gaseous combustion products.

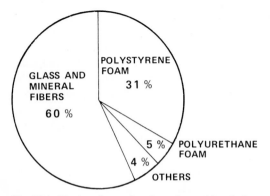

Fig. 13.1. Market share of various thermal insulation materials[3] in West Germany in 1983

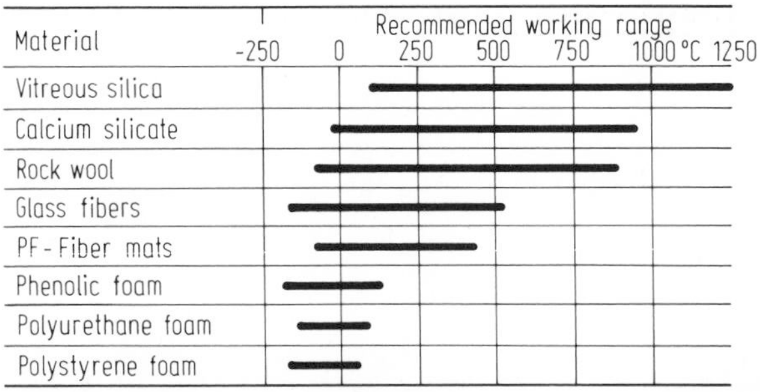

Material	Recommended working range −250	0	250	500	750	1000 °C 1250
Vitreous silica						
Calcium silicate						
Rock wool						
Glass fibers						
PF - Fiber mats						
Phenolic foam						
Polyurethane foam						
Polystyrene foam						

Fig. 13.2. Temperature range of application for some insulating materials[2]

Fig. 13.3. Heat insulation of buildings with mineral fiber mats. (Photo: Deutsche Rockwool Mineralwoll GmbH, D-4390 Gladbeck)

Fig. 13.4. Heat insulation of pipe lines with mineral fiber shells (Photo: Grünzweig & Hartmann und Glasfaser AG, D-6700 Ludwigshafen)

The flame retardancy, low smoke density of phenolic resins and different national standards and ratings are mentioned in the phenolic foam section and Chap. 6.5. Phenolic resin bound fiber composites possess an attractive upper temperature range as compared to organic cellular plastics (Fig. 13.2).

Fibrous insulation is applied to blankets and felts with a density of 10–250 kg/m^3 and can be clad on one side with bitumenous membranes or quilted on both sides. The blankets can also be clad with plastic or aluminium foil which acts as a vapour barrier. Water vapor permeation, which increases the conductivity, is a serious problem in the low temperature range. Blankets for insulation of pipes and containers can be reinforced with wire mesh. One piece pipe insulation is produced for pipes of 20–900 mm diameter.

The blankets may be installed around the house, on exterior walls, basement masonry, floor, partitions and roof. Apart from heat insulation, good sound insulation is also desired.

Table 13.1. Chemical composition of mineral fibers[5]. (^{x} = including also 5–12% iron oxide)

	SiO_2	Al_2O_3	CaO	MgO	B_2O_3	$Na_2O + K_2O$
Glass fiber	50–65	3–15	5–15	2–5	1–12	1–18
Slag fiber	30–35	10–20	40–45	2–8	–	–
Stone fiber^x	50–55	6–15	25–35	2–6	–	2–3
Basalt fiber^x	45–50	12–15	9–12	7–10	–	2–4

Prefabricated shells are used to insulate pipe lines for hot water, steam or oil etc., or in the form of blankets for industrial furnaces, reactors and containers. The limit of application is up to 450 °C. Although the binder is slowly degraded at temperatures above 250 °C, the performance of the insulation is not affected. Cold insulation includes refrigeration equipment for the storage of foodstuffs, gas liquefication plants, house-hold appliances, refrigerated cars, ships etc. Glass wool is mainly used in the lower temperature region and for domestic purposes, while mineral wool is proposed in areas of higher temperatures and for industrial application[4]. As far as the quantity (volume) is concerned, glass fiber insulation volume is greater than mineral wool. The market share of slag fibers is relatively small and decreasing.

13.1.1 Inorganic Fibers and Fiber Production

Mineral fibers are multi-component systems with the main components (Table 13.1) being SiO_2, Al_2O_3, CaO, MgO with mean eutectic points at 1,170, 1,220 and 1,345 °C[5]. The chemical composition of the melt determines the thermal resistance, the devitrification temperature and devitrification rate.

Mineral wool production process is shown in Fig. 13.5. Basalt on diabase rocks is melted by the addition of lime and foundry coke at approx. 1,500 °C in a cupola furnace. The higher the content of silica, the "longer" the melt performance. The upper limit of SiO_2 content is determined by the increased melting and spinning temperature[6].

The molten material then flows to four rotating spinning wheels (3,000–5,000 rpm) and is spun into thin fibers with diameters of between 3–7 μm by centrifugal force. The formulated phenol resin is added as bonding agent in the blowing chamber along with oil (~0.2%) to make the wool dust-free and water-repellent[7].

The melt yield is within the range of 70% calculated on the rock amount. The resulting material may contain at least 40% of non-fibrous material generally in the form of small pellets, called shot, depending upon the method of production and chemical composition. The overall fiber yield can be as low as 40%.

High-silica glass fibers[8], on the other hand, are almost shotfree (0.1%). The absence of shots frequently justifies the higher material costs because of the higher effectiveness in relation to weight. The production of glass fiber insulation materials is shown in Fig. 13.6 and is analogous to mineral fiber.

In collecting chambers, the fibers are drawn onto a conveyor belt to build up a mat of desired thickness and density. Then the wool is passed through an oven in which the phenol resin is completely cured. Several additional processes follow, such as cutting to size, shape forming, controlling and packaging.

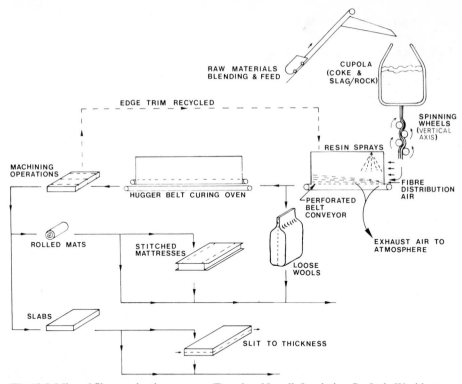

Fig. 13.5. Mineral fiber production process. (Drawing: Newalls Insulation Co. Ltd., Washington, GB)

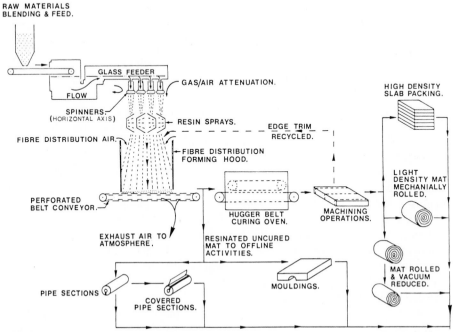

Fig. 13.6. Glass fiber production process. (Drawing: Newalls Insulation Co. Ltd., Washington, GB)

13.1.2 Resins and Formulation

Generally, an aqueous resol solution is sprayed onto a mass of hot fibers and then the mass is further heated until the resin is wholly cured. Since the highly diluted resol is spread around the fibers in form of extremely small drops and is subjected to relatively high temperatures of 200 °C and higher, a significant amount of low molecular weight resol components, mainly phenol, formaldehyde and saligenin, are volatilized. Because of this the economics of production are adversely affected. The emission of phenols and formaldehyde creates an environmental problem. Due to the huge amounts of air required in the blowing chamber and therefore low concentration of combustible substances, the treatment of the exhaust air has not yet been satisfactorily resolved.

The application efficiency of a resin binder solution is determined by the following:

$$E = \frac{W}{G \cdot V \cdot S} \times 100 \; (\%)$$

E figures for efficiency, W for the increase in weight of the fibrous mass after curing, G for the specific gravity of the binder solution, V for the volume of resin used and S for the percentage of solids in the resin as evaluated by standard methods. The achievable resin efficiency in large commercial plants today is within the range of 60–80%.

Appropriate resol resins are obtained by reacting phenol with an excess of formaldehyde at temperatures below 70 °C. The PF ratio is between 2.5 and 3.5. The remaining free formaldehyde up to 7% (Table 13.2) is used to bind urea. Not only does urea react with formaldehyde but it has also been shown[9] to undergo a co-condensation reaction with methylol phenols (Chap. 3.8). In general, alkaline earth hydroxides e.g. calcium or barium hydroxide, seldom sodium hydroxide, are used as catalysts. High-quality resins are normally ash-free, the catalyst is precipitated as sulfate or carbonate and removed by filtration. A high formaldehyde ratio favours high resin efficiency. Urea is added in most formulations up to 40% (calculated on dry weight) to reduce cost.

The resin is applied as a 10 to 15% aqueous solution; a high water dilutability is an important requirement. A satisfactory resin consists mainly of mono-nuclear polymethylolated compounds, the prevailing species being trimethylolphenol (Fig. 7.4). The amount of polynuclear compounds should be as low as possible. The effect of storage caused by condensation reactions is shown in Fig. 13.7. The effect of pH on the

Table 13.2. Phenol resol properties for the production of mineral- and glass fiber mats[11] (Fig. 7.4)

Dry solids content	%	40–50
pH	–	7.0
Viscosity at 20 °C	mPa · s	8–20
B-time at 130 °C	min	8–10
Dilutability with water	ratio	1:10
Content of free phenol	%	<1.0
Content of free formaldehyde	%	1–7
Storage life at 20 °C	days	14
Storage life at 10 °C	days	35

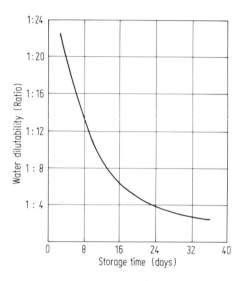

Fig. 13.7. Relationship of water solubility of the phenol resol of Table 13.2 during storage at 21 °C. (Drawing: Bakelite GmbH, D-5860 Iserlohn-Letmathe)

condensation rate of trimethylolphenol was reinvestigated recently[10] confirming a complex variation of the reaction rate indicating change in mechanism and a minimum rate at pH 4.

The following formulation may be used for the production of fibers mats:

100	pbw	phenol resol
7	pbw	20% ammonia solution
0.02	pbw	amino silane (γ-aminopropyltriethoxysilane)
0 –800	pbw	water

Ammonia serves to bind the free formaldehyde and for pH adjustment to a low alkaline value. Aminosilane acts as coupling agent to improve the humidity resistance and to increase the mechanical strength. In general the resin coat is adjusted through resin dilution. The resin content in mineral fiber mats is between 1–5%, mostly 3%. For glass fiber mats a higher resin content between 5–14%, mostly 7% is used. Urea is added to the phenol resin, either at the end of the reaction or immediately prior to use by the producer. Urea addition in large amounts reduces humidity, thermal and ageing resistance. The addition of lignin, calcium- or magnesium lignosulfonate[12, 13], dicyandiamide[14, 15] or melamine is recommended to improve PF-resin/urea blends as well as to reduce binder cost.

13.1.3 Properties of Fiber Mats

The mechanical strength depends upon the fiber length and diameter, the content of bonding agent and the proportion of shots. It is obvious that physical properties[16–18] depend to a great extent on material density (Table 13.3). These PF glass or mineral wool composites should be contrasted with higher PF resin containing glass fiber composites (Chap. 10) when, in fact, mechanical properties, rigidity, heat distortion temperature, etc. are of paramount importance.

Fibrous insulation materials are resistant to humidity, are biologically inert, chemically neutral and temperature resistant to a great extent. Formaldehyde emission during use is extremely low[19]. Emission figures determined by the perforator test in com-

Table 13.3. Physical properties of mineral fiber mats[17]

Property	Test method	Unit	Test value
Density	DIN 18165	kg/m^3	20–80
Behavior in fire	DIN 4102	–	fire-proof A 2
Thermal conductivity	DIN 4108	W/mK	0.040
Spec. thermal capacity C_p	–	kJ/kg °C	0.84
Noise reduction coefficient	DIN 52212		
125 Hz			0.17
500 Hz			0.76
1,000 Hz			0.86
4,000 Hz			1.00

parison to other building materials are compiled in Table 11.2, Chap. 11. Formaldehyde emission depends on curing conditions, phenol/urea ratio applied and binder content. At low urea content figures between 0.1 and 0.5 mg/100 g material can be expected.

The upper temperature resistance is primarily limited by the organic bonding agent. The thermal decomposition of the binder commences at approximately 250 °C. The devitrification range of glass fibers is between 650–850 °C depending on composition, of mineral fibers between 750 and 1,000 °C.

13.2 Phenolic Resin Foam

Phenolic resin foam has been known for several decades as a fire resistant material with low smoke emission. The major drawbacks in considering it for thermal insulation were mediocre insulation performance as well as marginal mechanical and handling characteristics as compared to rigid plastic insulation materials like polystyrene, polyurethane or polyisocyanurate.

Within the last few years significant improvements in insulation and mechanical properties have been achieved for phenolic foam without adversely affecting fire performance or smoke emission properties. Deficiencies such as friability and mechanical strength, punking, corrosion and high open cell content have been resolved to a great extent thus reestablishing the potential of phenolic foam in thermal insulation. Continued increased energy costs and the public awareness of the hazards of non-flame-retardant thermal insulation will undoubtedly lead to increased share of the insulation market to phenolic foam. The current West European (8,000 t/a), US (10,000 t/a in 1983), and USSR[19a] markets have been predicted to increase significantly by the end of the decade[20] because of promising developments in materials and new processing techniques[21].

A different application which requires friability and lower mechanical strength is phenolic foam for floral arrangements. The foam must be wholly open cell and completely absorb water due to the special surfactant wetting agent composition.

Early methods of foaming mentioned the expansion of novolak/HMTA mixtures. Phenolic foams are now dominated completely by the acid catalyzed reaction of a resol type resin with added blowing agent, and surfactant[22, 23].

13.2.1 Resins

Aqueous resols consisting of phenol and formaldehyde in a molar ratio (1 : 1.5–2.5) are generally used. The trend in resol resin preparation is toward a lower formaldehyde ratio to maintain low formaldehyde emission during the foaming process. Sodium hydroxide or alkaline earth hydroxides are used as catalysts. Typical resol resins used in foam preparation contain about 80% solids, volatile components consist mainly of water, formaldehyde, and phenol. After the reaction is completed within a temperature range of 60 to 90 °C, the percent solids and viscosity are adjusted by distillation. The resin viscosity is an important variable, among others, for foam density adjustment.

Foam density is further regulated by the type of surfactant, amount of blowing agent, temperature and acid-reactivity of the resin. The acid-reactivity is influenced by the storage time. The storage life of highly concentrated resols is limited, in general, to 2–4 months at room temperature; the acid-reactivity decreases with increasing storage time. At the same time, the viscosity increases considerably; maximum viscosity should be 10,000 mPa · s.

Favourable foam properties can be obtained in a 2-step process whereby o-cresol is pre-reacted with formaldehyde under basic conditions followed by the addition of an equimolar amount of phenol (compared to o-cresol) and formaldehyde[24]. These cresol/phenol resol resins with selectively "capped and grafted" surfactants yield phenolic foam with excellent mechanical and thermal properties.

Hybrid phenolic foam systems have been reported for resin combinations such as phenol resol resin and furfuryl alcohol[25], and phenol resol resin, furfuryl alcohol and melamine[26]. The former system was foamed via strong acid conditions whereas the latter under urethane catalysis conditions. The resulting hybrid foam products were deficient in either mechanical properties or fire resistance performance.

Catalysts

Crosslinking of the resol resin is promoted by the addition of a strong organic sulfonic acid such as p-toluene sulfonic acid, methane sulfonic acid or phenol sulfonic acid. Inorganic acids are rarely used with the exception of phosphoric acid which is sometimes combined with aromatic sulfonic acids such as p-toluene sulfonic acid or phenol sulfonic acid for improved flame resistance of the foam. Phenol sulfonic acid is allegedly incorporated into the macromolecular structure thereby reducing foam metal corrosion. Oligomeric sulfonic acids[27], sulfonated phenol novolaks, and resorcinol novolaks[28] have also been proposed as catalysts with anticipated minimal foam corrosion.

Blowing Agents

Liquids with a low boiling point and a corresponding low heat of vaporization are mixed with the aqueous phenol resol, and the catalyst is then added to a vessel equipped with a rapid agitator or by use of a three component mixing head. Within a short period of time the temperature rises due to the exothermic resol curing reaction and the blowing agent is volatilized.

The liquid blowing agent must volatilize at the moment when the viscosity or plasticity of the mixture increases. Foam expansion occurs as the exothermic crosslinking reaction continues with either the blowing agent contained within the cellular structure (closed cell) or chemical blowing agents which undergo decomposition and expel gaseous products. Chemical blowing agents have been mentioned periodically in the literature but are not the preferred blowing agent system for optimum foam properties.

Suitable blowing agents are trichlorofluoromethane, trichlorotrifluoroethane, dichloromethane and *n*-pentane. The quantity of blowing agent depends upon the desired density of the foam. Other factors are also considered in foam density: foaming temperature, resin reactivity, and acid catalyst content. Volatile fluorocarbons are preferred blowing agents because they possess low thermal conductivity values, low heats of vaporization, and are non-flammable[29].

Surfactants

The production of homogeneous, reproducible foam composites requires the use of a surfactant that is miscible or dispersible in water, non-hydrolyzable and moderately resistant to the acidic resol resin media. The surfactant or cell regulator plays a vital role by lowering surface tension of the resin formulation and providing an interface between the non polar blowing agent and the polar resin phase. The surfactant keeps the developing foam from collapsing or rupturing. Proper selection and/or design of the surfactant generally yields a phenolic foam with a uniform fine cellular structure. The non-ionic types are the most commonly used materials. The quantities to be added vary up to 5%. Too little surfactant fails to stabilize the foam and too much causes the cells to congeal and collapse. Polyoxyethylene sorbitan fatty acid esters, siloxane-oxyalkylene copolymers and castor oil-ethylene oxide adducts are preferred[22]. Surfactants also influence the amount of open or closed cells within the foam microstructure. The open cell foam relates to wettability and water absorption and is of great importance for the floral foam application. Recently special surfactants have been reported to yield fine cellular phenolic foams with high closed cell content, a low initial thermal conductivity factor (K) with corresponding low deviation in K over a period of time (see closed cell discussion). These surfactants are hydroxyl terminated ethylene and/or propylene oxide triols, diamines or castor oil materials that are hydroxyl capped, preferably with acetic anhydride and free radical grafted with maleate esters of high MW alcohols[30]. Some of these latter surfactants are equally useful in the preparation of polyisocyanurate foam[31].

13.2.2 Foaming Equipment

Discontinous, semi-continuous or continuous foaming can be conducted in the manufacture of phenolic foams. Conventional equipment such as high speed agitation components, metal or wooden molds with capacities to 4 m³ have been described for discontinuous foaming operations. Those equipment components which are exposed to acid catalysts are either stainless steel or of a rigid plastic composition.

Semi-continuous or continuous processing operations are mainly identified with the manufacture of slabstock with facings on both sides. Polyurethane machinery manufacturers mention in their literature low and high pressure dispensing machines for phenolic foam in addition to modified processing equipment[32, 33]. The intensive mixing head is fabricated of stainless steel for acid catalyst use. Either two or three inlets of the mixing head are employed depending on whether resin/surfactant is combined with blowing agent system or introduced as separate streams. One inlet is required for catalyst. More powerful pumps are necessary to transfer phenolic resins which are more viscous than urethane resins.

In a continuous process[34] foaming is conducted under pressure whereby a distribution device that traverses laterally the curing cavity introduces the foamable composition onto the lower facing material (Fig. 13.8). Side rails confine the expanding foam within the lateral width of the curing cavity. Pressure plates (the lower plates are fixed while the upper plates can be raised or lowered for foam thickness) are heated. Line rates are 12–15 m/min for thicknesses from 3.5 to 7.5 cm.

13.2.3 Foam Properties

Cellular Structure

The cellular structure of phenolic foam is extremely fine and spherical, unique as compared to most other plastic cellular products. Depending on reaction conditions and surfactant/formulation selectivity, an optimum of about 60% closed cells is obtained within a density range of 35–80 kg/m^3. Lower density foams my contain up to 90% open cells.

Interconnecting open cells lead to high thermal conductivity, higher water absorption, and water vapor permeability, all undesirable thermal insulation properties but satisfactory for acoustical or for floral foam application.

A majority of closed cells is necessary in cellular plastic materials if the subject materials are to be economically viable for thermal insulation. The closed cells entrap the expanding volatile or gaseous blowing agent. The gaseous phase (voids) within the closed cells occupies approximately 97% of the volume for foams in the 30–50 kg/m^3 density range. Further these cellular plastics materials maintain their thermal insulation superiority over air filled (or open cell) insulants due to their ability to retain low conductivity gases in their closed cells. K values of gaseous compounds vary between 0.0242 (air), 0.0207 (water vapor), and 0.0066 (CCl_3CF_3).

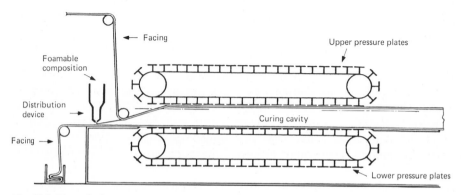

Fig. 13.8. Phenolic foam continuous line unit

The main objective in obtaining good thermal insulation K factors of ≤ 0.02 W/m/ C is a preponderant amount of closed cells with low K value gaseous blowing agent, preferably fluorocarbon, contained within the cells.

It is also necessary to maintain the K factor constant over an extended period of time or during the lifetime of the thermal insulating material. The newly established Thermal Insulation Manufacturers Association, Roofing Insulation Committee, in the U.S. (RIC/TIMA) is currently evaluating methods and procedures for determining long term insulation values of various cellular plastic products. Fundamental studies of cellular plastics aging as related to cell geometry and gas diffusion identify polymer distribution (cell strut to wall), polymer permeability, and foam density as the most important parameters for predicting aging in foam insulation[35, 36].

Data compiled for closed cell rigid polyurethane[29, 37, 38] identified the following important parameters:
– high initial fluorocarbon cell pressure or overpacking;
– use of a polymer matrix providing a high resistance to gaseous diffusion of the fluorocarbon;
– moderate foam density;
– low to moderate ageing temperatures;
– use of effective barriers or foam facings to gaseous diffusion;
– small cell diameter;
– molded surface skins to retard fluorocarbon diffusion.

Scanning electron micrographs[39] identified some small cellular defects or pin holes which are probably responsible for K drift. Early studies by Sefton and Carlson[40] demonstrated via scanning electron microscopy (SEM) and K factor data that the use of small amount of N-methyl pyrrolidone yielded closed cell phenolic foam. Later in a series of patents assigned to Koppers, Kifer and coworkers[41] showed that completely closed cell phenolic foam with good compressive strength and mechanical stability can be obtained from aqueous resols with silicone surfactant, a blend of

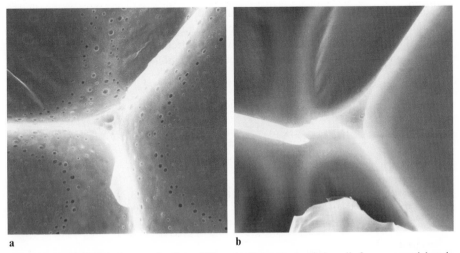

a   b

Fig. 13.9a and b. SEM micrographs show different cell structures of phenolic foams; **a** semiclosed; **b** closed cell foam. (Photo: Koppers Co. Inc., Monroeville, PA 15146)

fluorocarbon blowing agents, and mixture of aryl sulfonic acids such as *p*-toluenesulfonic acid and xylene sulfonic acid. No added N-methyl pyrrolidone was necessary for optimum foam properties. The use of low water containing components is suggested as reducing resolwater compatibility and thus the perforation of foam cell walls. The intensively mixed, creamed foam composition is "molded" at 65 °C with maximum temperature range of 95–100 °C and a pressure of 8 N/cm².

The effect of capping and grafting surfactants and the moderately constant K factor values obtained over a period of time, compared with a non grafted surfactant whose K factor decayed within several days, is disclosed[30]. Yet the K factor value[41] of newly commercialized phenolic foam is unusually constant with conventional surfactant[42, 43].

Inspection of electron photo micrographs of a typical "semi-closed" phenolic foam and newly commercialized foam (Fig. 13.9) visibly illustrates the closed cell structure of this product as compared to the micro-perforations present in semi-closed foams. Similar fine detailed perforations were illustrated by Lowe in a series of electron photo micrographs[39].

Mechanical Properties

Aside from open/close cell considerations, friability related to foam mechanical and handling properties and punking characteristics were also responsible for minimizing the use of phenolic foam in thermal insulation. The phenomenon of punking is discussed in Chap. 6. Friability is the capability of material being easily crumbled into a powder. The undesirable high friability of the early phenolic foam materials was accompanied by increased dustiness and low tensile strength. Improved cellular structure has been aptly demonstrated by the recent phenolic foam products with reduced friability and satisfactory mechanical properties (Table 13.4).

Fire Resistance Properties

A number of major fires stimulated public and official apprehension regarding the general fire performance of building materials and in particular synthetic materials. The major advantage of phenolic foam over polyurethane or polystyrene foam is in the area of fire resistance. Newly developed fine cellular phenolic foam blown with

Table 13.4. Physical properties of various phenolic foams[42, 44, 45]

	Exeltherm[a] Xtra	Cellobond k[b]	Celotex[c]
Density kg/m³	40	35	38
Compressive strength (N/cm²)	17.2	17.5	20.7
Service temp. °C	−196 to +150	−196 to +150	−196 to +150
Coefficient of linear thermal expansion (cm/cm °C)	33×10^{-6}	$30–40 \times 10^{-6}$	NA

[a] Koppers Co. Inc.
[b] BP Chemicals Ltd.
[c] The Celotex Corp.

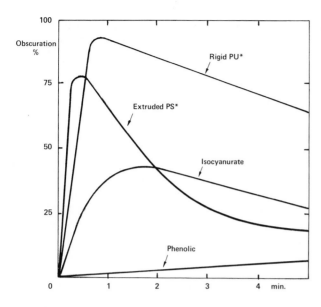

Fig. 13.10. Comparative smoke emissions of foam materials; measurements made in XP2 chamber. * Fire-retardant grades

Table 13.5. Critical oxygen index of various foams[39]

Polystyrene	19.5
Polyurethane	21.7
Polyurethane, flame retardant	25.0
Polyisocyanurate	29.0
Phenol-formaldehyde	32 –36

fluorocarbons with high closed cell content exhibits excellent fire resistance, no punking, low smoke density and low toxicity of the combustion gases.

Comparative smoke emissions of several foam systems are shown in Fig. 13.10. Flame retardant polystyrene and rigid polyurethane (PUR) are contrasted with polyisocyanurate and phenolic foam. The low smoke emission of phenolic foam is illustrated convincingly.

Flammability tests[46] apply for individual materials (e. g. case of ignition, surface flame spread, and smoke density) or for complete systems (e. g. fire endurance and surface flame spread). Some methods and comparative performance of materials are mentioned in Chaps. 6 and 10. Ease of ignition can be measured according to ASTM D 2863–77 critical oxygen index (COI) test designating the percentage of oxygen in a mixture of oxygen and nitrogen that will support the combustion of an ignited cellular plastic specimen. The COI of various foams compared to phenolic foam products are listed in Table 13.5.

The most widely accepted test for surface flame spread is the ASTM E 84 25 foot tunnel test. Flame spread classification is determined on a scale whereby asbestos cement board is zero and select grade red oak flooring is 100. Fuel contribution and smoke evolution are determined on a similar scale. ASTM E 84 values are presented in Table 13.6. This method also provides a measure of heat release based on the temperature rise in the gases produced.

Table 13.6. Flame spread of phenolic[42] and polyisocyanurate foam by ASTM E 84

	Density kg/m^3	Flame spread index	Fuel contribution	Smoke density
Phenolic	48	20–25	10	3–15
Polyisocyanurate	32	15	15	40

The use of cellular insulation products in US commercial buildings as roof decking requires Factory Mutual (FM) approval as well as Underwriters Laboratories (UL) classification. Successful testing of a new foam insulation product by FM and UL results in identification of the foam as FM Class I material and UL Class A material. Only foams with excellent thermal and fire resistance are able to comply with the rigorous requirements of FM and UL criteria.

Polyisocyanurate foam is used extensively as roof decking over steel decks because it can comply with FM Class I fire requirements. A newly developed closed cell phenolic foam[42] has been classified as FM Class I and is presently undergoing field performance for FM Class I acceptance.

Gas toxicity evaluation indicates that combustion gases from phenolic foams (CO_2, CO, H_2O) are no more toxic than those expected from wood.

For cellular plastic pipe insulation products, other fire performance tests have been developed besides ASTM E 84. These specific tests include Armstrong Vertical Pipe Chase test (provides data on vertical flame spread on pipe insulation) and Factory Mutual Full-Scale Simulated Pipe Chase test that evaluates the potential for a spreading fire in multiple parallel runs of insulated pipes that are horizontal[47]. No satisfactory phenolic foam pipe insulation with low, constant K factor has been developed to date.

In the U.K. the Rubber and Plastics Research Association of Great Britain (RAPRA) has evaluated the behavior of many cellular plastics materials in fire[48] by examining cellular structure, smoke and gases from fire, and fire performance in building applications. RAPRA corner tests of unprotected foam showed phenolic foam merely charred when exposed to a heat flux of more than 7 W/cm^2 whereas polyisocyanurate foam burned rapidly under similar test conditions.

The reported superior high temperature resistance, fire resistance, and very low smoke generation of phenolic foam may result in phenolic foam becoming an attractive material for the fast growing cellular insulation industry.

13.3 Bonded Textile Felts

Phenolic resin bonded felts from shredded organic fibers have become an increasingly important application for solid pulverized phenolic resins. These porous felt materials are common acoustical insulation products[49, 50] used in the automotive industry, for machinery housings, in living or office spaces[51]. As sound passes through the interconnecting fibers, sound energy is transformed into frictional, and vibrational heat and minimizes sound.

Table 13.7. Properties of powder resins for felt
bonding

Flow distance	mm	~ 18
Melting point	°C	85
Gel time	min	5.5
Bulk weight	g/l	350
Grain size <40 mm	%	~ 70
HMTA content	%	10

Textile cuttings or clean reclaimed material is shredded into fibers 15 to 35 mm
long. Depending on requirements, the fiber-phenolic resin powder ratio varies from
80:20 to 60:40. Additives like lubricants, sizes, fungicides, water repellants, and flame
retardants can be added. The mixing operation can be performed separately or during
web formation which is either a mechanical or aerodynamic process. Carding machines and aerodynamic web formers are used[52, 53].

The "spot welding" of the fibers is performed where fiber overlap occurs. Principally, the powder resin is uniformly scattered on the fleece web from a hopper which
covers the whole width of the web and is distributed by roll brushes, toothed rolls or
other apportioning devices. The bonding agent is also uniformly distributed in the interior of the fleece by special operations. The resin loaded fleece is carried by two wire
supports through a heated oven about 20 m long. The resin initially softens, immobi-
lizes the fibers and bonds them firmly together. For certain applications the binder resin
is totally cured in the oven. Curing is performed by hot air (180–200 °C) which is
blown vertically on the surface of the fleece or drawn through it. According to a special process, steam is blown onto the fleece prior to the hot air curing to fix the powder
resin. The temperature must be carefully controlled because high temperatures in the
interior of the fleece may easily cause ignition due to the flammability of the material.
After curing, the fleece material is quickly cooled to 40–50 °C, cut to size, stacked and
packaged. It is applied, e.g. in engine hood insulation, instrument panel insulation,
and transmission tunnel coverings.

On the other hand, for curved pressed parts, used as self-supporting elements, the
resin is only partially advanced for preliminary bonding. The web is cut into slabs
which are then consolidated in a hydraulic press and totally cured in appropriate
moulds at 170–185 °C for 1–3 min. These parts are overlaid with decorative
facings. Examples in the automotive industry include door panelling, roof linings, tunnel transmission side parts, and rear package trays. The specific gravity for
oven cured materials varies from 100 to 250–kg/m^3, for pressed parts from 400 to
900 kg/m^3. Physical properties are reported in the literature[51].

Pulverized novolak resins with a low free phenol content and special particle size
distribution are used (Table 13.7). They are mixed with generally 10% HMTA and
often with flame retardant inorganic materials. The fines content should be kept low
for clean working conditions and to prevent dust explosions (Chap. 5). Special resin
formulations are required for "odor-free" materials. Besides odor emanating from the
organic fibers, amines are generated via hydrolysis of HMTA in a high humidity
environment. High-cure rate HMTA-free resins (i. e. novolak/resol combinations) are
currently available.

13.4 References

1. Harrison, M. R., Pelanne, C. M.: Chem. Engng. 19, 62 (1977)
2. Probert, S. D.: Insulation, Sept. 1968, P. 190
3. Fachverband Schaumkunststoffe e.V., D-6000 Frankfurt/M., Am Hauptbahnhof 12
4. Harrison, M. R., Pelanne, C. M.: Cost-effective Thermal Insulation. Chem. Engng. Dec. 19, 62 (1977)
5. Klingholz, R., Eberle, H.: Mineralfasern. In: Ullmanns Encyclopädie d. techn. Chemie, Bd. 12, P. 537. München: Urban u. Schwarzenberg 1959
6. Gould, T. R.: Refractory Fibers. In: Encyclop. Chem. Technology, Vol. 17, 285. New York: Wiley 1968
7. Jungers Verkstads AB: Mineral Wool Plants, Technical Bulletin
8. Tooley, F. V.: Handbook of Glass Manufacture, Vol. II, Books for Industry Inc. New York 1974
9. Tomita, B.: ACS Polymer Preprints, 24 (2) 165 (1983)
10. Jones, R. T.: J. Polymer Sci., Polymer Chem. Ed. 21, 1801 (1983)
11. Bakelite GmbH: Bakelite-Harz M 425, Technical Bulletin
12. West Virginia Pulp and Paper Co.: DE-PS 1226926 (1962)
13. Fiberglass Ltd.: US-PS 1316911 (1971)
14. Owens Corning Fiberglass Corp.: US-PS 2,604,427
15. Westvaco Corp.: US-PS 3 463 747
16. Newalls Insulation Co Ltd.: Technical Bulletin
17. Grünzweig & Hartmann AG: Technical Bulletin
18. N. N.: Modern Plastics Int. July 1974, P. 14
19. Tieseler, H., Achenbach, G., Kuhn, P. J., Gollner, H., Johannes, K.-H.: Zbl. Arbeitsmed. 33, 222 (1983)
19a. Shutov, F. A.: Cell. Polym. 3, 95 (1984)
20. N. N.: Chemical and Engineering News, Dec. 19, 1983, P. 10
21. N. N.: Plastics Technology, April 1983, P. 97
22. Papa, A. J., Proops, W. R.: Phenolic Foams. In: K. C. Frish, J. H. Sounders (ed.): Plastic Foams, Vol. II. New York: Marcel Dekker 1973
23. Benning, C. J.: "Plastic Foams" Vol. 1, Wiley Interscience, N.Y. (1969)
24. Celotex: US-PS 3,953,645 (1976)
25. Thermoset A.G.: US-PS 4,350,776 (1982)
26. Thermocell Development LTD.: US-PS 4,390,642 (1983)
27. Colombo, A., Pilato, L. A.: ACS Polymer Preprints 24 (2) 209 (1983)
28. Ciba Geigy AG: US-PS 1,488,527
29. Ball, G. W., Hurd, R., Wallace, M. G.: J. Cell. Plast., March/April 1970, P. 66
30. The Celotex Corp.: US-PS 3,953,645 (1976), US-PS 4,133,931 (1979)
    The Celotex Corp.: US-PS 4,204,020 (1980), US-PS 4,205,135 (1980)
    The Celotex Corp.: US-PS 4,247,413 (1981), US-PS 4,365,024 (1982)
31. The Celotex Corp.: US-PS 4,365,024 (1982)
32. Polymer Technology International BV: Company Brochure
33. Kornylak Corp.: Company Brochure
34. Koppers Co. Inc.: US-PS 4,423,163 (1983)
35. Reitz, D. W., Schuetz, M. A., Glicksman, L. R.: J. Cell. Plast. 20, 104 (1984)
36. Schuetz, M. A., Glicksman, L. R.: J. Cell. Plast., 20, 114 (1984)
37. Norton, F. J.: J. Cell. Plast., Jan. (1967)
38. Schmidt, W.: Appl. Phys. 11 (4), 19 (1968)
39. Lowe, A. J. et al.: International Cellular Plastics Conference Nov. 1976, Montreal, Canada
40. Koppers Co. Inc.: US-PS 4,165,413 (1979)
41. Koppers Co. Inc.: BE 897,254; 897,255; 897,256; Nov. 3, 1983
42. Koppers Co. Inc.: Company Brochure; US-PS 4,444,912 (1984)
43. Kifer, E., Koppers Co. Inc.: Private Communication (1984)
44. BP Chemicals Ltd.: Company Brochure
45. Celotex: US-PS 4,204,020

46. Hilado, C.: "Flammability Handbook for Plastics" 2nd Edition, Technomic, Westport, Conn. (1974)
47. Fritz, T. W.: J. Therm. Insul. 6, 183 (1983)
48. Paul, K.: Chem. and Ind. 63, Jan. (1983)
49. Carlowitz, B.: Plastverarbeiter 31, 84 (1980)
50. Jokel, C. R.: Insulation, Acoustic. In: Kirk-Othmer: Encyclopedia of Chemical Technology, Vol. 13, 513, J. Wiley & Sons, New York, 1981
51. J. Borgers GmbH, D-4290 Bocholt: "Hard Pressed-Formed Parts (HPF) from shredded Fibermaterial, Bonded with Phenolic Resin". Company Brochure
52. Eisele, D.: Melliand Textilberichte 56, 916 (1975); 60, 135 (1979)
53. Hofer, H.-G.: Melliand Textilberichte No. 8 (1981) P. 641

14 Industrial Laminates and Paper Impregnation

The hydrophilic structure of non-cured phenolic resins characterizes them as suitable material for the impregnation of paper and cotton fabrics for use in the manufacture of electrical and decorative laminates, molded parts, filter papers and battery separators. Low molecular weight components, preferably mononuclear phenol alcohols penetrate into the capillary cavities of cellulose fibers and fill the cavities[1], whereas resins of higher molecular weight coat the fibers and make them water-repellent. During cure, a chemical reaction between cellulose and phenol alcohols probably occurs which contributes to the increased water- and chemical resistance.

14.1 Electrical Laminates

After a few years of slow growth and partial replacement of phenolic paper based electrical laminates by epoxy glass laminates, the recent shift in the industrial sector towards information items and services – computer manufacturing, telecommunications, new trends in printing, mass media and education – have created a strong demand for low cost printed circuit boards enabling economic mass production. New developments in phenolic paper boards offer improved and adequate performance in many areas at half the cost compared to epoxy glass laminates.

Fig. 14.1. Phenolic printed circuit board and electrical components of a radio. (Photo: Isola Werke AG, D-5160 Düren)

Table 14.1. Comparison of various national and international standards for industrial laminated thermosetting sheets[5]

National Standards			Hp 2061	Hp 2061.5	Hp 2061.6	Hp 2062.8	Hp 2062.9	Hp 2063	Hp 2064
West Germany	DIN 7735		S-PF-CP 1	S-PF-CP 2	S-PF-CP 3	S-PF-CP 4	–	S-PF-CP 4	–
Switzerland	VSM		P, PO	Pa	PP, PPO	–	PPPOO	PPP	–
France	NF		P 1	P 2	P 3	P 4	P 4	P 4	–
Great Britain	BS		X, XP	XX	XXP	XXXP	XXXPC	XXXPC/FR-2-	–
USA	NEMA-LI 1								
ISO/R 1642			PF CP1	PF CP 2	PF CP 3	PF CP 4	–	PF CP 4	–
Reinforcement			Paper						
Flexural strength d < 10 mm	N/mm^2	min.	150	130	130	80	60	80	130
Impact strength a_{k10}	kJ/m^2	min.	20	20	15	8	–	7	20
Impact strength, notched	kJ/m^2	min.	5	4	4	2.5	–	2.5	5
Tensile strength	N/mm^2	min.	120	100	100	70	60	70	100
Compressive strength	N/mm^2	min.	150	150	100	120	–	–	100
Fission force	N	min.	2000	2000	2000	2000	–	–	2000
Modulus of elasticity	N/mm^2	≈	$7 \cdot 10^3$	$7 \cdot 10^3$	$7 \cdot 10^3$	$7 \cdot 10^3$	$5 \cdot 10^3$	$7 \cdot 10^3$	$7 \cdot 10^3$
Resistance between plugs	Ω	min.	–	–	$5 \cdot 10^7$	10^{10}	10^{10}	10^{10}	–
Breakdown voltage, ∥, 90 ± 2 °C, 1 min, d = 25 mm	kV	min.	15	40	25	25	20	20	–
Breakdown voltage, ⊥, 90 ± 2 °C, 1 min, d = 3 mm	kV	min.	15	40	30	30	25	25	–
Dielectric loss tgδ, 50 Hz, 96[h]/105 °C		max.	–	0.05	0.08	0.08	–	–	–
Dielectric loss tgδ, 1 MHz, 24[h] H_2O		max.	–	–	–	–	0.06	0.05	–
Dielectric constant		≈	5	5	5	5	5	5	5
Tracking resistance	grade		KC 100	KC 100	KC 100	KC 100	KC 100	KC 100	KC 100
Electrolytic corrosion	grade	max.	–	–	–	–	–	AN 1.4	–
Glow resistance	grade		2 b	2 b	2 a	2 a	2 a	2 b	2 b
Thermal conductivity	W/m · K	≈	0.2	0.2	0.2	0.2	0.2	0.2	0.2
Coefficient of thermal expansion	10^{-6}		20–40	20–40	20–40	20–40	20–40	20–40	20–40
Limiting temperature	°C		120	120	120	120	90	120	120

Table 14.1 (continued)

National Standards:				Hgw 2082	Hgw 2082.5	Hgw 2083	Hgw 2072
West Germany	DIN 7735			S-PF-CC 1	S-PF-CC 2	S-PF-CC 3	S-PF-GC 1
Switzerland	VSM			C	–	CC	–
France	NF			F2, F3	–	F 2	PF 1
Great Britain	BS			C	CE	L	G-3
USA	NEMA-LI 1						
ISO/R 1642				PF CC 1	PF CC 2	PF CC 3	PF GC 1
Reinforcement				Cotton fabric			Glass fabric
Flexural strength d < 10 mm	N/mm^2	min.		130	115	150	200
Impact strength a_{k10}	kJ/m^2	min.		30	20	35	50
Impact strength, notched	kJ/m^2	min.		10	10	12	40
Tensile strength	N/mm^2	min.		80	60	100	100
Compressive strength	N/mm^2	min.		170	150	170	150
Fission force	N	min.		2500	2500	2500	2000
Modulus of elasticity	N/mm^2	≈		$7 \cdot 10^3$	$7 \cdot 10^3$	$7 \cdot 10^3$	$14 \cdot 10^3$
Resistance between plugs	Ω	min.		–	10^7	–	10^8
Breakdown voltage, ‖, 90 ± 2 °C, 1 min, d = 25 mm	kV	min.		8	20	8	20
Breakdown voltage, ⊥, 90 ± 2 °C, 1 min, d = 3mm	kV	min.		5	5	5	25
Dielectric loss tgδ, 50 Hz, 96[h]/105 °C		max.		–	–	–	–
Dielectric loss tgδ, 1 MHz, 24[h] H_2O		max.		–	–	–	–
Dielectric constant		≈		5	5	5	5
Tracking resistance	grade			KC 100	KC 100	KC 100	KC 100
Electrolytic corrosion	grade	max.		–	–	–	A/B 2
Glow resistance	grade			2 b	2 b	2 b	2 a
Thermal conductivity	W/m · K			0.2	0.2	0.2	0.3
Coefficient of thermal expansion	10^{-6}	≈		20–40	20–40	20–40	10–20
Limiting temperature	°C			110	110	110	130

Abbreviations: DIN: Deutsche Industrienorm, VSM: Verein Schweizerischer Maschinen-Industrieller, NF: Norme Française, BS: British Standards, NEMA: National Electrical Manufacturers Association, USA

Table 14.2. Characteristics and recommendations for the application of laminated thermosetting sheets according to NEMA-LI 1, 1971 (R 1976)

X: Primarily intended for mechanical applications where electrical properties are of secondary importance. Should be used with discretion when high-humidity conditions are encountered.

XP: Primarily intended for hot punching. More flexible and not as strong as Grade X. Intermediate between Grades X and XX in moisture-resistance and electrical properties.

XPC: Primarily intended for cold punching and shearing. More flexible and higher cold-flow but lower in flexural strength than Grade XP.

XX: Suitable for usual electrical applications. Good machinability.

XXP: Better than Grade XX in electrical and moisture-resisting properties and more suitable for hot punching. Intermediate between Grades XP and XX in punching and cold-flow characteristics.

XXX: Suitable for radio frequency work, for high-humidity applications. Has minimum cold-flow characteristics.

XXXP: Better in electrical properties than Grade XXX and more suitable for hot punching. Intermediate between Grades XXP and XX in punching characteristics. This grade is recommended for applications requiring high insulation resistance and low dielectric losses under severe humidity conditions.

XXXPC: Similar in electrical properties to Grade XXXP and suitable for punching at lower temperatures than Grade XXXP. With good punching practice at ambient temperature. This grade is recommended for applications requiring high insulation resistance and low dielectric losses under severe humidity conditions.

The standardization for paper reinforced base material in various countries is given in Table 14.1. Due to the rapid development of materials and the everchanging requirements of the market, the standards are not current. This is especially true with regard to processability, dimensional stability, flame and tracking resistance. Higher requirements are set forth than are stated in NEMA for FR-2. The description of quality marks and recommendations for the application according to NEMA specifications are compiled in Table 14.2.

14.1.1 Materials

The following description of the basic materials, paper, resins and production refers mainly to the more sophisticated copper clad laminates for printed circuit boards.

As an example, materials required to manufacture one square meter of an 1.5 mm thick, one side copper clad laminate (FR-2 or XXXPC) are:

295 g copper foil, 35 μm
  30 g thermosetting copper adhesive
1,170 g phenolic resin, dry weight
  970 g paper

All materials are subject to strict selection criteria and quality control to ensure trouble free production, favorable electrical and processing properties.

Paper

Bleached kraft paper as well as cotton linter paper is used. Cotton linter paper is easier to impregnate, offers qualitative advantages, especially with reference to electrical properties and punchability, but is relatively expensive. Kraft paper is cheaper and somewhat more favorable as far as the mechanical strength is concerned. Since cotton linter paper, which has been impregnated only once, meets the high requirements, while kraft paper often has to be impregnated twice, economic advantages are finally outweighed by plant and process conditions.

Resins

Phenol, cresol- and xylenol resols are used to manufacture high quality laminates. *p*-tert-butylphenol or nonylphenol enhance flexibility. The higher compatibility of cresol resins with natural oils results in mechanically tougher and less brittle resin films. Today, phenol resols modified with synthetic flexibilizers (i.e. polyether-, polyester-, polyurethane- or polybutadiene compounds) are utilized which are comparable to cresol- and xylenol resins and, in addition, offer economic advantages.

Resins of different MW are necessary regardless whether the impregnation is performed in one or two steps. Presently, the one step impregnation technique prevails. The first (preimpregnating) resin, a low molecular resol resin contains only a minor portion of di-nuclear phenols (<20%). It mainly consists of mono-, di-, and trimethylolphenol in addition to free phenol (<10%). Melamine resins are added to the preimpregnating resol to improve flame retardancy. Bromine compounds are in general not introduced in these laminates.

Hydrophobic modified resols of medium average MW are used as the "main" impregnating resin. The characteristic properties and differences of both resins are indicated in Table 14.3.

Calcium-, magnesium- or barium hydroxide are used as catalysts for the production of pre-impregnating resins which are precipitated at the end of the reaction by the addition of sulfuric acid. A low content of inorganic ions is essential for electrical applications. Only deionized water is used for the production of resins. The most frequently used catalyst for the main impregnating resins is ammonia or HMTA; primary, secondary and tertiary amines are also suitable.

Table 14.3. Properties of phenol resols used in the manufacture of copper-clad laminates according to NEMA FR-2, V-1[8)]

		1st Impregnation	2nd Impregnation
Resin content	%	45	60
Solvents		Water	Acetone/Methanol/Toluene
Viscosity	mPa·s	10–15	300–500
Gel time 130 °C	min	15–20	18–25
Gel time 150 °C	min	6– 8	6–10
Water solubility	ratio, min	1:5	1:0.1
Free phenol content	%	<10	<5
Free formaldehyde content	%	< 3	<0.1
Ash	%	< 0.1	<0.1
Storage stability	weeks	~4	~12

The resins for the second impregnation must be flexibilized to meet the requirements for punchability. A multitude of compounds has been recommended to improve the flexibility. The resins used at present time contain mostly phosphoric acid esters (triphenyl- or diphenylcresylphosphate), synthetic aliphatic prepolymers or tung oil. Tung oil, obtained from the tung tree, which superficially resembles an apple tree, is also known as China wood oil since the oil came originally and exclusively from China. Plantations are now also found in some areas of the U.S., Argentina, Paraguay and New Zealand. Tung oil is largely composed of glycerides of α-elaeostearic acid which has three conjugated double bonds[9]. Type F tung oil (Aleurites fordii) contains 79–82% elaeostearic acid, together with about 12% oleic acid, the residue being saturated fatty acids. The type M oil (Aleuritis montana) contains smaller quantities of elaeostearic acid (75–78%). Mechanisms of the reaction of phenol and phenol resols with olefinic double bonds are described in Chaps. 3.3.5 and 19.1. The advantages of tung oil modified resins are flexibility and toughness, i.e. favourable punching properties and good solvent resistance. A disadvantage, however, is the price instability of this natural material as well as the enhanced flammability. For many of these reasons tung oil-free resins using synthetic additives have been developed.

Similar oils, for instance oiticica oil[9, 10], follow price quotations of tung oil very closely. Oiticica oil contains up to 80% licamic acid (4-keto-octadeca 9,11,13-trienoic acid). Formulated resins are inferior to some extent in comparison to those based on tung oil so that they are not satisfactory alternatives. Recently extensive studies have been conducted to elucidate the structure of reaction products from m-cresol/tung oil acid catalyzed process[11].

Aromatic phosphoric acid esters improve the punchability as well as the flame resistance. Since they act as external flexibilizers which are not incorporated in the macromolecular system, solvent resistance is diminished. Aliphatic phosphoric acid esters are not suitable due to their ease of hydrolysis. To improve the flame resistance bromine containing additives, e.g. tetrabromobisphenol-A, octabromodiphenyl, pentabromophenol, pentabromodiphenyl ether[12], brominated epoxies, and others can be used. However, the laminate manufacturers prefer halogen-free laminates to a greater extent. This is achieved by using melamine resins for the preimpregnation and eventually the addition of inorganic fillers.

14.1.2 Production of Electrical Laminates

A plant for the production of industrial laminates consists of three units: the impregnation plant, the press plant and the finishing line. In general, horizontal driers are used for paper impregnation[13].

A one step impregnation technology allows a significant cost reduction with an increase of 60–80% in capacity compared to a two step impregnation process. The impregnation of the one step process can be accomplished with two impregnating troughs arranged consecutively with a breathing line (Fig. 14.2) with sufficient space between troughs.

On the other hand, the preimpregnating and main resin can be mixed and applied from one trough. Consequently either an emulsion can develop (e.g. with tung oil-modified resins) or a clear solution. During this combined operation a low stability of the emulsion or advancement of the resin solution can cause problems. The precise

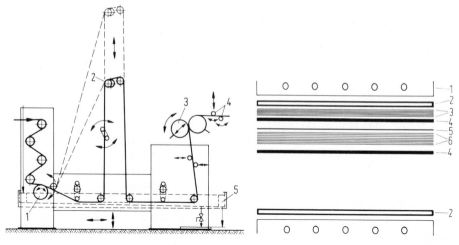

Fig. 14.2. Impregnation with prewetting. (Drawing: Impla, I-20060 Cassina de' Pecchi) *1* prewetting roll; *2* mobile extension rolls; *3* squeeze and doctor roll; *4* smoothing rolls; *5* heated bath

Fig. 14.3. Structure of the press composition. *1* upper heater plate; *2* steel plate; *3* equalizing sheets of kraft paper; *4* press sheet; *5* copper foil; *6* resin impregnated papers

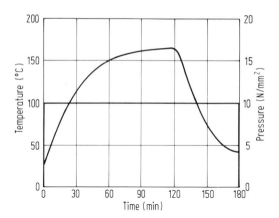

Fig. 14.4. Temperature-(t) and pressure-(p) cycle for the production of copper-clad laminates according to NEMA FR-2 and XXXPC

adjustment of the average molecular weight of both resins is of critical importance for the success of the one step impregnation as well as the choice of the flexibilizing polymer.

Solvent release and the controlled advancement of the resin is adjusted by the temperature in the drier and the impregnation speed. Finally the web is cooled in a cooling section or by cooled rolls, and rolled, or cut to size. The prepregs are laminated, after the required properties are confirmed (see Table 14.4), either immediately or after intermediate storage in air-conditioned storage rooms to avoid the absorption of water, in a multidaylight press with up to 20–22 openings. Approximately 8–12 laminates are produced simultaneously per opening (Fig. 14.3). Lower curing temperatures

($ < 160 $ °C) improve the punchability, but impede, however, other laminate properties. The cooling rate is important for the extent of bow. Slow and programmed cooling, especially in the upper temperature range, results in considerable improvement. Pressure stages are not customary; the pressure is between 10 and 12 N/mm^2 (Fig. 14.4). After cooling, a series of finishing operations follows[14].

Special requirements which mainly result from the processing of copper clad laminates, i.e. printing, etching, drying, punching, and soldering, are not covered in the applied specifications. The progressive automation in processing of printed circuit boards and the trend to miniaturization require FR-2 laminates with high quality standards combined with very specific properties. The ratio of low MW mono-nuclear polyfunctional species to high MW resins effects important properties such as low water absorption, high dimensional stability, printing and punching behavior. The low MW resin favors low water absorption and provides high dimensional stability, but adversely affects printing and punching. The opposite behavior in properties is achieved with high MW resin. Therefore, an optimum compromise is sought during formulation of both resins and their mix ratio.

14.2 Laminated Tubes and Rods

Laminated tubes and rods of phenolic resin impregnated papers or cotton fabrics are used as insulation and machine elements in electrical engineering, mechanical engineering and in the textile industry because of their high dielectric and mechanical strength and low specific weight.

Further examples are winding supports in transformers, insulating parts of quenching chambers, switch shafts, compressed gas capacitors, bobbins and guide rolls for paper and textile machines. The use of cotton fabrics results in high mechanical strength, which is necessary for slide bearings and bearing shells.

Quality requirements for tubes are established by national standards (NEMA, DIN). Natron kraft paper is preferred for the manufacture of laminated tubes, but with lower absorbency than paper recommended for flat laminates. The papers are coated with phenolic resin on one side, dried and rolled up before they are made into tubes in special machines[15]. The coated papers are wound around a preheated core by means of electrically heated (150–170 °C) counter pressure rolls. During this operation, the resin flows and binds the individual layers. The laminate and core are then placed in a drying oven for complete cure. After cure the core is removed and the tube is machined to size. The procedure requires thorough control of curing conditions to avoid blisters and delamination. The phenol resols are not flexibilized. Often *ortho*-cresol mixed with phenol is used to reduce the reaction rate. Thorough control of the volatile components content and resin flow are further prerequisites of a trouble free production.

In the manufacture of bobbins for the textile industry, which commonly spin at 4–12,000 rpm, special emphasis must be given to a perfect cylinder shape without differences in density to guarantee satisfactory spinning and balance. High quality bobbins with a wall thickness of 6 mm or more can be rotated at speeds to 20,000 m/min[16].

Fig. 14.5. Applications for cotton fabric reinforced laminates. (Photo: Isola Werke AG, D-5160 Düren)

14.3 Cotton Fabric Reinforced Laminates

Higher strength laminates and shaped parts are obtained by the use of cotton fabrics as reinforcing material instead of paper. Such shaped parts are used as construction material in mechanical engineering and as insulating materials in electrical engineering. They are distinguished by high specific strength, resistance to continuous heat to 110 °C, high wear resistance, very low water absorption and resistance to lubricants, solvents, acids and weak alkali. These materials possess very good machining properties and are used as cog-wheels, driving rolls, linings for sliding surfaces, guide ledges, switch base plates, wheel bearings and other applications.

Because the laminates are quite thick, the reactivity of the resins must be carefully balanced. High exothermic heat may result in carbonization in the center of the laminate or in some cases even in explosions. Cresol resols with high *ortho*- and *para*-cresol content are preferred because of their low curing rates. In most cases, these resins are not flexibilized to meet the high requirements of compression strength and delamination resistance. Cotton fabrics with weights between 130 and 200 g/m^2 are used. Fabrics for commercial application contain considerable proportions (4.5–5.5%) of size based on polyvinyl alcohol, in order to facilitate the manufacture of the fabric and its handling. Special electrolyte-free fabrics, which are washed with dilute NaOH prior to their use, are used for electrical applications. The phenolic resins contain methanol, ethanol, acetone or toluene as solvents. The impregnation is accomplished with solutions of approximately 50% solids. After a preimpregnation, the fabric is squeezed to ensure better penetration of the resin; then the desired amount of resin is applied by a second impregnation. The resin content of the prepreg is, in most cases, adjusted to 50% and the resin flow to 5%. In no case should the proportion of volatile components exceed 3%, especially if thicker laminates are required. Explosions may occur due to the gas-steam pressure which can not be controlled. The prepregs are pressed at a pressure of 10 N/mm^2 without pressure stages and at temperatures of 160–165 °C. The curing time is approximately 4 min/mm; a cure time of approximately 17 h and additional 8–10 h for cooling of laminates 250 mm thick must be considered.

14.4 Decorative Laminates

The main application for decorative laminates, also called high-pressure laminates (HPL), is the furniture industry. Decorative laminates as well as low-pressure lami-

Table 14.4. Structure of an HPL-laminate[19]

Number	Definition of paper	Type of paper	Resin	Resin content %
1	Overlay paper	α-Cellulose 20–50 g/m^2	Melamine	67
1	Decorative paper	α-Cellulose 80–160 g/m^2	Melamine	50
1	Barrier sheet	Kraft paper, bleached 80–160 g/m^2	Melamine	35
X	Core sheet	Kraft paper, unbleached 80–160 g/m^2	Phenol	35
1	Back paper	Kraft paper, unbleached 80–160 g/m^2	Melamine	50

nates, decorative papers, wood veneers, coatings and vinyl facings are used to laminate particle boards[17] and other wood materials used mainly for the manufacturing of kitchen cabinets. The hard and easy to clean surface is an important prerequisite. Further areas of application are furniture for children, laboratory furniture, wall elements in buildings, ships, boats and rail cars. The high surface hardness, scratch resistance, light colors and fastness to light of decorative laminates are attributed to the hard surface layer of melamine resin impregnated cellulose. The core sheet layers are of kraft paper impregnated with phenolic resin[18]. Sodium hydroxide catalyzed aqueous resols (60% dry solids) are almost exclusively used.

Conventional sizes range between 3.6×2.0 m, the thickness of the laminates is mostly 1.3 mm[19]. In general, the definition of electrical laminates is also applicable to HPL. The individual layers consisting of different kinds of paper are also impregnated with different kinds of thermosetting resins. The components of an HPL, 0.3–1.5 mm thick, are indicated in Table 14.4.

The production of decorative laminates is very similar to the production of electrical laminates. The speeds of impregnation are generally higher; they are 10–16 m/min for decorative paper facings and 30–50 m/min for phenol resin impregnated core plies. The volatiles content, resin content and flow are the most essential criteria of prepreg quality.

They are pressed in a multi-daylight press with 6–12 laminates per daylight. The pressure is between 8 and 12 N/mm^2, the temperature between 135–160 °C.

A continuous process for production of endless-laminates using a double belt press was recently developed[20]. The double belt press (Fig. 14.6) with a length of the pressure zone of 2 m and more is operated at 160–190 °C with a specific pressure of 5–50 bar. The setting time can be as low as 20 s, e. g. at a feeding speed to 20 m/min. A high cure rate of the resol resin used is required which is achieved by addition of accelerators.

The main problem of the HPL production is indicated in Table 14.4. The asymmetrical structure due to different qualities of paper and resin may lead to internal stress, bow and cracks because of the different thermal expansion and swelling when exposed to heat and moisture. In addition, the thermal expansion (approximately $45 \cdot 10^{-6} \text{K}^{-1}$) is different in length and cross direction due to the fiber orientation.

Fig. 14.6. Double belt press for the continuous production of decorative laminates. (Photo: Fagus-Grecon, D-3220 Alfeld-Hannover)

Table 14.5. National standards for decorative laminates

USA	NEMA	LD 1
West Germany	DIN	53 799
Great Britain	BS	3794
France	NF T	54-001
Japan	JIS-K	6903

The materials – plywood and particle boards or inorganic materials – also have different coefficients of thermal expansion.

Polyvinyl acetate dispersions, urea- and urea-melamine resins, phenol- and phenol-resorcinol resins are used as adhesives depending on the requirements and equipment. The adhesive coat is approximately 80–150 g/m^2. Adhesive bonding can be made at ambient or elevated temperatures up to 60 °C and pressure of 0.015 to 0.5 N/mm^2. Curing time is determined by the formulation and temperature.

Properties specified in national standards (Table 14.5), besides mechanical properties, are abrasion resistance, scratch resistance, light fastness and resistance to cigarette glow, hot pot bottoms and boiling water[15, 21].

In Europe, the post-forming of decorative laminates which was introduced in the USA, is increasing due to the tendency towards round edges in furniture manufacture (Fig. 14.7). Post forming is achieved by heating partially cured phenolic laminates just below the blistering point until pliable and then quickly shaped in a suitable mold. For convex forming the exterior plies (pattern sheet) are strained to tension, the back plies, however, to compression. The center remains free of strain. Kraft paper can be

Fig. 14.7. Postformed profiles of particle boards. (Photo: Resopal Werk H. Römmler GmbH, D-6114 Groß-Umstadt)

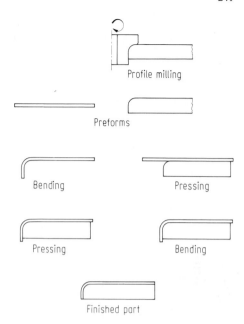

Profile milling

Preforms

Bending   Pressing

Pressing   Bending

Finished part

Fig. 14.8. Postforming working process[22]

stretched only to a limited extent. In the USA, it is still customary to produce core sheets of crepe paper. In Europe, this process is not practicable due to production problems. Flexibilized phenolic resins, which display some thermoplasticity are used in Europe. The moisture content of the laminate plays an important role in the deformation ability[22]. Water facilitates the elongation of the cellulose fiber, affects the sliding of the interfaces and acts as internal lubricant in the resin. The laminates are heated to temperatures between 140 and 220 °C by profiles or heated pipes. Additional heat can be applied by hot air. PVAc and UF-resins are used as glues.

Phenolic resin impregnated kraft papers (40–80 g/m² area weight) are used for overlay plywood and fiber boards for outdoor applications. The main objective is the improvement of weather and abrasion resistance and to reduce water absorption. More important examples are concrete form plywood and construction elements for railway cars and other transportation vehicles[17]. The resin penetration is relatively high, generally between 120–200% depending on paper weight. Release agents (stearates, about 1%) can be added to the impregnating resin. The prepregs, which may be pigmented, generally have a storage life between 4–6 months if stored under proper conditions, i.e. not exceeding 25 °C and 65% humidity. Plywood covering can be made during plywood manufacture or in a separate operation with the following conditions: pressure 1.5–3 N/mm², temperature 100–120 °C, molding time 5–10 min.

14.5 Filters

High porosity and filler-free papers, impregnated with phenolic resins, are used to produce oil, fuel and air filters for the automotive industry among others. This impreg-

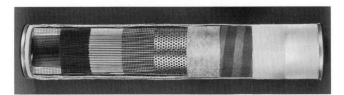

Fig. 14.9. Filter element of a water separator. (Photo: Faudi Feinbau GmbH, D-6370 Oberursel)

nation gives the paper the required strength and swelling resistance without appreciable reduction in porosity.

Generally, phenol novolak solutions in volatile solvents like methanol and acetone, with the required amount of HMTA, are used. Such solutions have a viscosity on 400–500 mPa·s at 65% solids content and a gel time of 7–10 min at 130 °C (flame retardant additives may be added to reduce burning rate). Prior to impregnation, the resin is diluted to approximately 10–18%. The resin content of the paper should be 20–30%. The temperature and impregnation speed are adjusted so that the resin is only partially cured. The webs are then cut to size, folded and heated up to 180 °C in a tunnel oven for complete resin cure.

Those filters must have high porosity, resistance to gasoline and oil, low swelling tendency and mechanical strength as well as sufficient aging resistance. The bursting pressure must be at least 0.2–0.3 N/mm². The value should not drop below 0.1 N/mm² after 24 h of service at 160 °C.

14.6 Battery Separators

Separators in batteries[23] maintain the distance between electrode plates and serve as electrical insulation between electrodes placed at narrow intervals (Fig. 14.10). They should not impede ion migration and current flow. Materials for separators are mainly paper and sintered PVC, rarely polyethylene fiber sheets. A lead-free plastic battery based on polyacetylene doped with I_2 or AsF_5 is under development.

Paper separators are made by impregnating paper of controlled porosity with PF-resins, curing, corrugating to form the rib and cutting to size. The resins have an average MW ranging between 130 to 300 with a phenol to aldehyde ratio preferably be-

Fig. 14.10. Automobile battery and battery separators made of phenolic resin impregnated paper. (Photo: Moll Akkumulatorenfabrik, D-8623 Staffelstein)

tween 1 : 1.8 and 1 : 2.3. Basic catalysts, for example alkali hydroxides, are used for the reaction[24].

Special kraft/cotton linter mixtures are suitable for the manufacture of separators. The fibers are protected from acid attack by coating them with phenolic resin. The separator must be resistant to sulfuric acid and must not evolve any by-products which affect the electrode reaction. The resin content is generally between 25–50%. The oxidation resistance is a very important criterion in view of the very high potential of the PbO_2 electrode besides high ion permeability. A test device and test performance for this property are mentioned in the literature[24]. Paper quality and resin properties are the critical parameters that influence oxidation resistance. Separators for automobile batteries, approximately 1.8 mm thick, or separators for heavy duty and stand-by batteries up to 6 mm thick, must also have sufficient mechanical strength. Sometimes, to stabilize the separators, they are covered with perforated plastics sheets or glass fleeces.

14.7 References

1. Schwaner, K., Komaromi, C.: Plaste u. Kautschuk 15, 413 (1968)
2. Knewstubb, N. W.: Plastics 27, 49 (1962)
3. Eisler, P.: The Technology of Printed Circuits, Chapman & Hall Ltd. 1956
4. Coombs, C. F.: Printed Circuits Handbook. New York: McGraw Hill 1967
5. Isola Werke AG: Herstellverfahren u. Anwend.-möglichkeiten gedruckter Schaltungen. Technical Bulletin
6. Herbert, J.: Electronic Production, October 1974, 35
7. Tillessen, R.: Feinwerktechnik & Meßtechnik 85, 8 (1977)
8. Bakelite GmbH: Bakelite Phenolharze-Bindemittel für Elektroschichtpreßstoffe. Technical Bulletin
9. Nylen, P., Sunderland, E.: Modern Surface Coatings. New York: Interscience 1965
10. Thomas, A.: Fette und Öle. In: Ullmanns Encyclopädie d. techn. Chem., Vol. 11, 4. Ed. Weinheim: Verlag Chemie 1976
11. Yoshimura, Y.: J. Appl. Polym. Sci., 28, 3859 (1983)
12. Dynamit Nobel AG: DE-OS 2 142 890
13. Caratsch AG: Technical Bulletin
14. G. Siempelkamp & Co: Press Lines and Finishing Lines for the Production of Technical Laminates. Technical Bulletin
15. Franz, A., Jüngerich, W., Schröder, W.: Schichtstoffe, Hartpapier u. Hartgewebe. In: R. Vieweg, E. Becker (ed.): Kunststoff-Handbuch Bd. 10 – Duroplaste. München: Carl Hanser 1968
16. Isola Werke AG: Spinnspulen aus Hartpapier CARTA für d. Chemiefaser Ind., Technical Bulletin
17. Enzesberger, W.: Holz als Roh- u. Werkstoff 27, 31 (1968)
18. Böhme, P., Karger, S.: Holztechnologie 10, 130 (1969)
19. Steinmetz, O. H.: Dekorative Schichtstoff-Platten. Stuttgart: DRW-Verlag
20. Fagus-Grecon, D-3220 Alfeld-Hannover: Company Brochure
21. Resopal Werk H. Römmler GmbH: Techn. Daten Resopal-dekorative Schichtpreßstoffplatten, Technical Bulletin
22. Menge, W.: Verfahrenstechnik des Verarbeitens dekorativer Schichtstoffe im Post-forming-Verfahren zum Fertigen von Möbelelementen. In: Verbund v. Holzwerkstoff u. Kunststoff in d. Möbelindustrie. Düsseldorf: VDI-Verlag 1977
23. Berndt, D.: Galvanische Elemente. In: Ullmanns Encyclopädie d. techn. Chem., 4. Ed. Vol. 12. Weinheim: Verlag Chemie 1976
24. Monsanto Co.: US-PS 3,926,679 (1975)

15 Coatings

All structural metals experience some corrosion in a natural environment. Materials that are especially susceptible are iron and steel whose corrosion is accelerated in the presence of oxygen and water[1].

High performance organic coatings provide the passivation, galvanic protection, and barrier properties that are necessary to inhibit corrosion. High metal adhesion, and low water vapor/oxygen transmission (Chaps. 19 and 20) are the key factors for the effectiveness of phenolic coatings. Additional advantages are the excellent chemical and abrasion resistance and moderately high temperature characteristics. Since unmodified phenolic resins result in brittle coatings due to structural rigidity of cured resins, considerable technical effort has been expended in flexibilizing the rigid resin system without adversely affecting the previously mentioned features. Hence, phenolic coatings resins are always used in combination with more flexible, hydrophobic resins like epoxy, alkyd, or natural resins, maleinized oils and polyvinyl butyral. Because of coloration or the propensity toward discoloration, phenolics are mainly employed as primers and undercoating materials. If solubility in hydrocarbons and aromatics is required, it is easily achieved by the use of alkyl phenol resins. Cure is performed by heating to a temperature range of 160–200 °C or by room temperature cross linking reaction promoted by acids or prereaction with drying oils.

Extensive reviews summarizing the chemical modification of phenolic resins for coating applications have been published[2-4]. The most important application areas for phenolic coating resins are automotive primers, coatings for metal containers, anticorrosive marine paints, and printing inks.

The adoption of Rule 66 by Los Angeles County (USA) and the Environmental Protection Agency (EPA) of USA in the late 60's ushered in a dramatic change in coatings technology. It required that most coatings systems would possess a significant reduction in solvent emissions and contain only limited amounts of certain photochemically active solvents. Exemptions are granted for coatings with a volatile portion of 80% water and 20% of a non-photochemically reactive solvent. These restrictions were instrumental in the development of powder coatings, reactive diluent coatings, high-solids coatings, radiation cured coatings[5], vapor cured coatings[6], as well as the easily recognizable water borne systems such as emulsions, dispersions or water reducible coatings.

Further governmental legislation such as Clean Air Act, Superfund, and Toxic Substances Control Act (Chap. 6) has required additional technical innovations in the coatings industry. Changes promoted by environmental, toxicological and energy factors have dramatically changed many conventional solvent based low to medium solids coatings systems into wholly "water based" systems[7, 8].

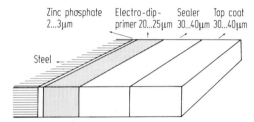

Fig. 15.1. Metal pretreatment and coating systems for automotive body corrosion protection

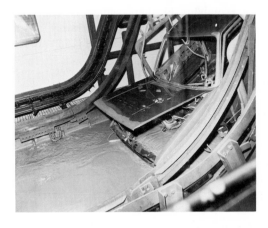

Fig. 15.2. Electrophoretic dip priming. (Photo: Volkswagenwerk AG, D-3180 Wolfsburg)

15.1 Automotive Coatings

Automotive coatings consist of an undercoat and a topcoat with a "tiecoat" between the two layers as a barrier against resin migration to avoid discoloration and improve intercoat adhesion[9]. The primary material to be coated is cold rolled steel whose surface is pretreated in several operations such as alkali cleaning, acid etching and chromate rinse. The formation of a 2–3 μm thick crystalline surface layer consisting of hydrated zinc iron phosphate is very effective.

The automotive undercoat is applied by electrodeposition (ED) or a variation of the "dip" operation. Electrocoating process has been popularized[10] since the late 60's and is now operated by all the major automobile manufacturers in the world. The fundamental property which electrocoating paints must have is the capability of being readily converted from a highly hydrophilic (water soluble or at least water dispersible) state to a highly hydrophobic state after deposition and curing[11].

The modern automotive enamel topcoat is in most cases an acrylic resin[12] containing hydroxyl groups (2-hydroxyethylacrylate) and is crosslinked with MF resins under moderate conditions, or an alkyd-melamine resin combination which is often used in Europe.

Latent acid catalysts[13, 14] have been suggested for these melamine resin systems. At ambient temperatures these latent acid catalysts are inactive and provide excellent storage stability for water borne or high solids coatings. At an elevated temperature but lower than usual, the catalyst rearranges or decomposes into the active acid to cure the resin system.

15.1.1 Water-Borne Coatings and Electrodeposition

Water-borne or water-reducible coatings[15] are increasing in importance due to ecological and worker safety reasons. They are formulated as maintenance and industrial
paints for interior and exterior metal substrates. Water solubility of the phenolic
moiety is frequently achieved by introducing carboxy-, sulfone- or sulfomethyl (see
also Sect. 21.5) groups followed by ion formation with ammonia, amines or inorganic
bases.

Low molecular weight resols are water soluble while higher MW resols require
higher amounts of inorganic bases for water solubility. Neither is suitable for coatings.
Many water-reducible phenolic coating resins are described in literature[16-27].
Phenolic resins incorporating salicylic acid (15.1)[20] or diphenolic acid (15.2) are reported.

(15.1)

Salicylic acid

(15.2)

4,4-Bis(4-hydroxyphenyl)
valeric acid
("Diphenolic acid")

The early electrocoating paints were deposited via anionic conditions and consisted
of such materials as natural drying oils, or polybutadiene resins[22] reacted with maleic
anhydride, epoxy ester, alkyd and acrylic resins. These vehicles were crosslinked with
phenolic resins for better corrosion resistance and throwing power. The ability of the
paint to coat completely recessed and shielded areas is known as throwing power. Either novolak or resol type resins as well as BPA resols etherified with methanol or butanol are effective. Additional flexibility is provided by alkylphenols. Aside from some
electrochemical disadvantages, a major coating disadvantage is the poor chip resistance culminating in a rapid spread of corrosion from small areas of damage on exterior surfaces of the vehicle body.

The development of cationic electrodeposited resins has rapidly displaced the anionic systems and is now the preferred process[28]. These cathodic coatings have up to
four times greater corrosion resistance than the anionic electrocoat. Cathodic resins
are solubilized by the presence of tertiary ammonium cations along the backbone of
the polymer molecule with acetate as a typical counter ion. The most convenient synthetic method of the protonated amine materials is the reaction of the oxirane functionality of an epoxy resin with a secondary amine. These cationic amine groups can
exist either as end groups in the polymer or randomly in the polymer chain.

The process negatively charges the metal part which then is dipped in a tank of positively charged, water based epoxy resins. As the film builds up the submerged part is insulated from the charged epoxy polymer. An extremely uniform coating covering the entire part including the less accessible interior recesses is obtained. Not only are emissions reduced, but the uniformity of the coat allows manufacturers to use less material, a saving that pays for the expensive application equipment.

Oven cure promotes crosslinking of the polymer coating. Suitable crosslinking agents for cationic electrodeposition systems include MF, UF, blocked isocyanates, and phenolic resins. The latter are reported to be compatible with epoxy-urethane systems and produce well cured coatings[29].

A recent patent[30] describes a phenolic system for cationic electrodeposited coatings. It is a styrenated phenolic/epoxy resin system.

Future developments in ED would welcome a reduction in energy used in the curing step. A sulfonic acid catalyst has been identified for the anionic ED process and reduces the cure temperature from 175 to 150 °C. Film properties of the catalyzed resin are similar to thermally-cured uncatalyzed resin.

15.2 Coatings for Metal Containers

The choice and requirements of a container coating[31] are dictated by the product to be packed, the container construction and the manufacturing methods used in fabricating or filling the container. The container market in the USA is divided into several segments (Table 15.1).

Metal containers are made of steel (cold-rolled steel, blackplate), chromium plated steel, tinplate, aluminum or aluminum alloys. The sheet stock is generally coated by coil coating in highly automated units (Fig. 15.3).

Fabrication into the final product may involve extreme deformation. Therefore, one of the major prerequisites of a coating is the flexibility of the film. In the fabrication of the can, sheet steel for drawn- and redrawn and ironed can manufacture is coated with a phenolic/epoxy resin system. As far as food containers are concerned, it is essential that the coating not affects taste or odor of the enclosed ingredients. The use of coatings for food packaging is subject to official regulations, such as in those issued by the Food and Drug Administration (FDA) and US Department of Agriculture (USDA) in USA.

The coating must be resistant to sterilization because both the content and the container must be sterilized. Beer cans are sterilized at 63 °C/45 min., food products and cans usually at 120 °C/90 min. The problems of food packaging are extremely varied.

Table 15.1. Container market USA (1983)

Beverage cans	60%
Food cans	25%
Drums and pails	10%
Others	5%

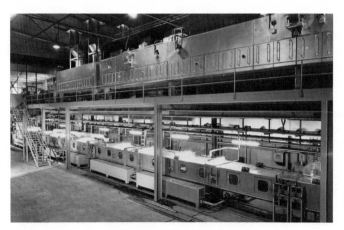

Fig. 15.3. Coil coating line, above, pretreatment section, below, coating and curing section. (Photo: Eisenmann KG, D-7030 Böblingen)

Most of the food stuffs with few exceptions are of acidic nature. Proteins containing sulfur cause black stains on the metal because of the formation of iron- and tin sulfide, as in fish or pea cans. Tomatoes and spinach also cause discoloration of the coating. Zinc oxide can be added to bind the sulfur by the formation of insoluble white zinc sulfide.

Phenolic coatings offer excellent resistance to corrosion, chemicals and sulfur staining compounds. They are used for pails, drums, collapsible tubes, aerosol cans and for the interior/exterior food containers. If direct contact must be avoided, a thin phenolic prime coat and a vinyl lining may be used.

One step cresols, sometimes also based on phenol, are used. Ammonia catalyzed resols are typically yellow in coloration (gold lacquers). The low flexibility is a disadvantage. Lighter or almost colorless films can be obtained by the use of etherified phenol- or BPA-resols.

Bisphenol-A is employed when taste is a critical factor, alkylphenol resins are occasionally used to improve the flexibility. Curing is effected at 180–220 °C for 15–20 min, temperatures up to 300 °C have been employed. In general, the phenolic resin must be flexibilized with other polymers, for instance, epoxy-, polyvinylbutyral- or alkyd resins.

A standard can coating formulation is described in Table 15.2.

Table 15.2. Epoxy/phenolic can coating formulation, solids content 40%[32)]

25.6 pbw	Epoxy resin (Rütapox 0197)
14.4 pbw	Phenolic resin (Bakelite 100)
0.3 pbw	Silicone oil (leveling agent)
20.0 pbw	Ethylene glycol
39.5 pbw	Ethylene glycol acetate
0.2 pbw	Phosphoric acid

A leveling agent (silicone) is included in the can coating formulation to promote a smooth coating and a lubricant (wax) to assist in the passage of the coated metal through metal forming equipment.

Fast curing systems with epoxy/phenolic combined resins are formulated with phosphoric acid, or dodecylbenzenesulfonic acid, both FDA approved. The higher temperature that occurs in rapid acid cure systems sometimes promotes a homopolymerization rather than a copolymerization and results in brittle film coating. Microvoids caused by traces of basic catalysts which neutralize the phosphoric acid, often are the cause of mediocre sterilization resistance. Prereaction of the resin at 60–80 °C generally results in a considerable improvement of coating performance.

The use of aluminum which has been replacing tin plate in recent years and its behavior on resin curing conditions must also be considered. Thus problems can be encountered if the cure temperature is too low. The brittle aluminium will break along the grain if it is not baked at 180 °C or above. Factors such as elevated temperature, cure conditions and temperature effects on the annealing of aluminum have to be evaluated.

New and emerging technology that represents new coatings opportunities in the metal container market area include high solids coatings, emulsion/latices, but largely a great emphasis on water based or water reducible systems[33]. Some of the performance requirements for water reducible can coatings are:
- water reducible with volatile organic content (VOC) $\leq$ 520 g/l;
- package stability of a minimum of 6 months;
- good adhesion to metal substrate;
- application via conventional equipment;
- cure latitude;
- minimum or no effect on flavor, odor, appearance;
- FDA compliance of cured film;
- applied cost not exceed conventional solvent based coatings.

15.3 Marine Paints

Marine paints[34] are important for all phases of marine construction, vessels and currently off-shore oil platforms. Marine coatings must prevent fouling, minimize surface roughness and more importantly prevent electrochemical corrosion. Besides these requirements, coating impact strength is also important. Coating of underwater sections consists of three basic coats:
- a primer coating;
- an anticorrosive paint;
- an antifouling paint.

Hot rolled steel for ship plating is first cleaned by shotblasting to remove the iron oxide layer. Then a quick drying shop primer – the most successful ones are based on phenolic resin/polyvinylbutyral combination, zinc dust or aluminium/epoxy resin combination, or zinc silicate – is applied in a thickness range of 10–15 μm[34]. When ships are repaired, the painting operation begins with the cleaning of the steel surface by brushing, flame or sand blasting. Then it is coated with a primer[35] based on polyvinylbutyral/phenolic resin/phosphoric acid to a dry film thickness of 10–15 μm. The anticorrosive paint for the second coat may consist of a variety of paint systems such as oil and rosin modified phenolic resins, chlorinated rubber, vinyl chloride/vinyl acetate copolymers, coal tar/epoxide resin combinations or zinc silicate. These paints are normally applied by roller either in dry dock or at sea in the open air. The coating is about 100–120 μm thick. The antifouling paint contains a biocide additive such as copper (I) oxide or organic tin or lead compounds, a bonding agent, fillers and solvents.

Copper (I) oxide is normally placed in a partially soluble matrix composed of resin and an oil based polymer[36]. Insoluble matrix coatings or vinyl resins, often called contact coatings, require higher biocide loadings. The use of phenolics has been recommended[37]. Recent studies have identified some addition copolymers with organotin substituents as attractive anti-fouling polymeric materials with longterm biocidal activity[38].

15.3.1 Shop Primers

A 10–15 μm thick coating of shop primer protects steel plates from corrosion in the plant or outdoors for a maximum of 12 months while the steel plate is being fabricated. Furthermore, it improves the adhesion of the other coatings. The coating formulation must not affect the quality of the welds. Some active pigments[27, 39, 40] such as chromates are avoided for ecological reasons. The generation of toxic vapors attributable to the coating during plate cutting and welding must also be avoided. No slag inclusions, gas bubbles or ash should occur in the welded seams. A shop primer's composition is comparable to a wash primer which is described in the next section (Table 15.3). The essential components are polyvinylbutyral, phenolic resin and micronized iron oxide.

15.3.2 Wash Primers

The term is a misnomer and somewhat misleading. A wash primer is applied by brush or a spraying operation. Up to World War II the only primer for the pretreatment of the cleaned steel surface of ships bottoms was phosphoric acid/red lead combination.

Table 15.3. Formulation of a wash primer[32]

		pbw
Resin	Phenol resol resin (Bakelite LG 721)	12.7
	Polyvinylbutyral (Mowital B 30 H)	12.1
	Epoxide resin (Rütapox 0191)	0.9
	Zinctetraoxychromate	8.6
	Talc, micronized	5.7
	Aerosil	0.3
	Wetting agent	0.4
	Isopropanol	42.9
	n-Butanol	16.4
Hardener	Phosphoric acid, 85%	6.0
	Isopropanol	29.0

This unsatisfactory primer was easily displaced by modern primers for steel surfaces based on polyvinylbutyral, phenolic resins and alcoholic phosphoric acid solutions. An example of a two component primer formulation is shown in Table 15.3. Phenol is mainly used in resin production; occasionally with addition of cresols or *p*-alkyl-phenols to improve flexibility.

15.3.3 Oil-Modified Phenolic Resin Paints

The most important applications for oil-modified phenolics are anticorrosive under-coats for ship painting and boat paints. Similar multicoat systems are customary for other transportation vehicles. Paints for railway cars may consist of an epoxy primer, an urethane oil/alkyd-modified phenolic undercoat and an urethane oil/alkyd based finish[41]. Alkyl- and arylphenolic resins can be cooked with drying oils as a result of reduced self reactivity[2, 3]. Tung oil is preferred; sometimes linseed oil or castor oil is used. The proportion of phenolic resin is between 25% (resols) and 100% (novolaks) in relation to the oil, depending on self condensation ability. The reaction of novolak resin made of *p*-tert-butylphenol, *p*-octylphenol or *p*-phenylphenol is conducted in the following manner to prevent gelation. One half of the resin is dissolved in the oil and heated to 190 °C for 60 min. Then, the rest of the resin is added and cooked at 230–240 °C until foaming ceases and then for an additional 30 min, until the reaction is completed. After cooling, the modified resin is diluted with white spirit and aromatic solvents. In order to accelerate air drying, cobalt- or lead driers are added, supple-mented by additives to prevent wrinkling and skin formation ("orange peel"). Tack-free coatings are obtained within 6–16 h at ambient temperature depending on tung oil content and solvents.

15.4 Printing Inks

Prior to 20th century, linseed oil in combination with carbon black and other pigments was exclusively used to make printing inks. New printing procedures and new mate-rials to be printed made it necessary to supplement the linseed oil by more effective synthetic polymers. The different printing procedures are divided into four main types: relief, planographic, intaglio and screen printing[42]. There is a definite trend to roller offset and rotation intaglio printing at the present time. Relief printing is pres-ently of no importance.

The most important application for phenolic resins as binder for printing inks are the publication gravure and offset inks which are used to a large extent for the printing of magazines and catalogues. Because of the extremely high speeds of the paper web (up to 800 m/min), the drying rate becomes the critical factor of the printing ink per-formance. Phenolic resin-modified colophonium resins release solvents completely and rapidly. They are mixed primarily with other resins due to cost.

Publication gravure inks may therefore contain 50–60% toluene and approximately 15% of a colophonium-modified phenolic resin, 15% calcium resinate, varying pro-portions of hydrocarbon resin and approximately 5–10% pigments. The resins should effect a fine and uniform distribution of the pigments during the ink preparation on a three-roll mill and also during the printing process. The solvent must be released rapidly and form a hard film. The modified colophonium resins (type B inks in the USA) are superior to the resinates (type A) and hydrocarbon resins as far as rate of solvent loss and film hardness is concerned. Cresols, xylenols or alkylphenols[43] are used as phenolic components. The compatibility with solvents and mineral oils is important in the choice of these components. In France, the use of toluene is not allowed, white spirit is preferred. Apart from solvent release, melting point, viscosity, yellowing and odor are further criteria.

15.4.1 Rosin-Modified Phenolic Resins

Rosin-modified phenolic resins are used in printing inks, in oil lacquers and as additives to alkyd paints because of their good compatibility with natural oils ("art copals") in which they improve the drying and gloss. However, the designation "phenolic resin-modified rosin" would be more accurate because of the low phenolic resin content, generally about 20%. Colophonium consists mostly of rosin acids, or special diterpenes (such as abietic-, laevo-pimaric-, neo-abietic acid etc.), and varying amounts of neutral components (rosin acid esters and several other diterpenes[44, 45]. The reaction mechanism with phenolic resins and the structure of the formed products, however, is not fully understood (15.3). The molecular structure and MP (between 100–140 °C) facilitate the solubility in aromatics and white spirit[2, 3]. Low molecular weight, "oil-reactive" alkylphenol resols are most frequently used to modify colophonium. They transform the monofunctional rosin acids

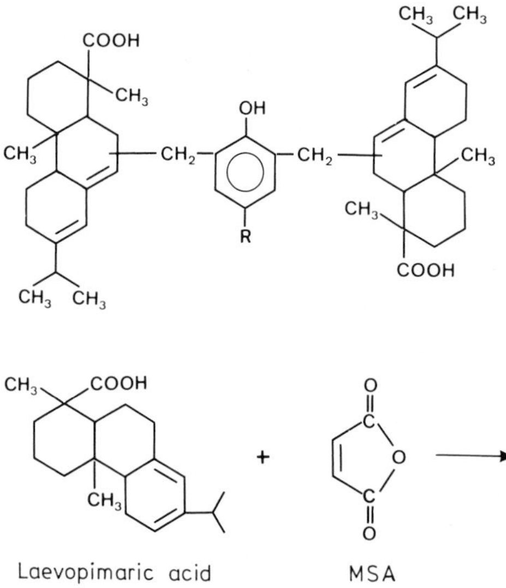

(15.3)

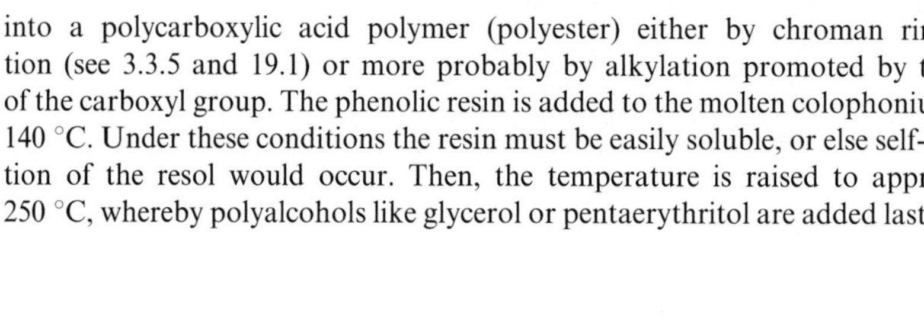

Laevopimaric acid   MSA

(15.4)

into a polycarboxylic acid polymer (polyester) either by chroman ring formation (see 3.3.5 and 19.1) or more probably by alkylation promoted by the acidity of the carboxyl group. The phenolic resin is added to the molten colophonium at 110–140 °C. Under these conditions the resin must be easily soluble, or else self-condensation of the resol would occur. Then, the temperature is raised to approximately 250 °C, whereby polyalcohols like glycerol or pentaerythritol are added last for esterification and to increase the molecular weight. Decarboxylation commences above 250 °C. In individual cases, cooking with novolaks at relatively high temperatures is performed.

Apart from the joint condensation of all components, rosin acids can also be prereacted with formaldehyde (Prins reaction mechanism, 2.17), yielding hydroxyl group containing compounds. The softening range of colophonium is raised by approximately 45 °C to 105 °C by such reactions. Maleic acid anhydride may also be slowly added to rosin acids at temperatures above 125 °C (1,4-dipolar addition mechanism, 15.4). The maleate resins which are formed by this process have considerable economic importance. Phenolic resin esters can be prepared by reaction with anhydrides.

15.5 Other Applications

The wide assortment of chemical equipment and the plant facility itself are not only endangered by environmental influences and high temperatures, but also by aggressive chemical compounds. Such chemical equipment components are containers, reactors, pipe lines and heat exchangers. Resistant or high-performance coatings are specialty products that provide long term protection under severe corrosion conditions to equipment or facility.

After the usual pretreatment of the steel to a surface roughness of 30–60 μm, differently pigmented heat curing phenolic coatings are applied in several operations either by dipping or pneumatic or electrostatic spraying. The thickness of one coat is 30–40 μm. After predrying for approximately one hour at room temperature, the coat is heated slowly to 120 °C with a holding time of 15–30 min. This process is repeated until the desired thickness of the coat (up to 250 μm) is obtained. It is then baked at 180–200 °C for two hours[46].

Non-plasticized phenol resols can be used as sole bonding agents as well as in combination with polyvinylbutyral, epoxy resins or furane resins. The alkali resistance can

Fig. 15.4. Phenolic resin coated heat exchangers before cure, drying oven in background. (Photo: Secova Ges. für Korrosionsschutz mbH, D-5040 Brühl)

be considerably improved by alkylation of the phenolic hydroxyl group with allyl alcohol, allyl chloride, dichloropropanol or epoxide compounds.

Cresol-, xylenol-, and phenol resols are of considerable importance in electrical engineering, either alone or in combination with flexibilizing resins (PVB, PVF, alkyds) as impregnating-, dynamo sheet- and wire varnishes. The requirements of this application are strong adhesion to copper, brass and aluminum, high electrical insulating properties, temperature resistance and resistance to humidity and transformer oil.

Protective coatings are used extensively in nuclear power plants for corrosion protection, maintenance and to facilitate the removal of radioactive soils (decontamination). The high resistance of phenolic resins to high-energy radiation was previously discussed in Chap. 8.3. Decontamination of concrete is difficult and costly to perform. Epoxy, epoxy-phenolic or phenolic surfacings are used instead of vinyl paints in critical areas[47].

15.6 References

1. Dickie, R.A., Smith, A.G.: Chem. Tech. 1980, P.31
2. Hultzsch, K.: Phenol-Lackharze u. abgewandelte Phenolharze. In: Vieweg, R., Becker, E. (ed.): Kunststoff-Handbuch Bd.10, Duroplaste. München: Carl Hanser 1968
3. Kittel, H. (ed.): Lehrbuch d. Lacke u. Beschichtungen Bd.1, Teil 1, Stuttgart, Berlin: W.A. Columb.
4. Richardson, S.H., Wertz, W.J.: "Treatise on Coatings Vol.1, Part III, Meyers and Long Ed. Chapt.3 – Phenolic Resins for Coatings" Marcel Dekker N.Y. 1972
5. Chemical Week, April 13, 1983, P.20
6. Chemical Week, January 12, 1983, P.52
7. Robinson, P.V.: J. Coat. Tech. 53, 23 (1981)
8. Fry, J.S., Stregowski, R.J.: Polymer Preprints 24 (2), 203 (1983)
9. McBane, B.N.: Automotive Coatings, In: R.R. Myers, J.S. Long (ed.): Treatise on Coatings, Vol.4, Formulations. New York: Marcel Dekker 1975
10. Schenck, H.U., Spoar, H., Marx, M: Progr. Org. Coat. 7, 1 (1979)
11. Rousse, R.E.: J. Oil & Col. Chem. Assoc. 55, 373 (1972)
12. Kline & Co. C.H. Fairfiled: The Kline Guide to the Paint Industry (1975)
13. Chattha, M.S., Bauer, D.R.: Ind. Eng. Chem. Prod. Res. Dev. 22, 440 (1983)
14. Hart, D.J.: J. Coat. Techn. 55, 87 (1983)
15. McEvan, J.H.: The Role of Water in Water-Reducible Paint Formulations. J. Paint Technology 45, 33 (1973)
16. Vianova Kunstharz AG: ÖS-PS 180 407; ÖS-PS 299 543
17. Reichhold-Albert-Chemie AG: US-PS 1 254 528; US-PS 1 254 529; DE-OS 1 669 286; DE-OS 2 046 458
18. Dtsch. Farben-Z. 27, 370 (1973)
19. Kansai Paint Co. Ltd.: DE-OS 2 237 830
20. Vianova Kunstharz AG: DE-PS 113 775
21. Daimer, W.: Dtsch. Farben-Z. 27, 358 (1973)
22. Chemische Werke Hüls: US-PS 3 546 184
23. Kita, R., Kimi, A.: J. Coatings Technology 48, 53 (1976)
24. Ford Motor Co: DE-OS 1 794 354
25. BASF AG: DE-OS 1 949 294
26. Rifi, M.R.: J. Paint Technology 45, 73 (1973)
27. Anisfeld, J.: Farbe u. Lack 81, 1024 (1975)
28. Kardomenos, P.I., Nordstrom, J.D.: J. Coat. Techn. 54, 33 (1982)
29. US-PS 3,963,663 (1976)

30. C.A. 99 39904s (1983)
31. Gerhardt, G. W., Seagren, G. W.: Coatings for Metal Containers. In: R. R. Myers, J. S. Long (ed.): Treatise on Coatings, Vol. 4, Formulations. New York: Marcel Dekker 1975
32. Bakelite GmbH: Phenolharze für d. Lackindustrie. Technical Bulletin
33. Demmer, C. G., Moss, N. S.: J. Oil Col. Chem. Assoc. 65, 249 (1982)
34. James, D. H.: Marine Paints. In: R. R. Myers, J. S. Long (ed.): Treatise on Coatings, Vol. 4, Formulation, New York: Marcel Dekker 1975
35. van Oeteren, K. A.: Korrosion u. Korrosionsschutz. Hannover: C. R. Vincentz 1967
36. Bufkin, G., Bounds, R., Thames, S. H.: Paint and Varnish Production, Feb. 1974, P. 25
37. Pitre, A. S., Saroyan, J. R.: US-PS 2 579 610
38. Rzaev, Z. M. O.: Chem. Tech., P. 58 (1979)
39. Hare, C. H.: Paint and Varnish Productions July 1974, P. 19
40. Ruff, J.: Dtsch. Farben-Z. 25, 199 (1971)
41. Timmins, F. D.: J. Oil & Col. Chem. Assoc. 60, 291 (1977)
42. Sixtus, H.: Druckfarben. In: Ullmanns Encyclopädie d. techn. Chem. 4. Aufl. Bd. 10, Weinheim: Verlag Chemie 1976
43. Hoechst AG: US-PS 1,428,285
44. Barendrecht, W., Less, L. J.: Harze, natürliche. In: Ullmanns Encyclopädie d. techn. Chem. Vol. 12, 4. Ed., Weinheim: Verlag Chemie 1976
45. Nylen, P., Sunderland, E.: Modern Surface Coatings, New York: Interscience 1965
46. Ernst, H., Mushak, J.: Plaste u. Kautschuk 24, 284 (1977)
47. Gad, W.: Kunststoffe im Bau 11, 5 (1976)

16 Foundry Resins

The production of shaped parts from molten metal can be performed either in permanent molds or in lost molds[1, 2, 3]. Permanent molds made of metal, graphite or ceramics are used to cast low melting non-ferrous metals. Lost molds consisting of fire-proof sands, an inorganic or organic bonding agent and sometimes various additives, are used for iron and other metals. Molten metal is poured into the mold cavity and solidifies to the desired shape. The mold becomes brittle under the influence of the molten metal temperature and can be readily removed from the casting. One mold (and core) must be used for each casting.

Fig. 16.1. Phenol resin bonded sand mold and core by shell process. Cylinder block for an automobile motor. (Photo: Halbergerhütte, D-6700 Ludwigshafen)

16.1 Mold- and Core-Making Processes

Molds for casting operations and cores, separate shapes which are placed in the mold to provide castings with contours, cavities and passages, are composed mainly of sand and a suitable binder. The binder may be of organic or inorganic composition. Natural inorganic binders are clays (montmorillonite, glauconite, kaolinite); synthetic ones are, for instance, sodium silicate, cement or gypsum. Mainly phenol-, furane- and urea resins are used as organic binders. Various oils and carbohydrate binders are of minor importance. However, there also are physical forming processes for sand molds (magnetic molding process and vacuum process).

16.1.1 Inorganic Binders

Green sand molds are made of sand, clay, water and sometimes organic additives and are used without further conditioning or drying (green). They are widely used as the least expensive method of producing a mold, but their dimensional tolerance and stability is low. Ferrous and non-ferrous castings can be produced in rapid production cycles[4]. Higher strength and dimensional stability can be obtained by drying in dry climates (skindried molds) or by oven drying (dry sand molds).

Sodium silicate is a further inorganic binder (CO_2 process) with key features being low cost, cleanliness and low toxicity. The disadvantages of this process are relatively low strength, limited storage and low breakdown capability of the mold. Many organic additives have been recommended to accelerate hardening, enhance the breakdown capability and reduce water absorption. Cement bonded sands can be used for the production of very large castings, but the low hardening rate has limited their wider use. Quick setting cements and accelerating additives have been recently recommended[5].

16.1.2 Organic Binders

Phenolic resin bonded cores and molds possess a number of favorable properties which are necessary for economic mass production of high quality castings. These are:
- short hardening time;
- high dimensional stability and hardness of cores and shells;
- good storage of cores and shells;
- high casting surface quality;
- good breakdown capability of cores and shells;
- low cleaning cost.

Other thermosetting resins currently used are furane-type resins based on furfuryl alcohol (including FA/UF- and FA/PF/UF-resin blends) and, by a considerable margin, polyurethanes. The most important foundry processes using synthetic polymer materials are the no-bake, shell molding, hot-box and cold-box processes. Newer publications indicate, that one of the most important considerations in the selection of a process is its energy consumption[6]. Therefore, today there is a definite trend towards the cold-set processes. Considering the various binders, phenol resins and furane resins are gaining further acceptance at the expense of the inorganic binders[7, 8, 9]. The organic materials limit but satisfy the requirements of work safety and environmental protection to a greater extent than the corresponding inorganic types. Further factors to be considered are[6, 7]:
- type of metal to be cast;
- number of pieces and production rate;
- volume of the casting;
- dimensional tolerances;
- availability of skilled personnel;
- extent of capital investment;
- types of sand available.

16.1.3 Requirements of Foundry Sands

Silica sands of high-purity are used for most sand molding operations. While some nations like West Germany, Belgium, The Netherlands and France have very good quality sand (99,8% SiO_2), other countries, like Italy or Japan must import sand for the production of high-quality iron and steel castings[7]. The chemical purity as well as the grain morphology affects the economy and strength considerably. Sands with poor surface quality need greater amounts of binder to obtain adequate strength, thus increasing cost and the amount of gases evolved during the casting process causing casting defects and higher environmental pollution[10, 11].

It is obvious that considerably larger amounts of acidic hardeners are needed when sands with higher portions of basic oxides (Na_2O, K_2O, CaO, MgO) are used in the cold-set process to obtain comparable hardening rates. This is also the case with a higher humidity content. Advantages of silica sand as mold and core material are low cost, ready availability in a variety of grades, consistency in composition, compatibility with all types of chemical binders and adequate thermal and chemical resistance in contact with metal melts.

On the other hand, there also are some disadvantages; for instance, relatively high thermal expansion results in expansion defects, when large castings are involved. Silica sand can react with certain molten metal oxides. Frequently, problems arise during the casting of austenitic manganese steels. The high and abrupt expansion of quartz (silica) is caused by inversions of the crystalline structure depending on temperature and pressure.

$$\text{α-quartz} \underset{575\,°C}{\rightleftharpoons} \text{β-quartz} \underset{870\,°C}{\rightleftharpoons} \text{β-tridymit} \underset{1470\,°C}{\rightleftharpoons} \text{β-cristobalit} \underset{1705\,°C}{\rightleftharpoons} \text{melt}$$

trigonal   hexagonal   hexagonal   cubic

These expansion-type defects can be avoided by using other inorganic materials like chromite, zircon or olivine. These sands, of course, are considerably more expensive and some of them are available to a limited extent, so that their use requires a thorough economic analysis. The thermal expansion of chromite, zircon and olivine (Table 16.1) is uniform and clearly lower as compared to quartz. They are used in approximately the same order as listed below.

A further, very important factor is grain shape and size distribution. When liquid metal comes into contact with the mold, steam and different gases of decomposed organic binders are generated. The mold must be porous enough to allow escape of gases and prevent an excessive pressure build up in the cavity which possibly might result in casting defects[12] and damage to the mold. On the other hand, the generation of gases has some advantages, for instance preventing metal penetration and yielding a

Table 16.1. Properties of different sand materials

Characterization	Sand material			
	Silica	Chromite	Zircon	Olivine
Chemical composition	SiO_2	$Cr_2O_3 \cdot FeO$	$ZrSiO_4$	$(Mg, Fe)_2 SiO_4$
Specific gravity g/cm^3	2.64	4.5	4.6	3.2
Expansion at 900 °C %	1.56	0.65	0.25	1.02
Volume price relation, silica = 1	1	9–11	60–70	6–7

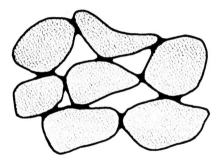

Fig. 16.2. Distribution of resin in bonded sand

good casting surface and excellent delineation to the casting. However, a balance between gas evolution and permeability must be maintained[1].

If the sand grains have a perfectly round shape and uniform diameter, the total volume of pores per unit volume is independent of the diameter of the grains. The pressure increase due to friction dependent upon the void diameter must be considered.
Although large pores are desirable to permit the escape of gases, they can permit the
penetration of the liquid metal into the voids. Fine sands and multimesh sands offer
better resistance to erosion and better surface quality of the casting, but require more
bonding material. All of these factors must be considered and invariably a compromise must be made.

Due to the highly increased cost for sand, transport and disposal, sand reclamation
becomes increasingly important. Reclamation plants provide the following regeneration processes:
– mechanical regeneration (breaking, grinding and dust removal);
– pneumatic regeneration (impact screening);
– wet regeneration (washing process for inorganic binders);
– thermal regeneration (incineration process).

Problems associated with reclaimed synthetic resin bonded sands have been thoroughly studied. Phenol levels observed in reclaimed sand eluted with water, are defi-
nitely lower than those determined in household sewage[13] (Chap. 6.3).

16.2 Shell Molding Process

In the shell molding process resin coated sand is placed into a heated metal pattern
(250–280 °C) by means of a dump-box or blowing with special machines into hot,
closed molds. Prior to this, the mold is sprayed with a silicone release agent. The resin
melts on the hot surface and binds the sand grains. A solid shell is formed; the thickness depends upon the time of direct contact, the temperature of the mold and the
hardening rate of the resin. As soon as the shaped molding has reached a thickness
of 4–7 mm, usually after 20–30 s, the unhardened surplus mixture is removed and fed
back into the chute. The shell is only hardened on one side. The reverse side is then
completely cured by infrared radiation or in a tunnel oven. This cured shell constitutes
one half of the mold. The mold is assembled by clamping both shell halves together
or gluing them together with thermosetting resins (bonding time 20–40 s), after the re-

quired cores are set. Shell molds can often be poured flat without further support depending on the casting volume and metalostatic pressure.

This process was invented by J. Croning around 1944[14] (Croning process). The process distinguishes itself by a very accurate reproduction of all the contours of the model; very narrow tolerances can be maintained during the casting operation[15] and is of particular importance to the automotive industry because of its high tolerances in comparison to other processes.

Precoated Shell Sand

A large series production of complicated cores with automatic core blowing machines necessitates a free-flowing resin coated sand with a controlled grain size distribution and uniform resin coat. While a number of foundries carry out their own precoating operation, many others purchase precoated sand from outside suppliers. The "warm" and "hot" coating processes differ from each other by the physical state of the resin[16] i.e. whether dissolved or granulated novolak is used.

In the hot coating process, emission of solvents is prevented by the use of granulated phenol resins and an aqueous HMTA-solution. The novolak resin with a MP of 70–75 °C must have a low melt viscosity so that a quick and uniform coat can be obtained.

Shell Sand Properties

To check the quality of the coated sand, the melting point, cold-tensile strength and hot-tensile strength are tested. Since the price of the resin coated sand is influenced by the resin cost, only enough bonding agent as is necessary is added. A larger amount of resin is required for the production of shells (3–4%) which are subjected to higher stress during casting than cores. In general, a resin content of 1.8–2.5% is sufficient for cores (Fig. 16.3). Further properties must be tested if complicated cores and shells are involved, for instance, gas volume, shell permeability, surface hardness, peel back and thermal shock resistance. The stick point of the coated sand, meaning

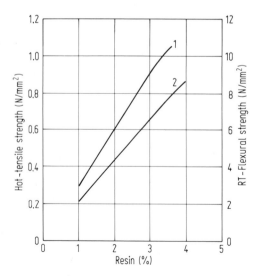

Fig. 16.3. "Hot coated" shell sand, hot-tensile strength (1) and RT-flexural strength (2) depending on resin content. HMTA 12%, calcium stearate 6%, based on resin weight. Cure at 280 °C for 3 min

the softening temperature of the resin, depends on the degree of advancement and reactivity of the resin. A broad melting range may indicate slow cure, low tensile strength and tendency to peel back. The cold-tensile strength reflects the ability of the shell and core to be handled without damage. The hot-tensile strength is an important factor of shell dimensional stability during casting. Strength and performance are improved significantly by modifying novolak resins with salicylic acid[17] or resol resin. Strength and temperature resistance are also controlled by the proportion of HMTA. A high proportion of HMTA, but not more than 18%, effects a high crosslink density and thereby higher thermal resistance. But the cores and shells become more brittle. The best results are obtained at a 10–13% HMTA level.

To prevent some frequently occurring mold defects in the shell casting process, such as mold cracking, peel back, soft mold or low hot-tensile strength, various additives can be added[17-19]. The thermal expansion of the sand (Table 16.1) during casting may lead to mold cracking. Apart from the use of sands with lower thermal expansion, cracking can be prevented by use of thermoplastic additives. The most common additive is a modified naturally occurring wood resin, called Vinsol, which is a complex mixture of substituted phenols, resin derivatives and hydrocarbon resins[19]. Vinsol has a softening point of 112 °C (ring and ball). By the addition of 0.25–0.5% Vinsol, in relation to the sand, thermal shock resistance can be increased and the metal penetration reduced. Excessive addition, however, leads to reduced hot-tensile strength.

Peel back, a serious mold defect frequently occurring during the drain cycle in the dump-box molding, is caused by several factors. On one hand, peeling from excessive weight may occur as a result of too high resin content. Another cause can be a very wide melting range of the resin due to an unfavorable MWD, excess release agent, insufficent solvent removal or presence of oily contaminants. Even though pattern design, pattern temperature and contact time can contribute to the phenomenon of peel back.

16.3 Hot-Box Process

In principle, the hot-box process is similar to shell molding. A wet resin/sand mixture is blown into the heated core box. A non-free-flowing sand is obtained from clay-free sand, resin and catalyst. It is more difficult to blow the wet mixture into a confined space than the dry, precoated shell sand, and thus, more difficult to produce accurately and consistently cores of complicated configuration and thin sections. An important advantage to shell molding is the essentially shorter curing time. By the use of a catalyst, a short heat surge is sufficient to start the curing while the final curing is performed outside the mold. After a few seconds, a hard surface layer is formed at the hot walls (180–280 °C) so that the core can be stripped out of the core box without any danger.

The surface quality can be improved by coating with a refractory wash. The hot-box process is applicable to mass production of small and medium volume castings. Examples in the automotive industry are transmission-, case-, crankcase-, body-, cylinder head- and gear box casing cores[20]. Phenol resols modified with urea resins (UF), furfuryl alcohol resins (FA) and FA/UF-resin blends are used together with catalysts. Aqueous solutions of ammonia salts of strong organic or inorganic acids are

Table 16.2. Properties of urea-modified phenol "hot-box" resins[16)]

Dry solids content	70 ± 1%
Viscosity at 20 °C	1,600 ± 200 mPa · s
Nitrogen content	7–8%
Water content	15–17%
Free phenol content	3–4%
Free formaldehyde content	< 0.5%

Table 16.3. Hot-box core sand formulation[16)]

Silica sand H 32, dry	100 parts
Hardener (ammonium chloride)	0.4 parts
Phenolic resin (urea modified)	2.0 parts
Release agent	0.2 parts
Iron oxide	0–1.5

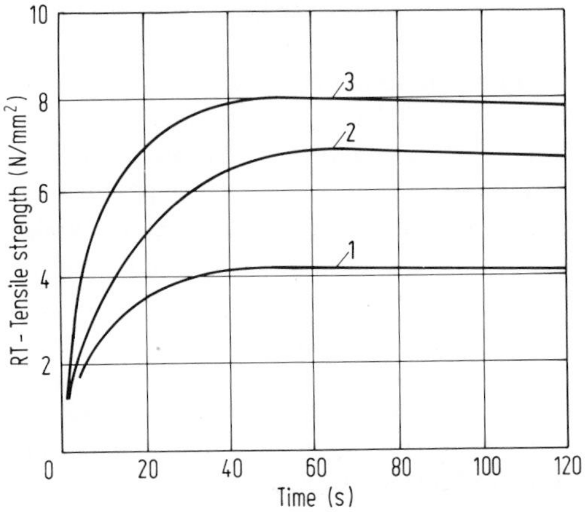

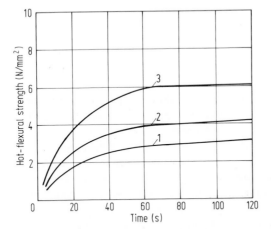

Fig. 16.4. RT-tensile strength of hot-box sands depending on curing time and resin content. Formulation: 100 parts silica sand H 33, 2.0 parts phenolic resin, 0.46 parts catalyst. Testing bars made at 220 °C and a pressure of 6 bar.
1 = 1.2% resin portion; *2* = 1.6% resin portion; *3* = 2.0% resin portion

Fig. 16.5. Hot-flexural strength of hot-box sands depending on curing time and resin content. Designation and conditions as in Fig. 16.4

used as catalysts. Ammonia reacts with formaldehyde contained in the resin whereby the acid is liberated and the reaction started. The hardening process starts while hardener and resin are being mixed, so that the mixture must be processed within a limited time which depends upon the temperature.

The cure rate of resols is insufficient for the short cycle times which are customary with UF/FA-resins. The reactivity can be increased by the addition of UF-resins.

However, higher urea levels would be necessary for phenolics in comparison to FA-resins to obtain comparable cure rates at low temperatures. UF-resins are less thermally stable than phenolic resins and decompose considerably faster during the high temperature operation. The gas shock caused thereby often leads to pinhole formation in the casting. The PF hot-box resins available on the market today contain approximately 7% nitrogen. To avoid pinholes, ferric oxide is added in the foundries. However, higher quantities of ferric oxide can lead to efflorences adhering to the mold[21]. Therefore, efforts are made to reduce the nitrogen content to at least 3% and still maintain the reactivity.

Appropriate PF-resins, made at a P/F molar ratio of between 1 : 1.4–1.8 and sodium hydroxide as catalyst have the following key properties (Table 16.2). The proportion of urea or UF-resin is indicated by the nitrogen content. Recently, PF-resins modified with furfuryl alcohol or furane resins are gaining increased market acceptance. Because of low viscosity and better wetting, higher core strength can be obtained. In order to improve the flow of the mixture and to facilitate the removal of the cores from the mold, lubricants and release agents like linseed oil or stearates can be added. The preparation of the mixture (Table 16.3) is performed discontinuously in a mixer.

16.4 No-Bake Process

The no-bake technique is very economical because the application of heat is not necessary. It is especially suitable for the casting of large pieces of gray-cast iron and cast steel for machine and vehicle construction[22]. Tool costs are considerably lower than those for the hot processes.

The most frequently used binders are those of the furane type: FA/UF-resin blends, UF/PF-resin blends and FA/PF/UF copolymers. Polyurethanes are also used in the USA. UF-resins offer high reactivity and thorough-cure, low cost and adequate bond strength. The high formaldehyde evolution of UF-resins during processing is a disadvantage. The high nitrogen content (20–25%) and the low temperature resistance lead to the formation of pinholes and metal penetration. Furane polymers are less dependent upon temperature and sand quality variations, and the fume evolution is low. The bond strength developed is high. However, they are very expensive. Phenol resins are attractively priced, provide acceptable strength and thorough-cure. Fume emission is often distressful. In order to reduce it, resins with an especially low content of free phenol and formaldehyde were developed. The high-temperature decomposition reaction and the nature of the gaseous decomposition products are described in Chap. 8.1. Nitrogen-free, fast setting phenol resins were introduced recently. A compromise between removal time, casting performance and price is achieved by the combination of three resin components, PF, UF, and FA.

The use of organic silane coupling agents in no-bake formulations has become common practice[23, 24] to improve the bond between sand and resin. An increase in strength of up to 40% can be obtained by the addition of γ-aminopropyltriethoxysilane or similar silanes between 0.1 and 1% in relation to the resin.

The phenol resins are characterized according to the content of FA, nitrogen, water and free formaldehyde. The content of free phenol and formaldehyde is considerably

Table 16.4. No-bake sand formulation[16)]

Silica sand, dry	100.0 pbw
p-Toluene sulfonic acid 65%	0.6 pbw
Phenol/FA/urea-resin	1.2 pbw
Iron oxide	0–1 pbw

reduced in the newer resins. At present, they contain approximately 3–4% phenol and less than 0.5% formaldehyde. The resin addition is generally between 1.2 and 2%.

For the cold-setting furane and resol resins[25)] inorganic acids (e.g. phosphoric acid) may be used; however, strong organic acids (phenol sulfonic acid, *p*-toluene sulfonic acid) which are enhanced in acidity by residual amount of sulfuric acid, are preferred. The most important criteria to differentiate the organic acids available on the market (65–85% acid content) are the content of sulfuric acid and water, viscosity and the price. A high water content inhibits the cure.

A great advantage of the no-bake technique is the capability of varying the setting and curing time by varying the catalyst addition rates (40–100% related to the resin). The cure rate depends entirely upon the temperature of the sand. When the temperature of the sand drops below 8–10 °C, very low-strength bonded sand is obtained. Further addition of acid as an attempt to speed up the reaction rate reduces the strength significantly[26)].

The preparation of the mixture (Table 16.4) can be performed in static mixers or in continuously operating high-speed mixers[27, 28)] which permit the use of extremely fast curing resins. Molding rates approaching those of green sand are possible with appropriate automation (carousel).

The acid catalyst is added first to prevent local "burn-red" of the resin under high local concentration of acid. Then, the bonding agent is added and further mixed until the material is uniform. At high nitrogen content iron oxide may be added to prevent the formation of pinholes. The maintenance of a certain temperature of the mixture

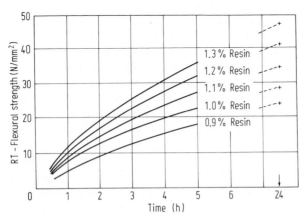

Fig. 16.6. No-bake core sand. Flexural strength at 20 °C depending on temperature and curing time at 20 °C

(between 15–30 °C) is important for a trouble-free production due to the strong dependence of reaction with temperature. A flowable, wet sand is obtained which is immediately filled into the molds and is compressed manually or by vibration.

The core boxes can be of nonmetal materials. Since the curing starts as soon as the acid is mixed with the bonding agent, the molding material has only a limited working time and must be processed immediately. Deformation of the preform during a critical stage of resin advancement results in reduced strength.

The strip time is the time from discharging the mixed sand into the core box until the time when the core can be safely removed. The strength increases as curing progresses until casting (Fig. 16.6).

Naturally, it is very desirable that the strength is as high as possible with regard to work time/strip time. Conventional single-trough continuous mixers may be used with sands having strip times of 20–40 min. The work time with most resins is at least 5–8 min.

16.5 Cold-Box Process

The cold-box process[29, 30] is based on the rapid reaction of poly-functional isocyanates with polyols. The reaction, which is rapid even in the cold, leads to the formation of polyurethanes (see Sect. 3.7). The important characteristic of this process is the use of gaseous organic amines as catalyst[31]. Their application requires specially designed gassing and exhaust systems. "High-*ortho*" novolak resins (see Sect. 3.4.3) are used as the hydroxyl component[31, 32]. They may be prepared in a water-free reaction medium in the presence of salts of divalent metals as catalysts. The phenolic resin/diisocyanate combination, e.g. 4,4'-isocyanatodiphenylmethane, is used in water-free solution.

High-boiling esters or ketones are used as solvents. The amount of binder normally used in this process is between 1 and 2%, for iron castings between 1.2 and 1.4%. Phenolic resin and isocyanate are used in the ratio 1:1. Since the bonding system is very sensitive to moisture, only washed and dry sands can be used. The humidity content should not exceed 0.1%.

The processing of a mixture of, for example, 100 kg silica sand H 31, 0.8 kg phenolic resin (65% dry resin) and 0.8 kg of diisocyanate (87–88%) is as follows. Initially, the sand is mixed with the phenolic resin solution in a continuous or batch mixer. Then the diisocyanate solution is added. A good flowing sand is obtained which has a work time up to 2.5 h. Core shooting machines are equipped with a gas inlet. After the core is shot, a catalyst, triethylamine or dimethylethylamine, is atomized into the tightly closed core box in an air or CO_2 stream. These amine vapors are highly flammable and, mixed with air within certain limits, tend to explode. Therefore, safety precautions must be taken. The use of CO_2 instead of air is preferred. An extremely fast cure is obtained by the sudden high amine concentration. The calculated consumption of amine is about 0.05% of the sand weight[30].

The cold-box process has some advantages, i.e. high core quality and cure rate, which renders them equally suitable for either mass production or short runs and jobbing work. Prototype tools can be made from inexpensive material like wood. However, stringent health and safety precautions must be considered because isocyanates and amines are highly toxic. The amines cause a strong cauterizing effect on the skin and the eyes. After the curing process, the amines must be displaced from the core and absorbed in phosphoric acid or incinerated. Phenolic resin-isocyanate binders are more popular in the U.S. with increasing application expected in Europe.

16.6 SO₂ Process

The most recent of the cold-setting mold and core making processes is the SO_2 process invented by Richard and assigned to Sapic, France, in 1971. The process is also referred to as Hardox[33].

It is basically a gas cured furane- or furfuryl alcohol/phenol resin system which relies on the transmission of SO_2 gas through the coated uncured sand in a presence of an organic peroxide to form sulfuric acid in situ[34, 35] (16.1).

$$2\ SO_2\ +\ HOO-\underset{\overset{|}{CH_2CH_3}}{\overset{\overset{CH_3}{|}}{C}}-OOH\ +\ H_2O\ \longrightarrow\ 2\ H_2SO_4\ +\ \underset{\overset{|}{CH_2CH_3}}{\overset{\overset{CH_3}{|}}{C}}=O \qquad (16.1)$$

High-purity dry silica sand with a low acid value and suitable grain size distribution is used. SO_2 is supplied as liquified gas. An aminosilane based adhesion promoter is used to improve the adhesion of the cured resin to the sand grains. From the great number of peroxides, methyl ethyl ketone peroxide (MEKP) and hydrogen peroxide with phlegmating additives, e.g. phthalates are used. Addition rates vary between 25–50% of the binder weight. The binder content varies from 0.7 to 1.5% and is dependent on requirements and sand quality. The furane resins used may contain 15–30% phenolic resin. Further modification is recommended, e.g. with epoxide or urea resins, to adjust selected properties or to reduce cost. Phenol resin addition avoids excessive crosslinking and brittleness at high SO_2 loadings. The urea resin improves economics, however, its content should not exceed 5% to avoid a decrease in curing rate. The role of the epoxy resin is quite specific. It avoids the formation of a coated film on the mould that occours with furane resins after repeated runs. The Hardox process also requires the use of silane coupling agent. The mixture shelf life is significantly longer (10–20 h) compared to the hot-box (2–4 h) and cold-box process (1–2.5 h). Generally all machines designed for gas-blowing can be used for mould and core production.

The SO_2 consumption is normally below 4 kg per ton of cured sand. The exhaust air treatment is performed in a similar manner as in the cold-box process with diluted sodium hydroxide using the countercurrent principle. After an oxidizing treatment of the resulting sodium sulfite/sodium sulfate mixture with hydrogen peroxide the final product is sodium sulfate which can be safely disposed.

The SO_2 process is used for mass production of iron castings with flake and spheroidal graphite and mass production of aluminum castings in permanent moulds. The advantages over other processes are close dimensional tolerances, improved physical properties, excellent break-down, long shelf life, acceptable cost and availability[36, 37].

16.7 Ingot Mold Hot Tops

To produce steel ingots, the melt is poured into large ingot molds consisting of three sections made of refractories (bottom) and steel. In order to obtain uniform solidifi-

Table 16.5. Materials for the production
of hot tops for ingot molds

Sand (fine)	35–40%
Asbestos and mineral wool	15–20%
Paper	5– 6%
Phenolic resin	4– 5%
Water	25–35%

cation, the cap sides of the ingot mold are insulated with phenolic resin bonded insulation boards called hot tops.

The hot top formulation consists of the fillers sand, asbestos, mineral wool and paper which are mixed to a fibrous pulp with water and bonding resin in a mixer similar to those used by the paper industry (Table 16.5).

The mixture is pumped into storage bins which are installed over molding tables. The molds consist of wooden frames with a sieve plate at the bottom. The material is poured into the mold, spread uniformly with a trowel and dehydrated by vacuum. The boards are removed and cured on a sheet metal or wire support in the oven. The oven temperature is approximately 180–190 °C. After cooling, the boards are cut or ground to the proper measurements. Powder resin (novolak type), blended with HMTA or paraformaldehyde, is used as binder.

Recently, phenolic resins have been partially replaced by urea resins or starch due to costs. This, however, reduces flexural strength and resistance to humidity.

16.8 References

1. Metals Handbook, 8th Ed., Vol. 5, Forging and Casting. Amer. Soc. Metals, Metals Park (1970)
2. Höhner, K. E.: Gießereiwesen und Gußeisen, Temperguß und Stahlguß. In: Ullmanns Encyclopädie d. techn. Chem., 4. Ed., Vol. 12, Weinheim: Verlag Chemie 1976
3. Gießerei-Lexikon. Brunhuber, E., Ed. Fachbuchverlag Schiele & Schön GmbH, Berlin 1983 (12. Ed.)
4. Middleton, J. M., Met, A.: Foundry Trade, April, 1976, 463
5. Kleinheyer, U.: Gießerei 63, 487 (1976)
6. Smith, J. R., Rowley, S., Young, J. S.: Can Hot Processes survive? In: Chemical Binders in Foundries, BICRA, Birmingham (1976)
7. Hatton, W. L. S., Cross, S.: Chemical Binders – the Supplier's Point of View. In: Chemical Binders in Foundries, BICRA, Birmingham (1976)
8. Howell, R. C.: Can Chemically-Bonded Sand Replace Green Sand? In: Chemical Binders in Foundries, BICRA, Birmingham (1976)
9. Kestler, J.: Modern Plastics Intern., Oct. 1976, P. 20
10. Goettmann, F. P.: Production of Sands for the Foundry Industry. AFS-Transactions 10, 15 (1975)
11. Middleton, J. M.: Alternatives to Silica for Mould Production. In: Chemical Binders in Foundries, BICRA, Birmingham (1976)
12. Levelink, H. G., Julien, F. P. M., De Man, H. C. J.: AFS Int. Cast Metals J. 6, 56 (1981)
13. Bradke, H. J., Klein, T.: Deponieverhalten und Verwertung von Gießereisanden, Teil I und II: Laborauslaugungen von Gießereisanden zur Ermittlung der eluierbaren Stoffe und des voraussichtlichen Deponieverhaltens. IWL, 5 Köln 51 (1975)

14. Croning, J.: DE-PS 810174, 83293, 832936, 832937
15. Meyers, W.: Gießerei 56, 101 (1969)
16. Bakelite GmbH: Bakelite-Resins for the Foundry Industry, Technical Bulletin
17. Hooker Chemical Corp.: DE-PS 1 095 516 (1960)
18. The Borden Chemical Corp.: DE-OS 1 808 673
19. Hercules Inc.: Vinsol Resin, Technical Bulletin
20. Awbery, M.: Automobile Castings by the Hot-box Process. In: Chemical Binders in Foundries, BICRA, Birmingham (1976)
21. Berndt, H., Unger, D., Räde, D.: Gießerei 59, 61 (1972)
22. Beale, D. W.: The Practical Application of Cold-set Processes to Larger Castings. In: Chemical Binders in Foundries, BICRA, Birmingham (1976)
23. Union Carbide Corp.: Technical Bulletin
24. Xuqi, D. et al.: AFS Int. Cast Metals J. 6, 54 (1981)
25. Lebach, H.: Angew. Chem. 22, 1598 (1909)
26. Tilch, W.: Prüfung kaltselbsthärtender Formstoffe, Gießereitechnik 22, 113 (1976)
27. Dhonau, H.: FASCOLD Cold-set Process. In: Chemical Binders in Foundries, BICRA, Birmingham (1976)
28. Christel, M.: Gießereitechnik 22, 158 (1976)
29. Kögler, H.: Gießerei 64, 95 (1977)
30. Truchelut, J.: Ashland Cold-box Process. In: Chemical Binders in Foundries, BICRA, Birmingham (1976)
31. Ashland Oil Inc.: DE-PS 1 583 521 (1967)
32. Ashland Oil Inc.: DE-AS 2 011 365: DE-AS 2 349 200
33. Société d'Application de Procédés Industriels et Chimiques S.A. (SAPIC): DE-PS 2239835
34. Gardziella, A.: Gießerei 69, 81 (1982)
35. Ellinghaus, W.: Gießerei 68, 135 (1981)
36. Briggs, K., Armistead, B.: Foundry Trade Journal, May 6, P. 641 (1982)
37. Gardziella, A., Kwasniok A.: Gießerei 71, 12 (1984)

17 Abrasive Materials

In 1907, a patent was assigned to L. H. Baekeland covering the manufacture of grinding devices with phenolic resins which were described as bonding agents for abrasive materials.

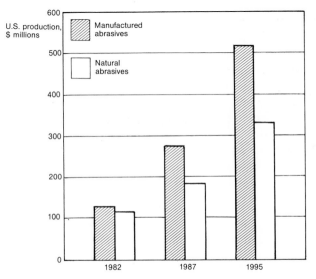

Fig. 17.1. Projected[1] growth of abrasives, U.S.

Many decades later, a recent market research study has indicated both natural (diamond, others) and manufactured (Al_2O_3, SiC) abrasives are expected to undergo strong growth into the 1990's (Fig. 17.1) with a corresponding higher growth for the synthetic based abrasives[1].

17.1 Grinding Wheels

Bonds normally used to manufacture grinding wheels[2-6] are inorganic such as vitrified, silicate and oxychloride bond or organic such as phenolic, rubber, shellac and others like polyurethane, polyester and epoxy bond.

Ceramic wheels which constitute more than 50% of all wheels, exhibit high porosity and allow a cool grinding in dry- and wet operation. However, they are very brittle and follow the movements of the machine very exactly. Therefore, they are used for profile and precision grinding, normally to a circumferential speed of 45 to 60 m/s.

The resinoid bond is stronger, more resilient and thermally shock resistant than the vitrious bond. The essential advantage of phenol resins in comparison to rubber or shellac is high-temperature resistance. The actual temperatures developed during grinding vary considerably with the kind of application. For grinding steel temperatures of at least 1,000 °C (possibly approaching 2,000 °C) on the surface layer are common by use of high-power operated machines. Therefore, it is easily understandable that laboratory performance evaluations in those cases can only serve as a guide. Actual thermal shock and mechanical stress is difficult to simulate.

Resinoid wheels are much less sensitive to impact, push and side pressure than ceramic wheels. The higher stability permits higher rotation speeds and thereby better grinding performance. The simpler manufacturing process has contributed to a relatively high market share of over 40%.

17.1.1 Composition of Grinding Wheels

A grinding wheel consists of three basic elements[2-6]: the abrasive grain, fillers and a bonding agent. Wheels differ by hardness (hard and soft) and grade. The wheel grade indicates the ratio of abrasive grain, bond volume and porosity, which is dictated by formulation and manufacturing. The bond volume identifies the hardness of the wheel, and designates the sloughing resistance of the abrasive grain rather than the hardness of the grain. Increasing hardness is indicated by the letter A to Z according to the scale developed by the Standards Committee of the Grinding Wheel Institute (U.S.)[4]. Distinction is also made between dry and wet grinding. Wet grinding is applied to heat sensitive materials. The grinding heat is removed by cooling liquids consisting of an oil in water emulsion and an anticorrosion additive. Mineral oils also are adequate.

Abrasive Materials

Corundum, the naturally occurring crystalline form of aluminum oxide (alumina, Al_2O_3) is very seldom used in industrial abrasives today. Synthetic, fused alumina is the preferred abrasive material. High-grade bauxite (hydrated aluminum oxide), purer than the grades for the production of aluminum metal, is the basic material. There are different types of fused alumina which differ from each other by their composition, mechanical properties and toughness: regular, high-titania, fine crystal and white alumina[2, 4, 6].

Silicon carbide (carborundum) which is obtained by reacting silica sand with coke in an electric furnace at approximately 2,000 °C, is harder but not as tough as aluminum oxide. Two types are produced, which differ in purity and color: "regular" (black) and "green" material. Pure silicon carbide forms colorless crystals. Fused alumina is preferred for the grinding of high-tensile materials like steel. Silicon carbide is more suitable for grinding of hard and brittle materials such as cast iron, stone, glass, ceramics and "hard metals". Boron carbide is extremely hard ($\sim$2,800 Knoop hardness) but also very brittle. It is used, like diamond, for special purposes only, such as the treatment of tungsten carbide. Broken glass and emery are used to a large extent for the production of coated abrasives. In general, the abrasive grain size may vary between 20 μm to 3 mm.

Table 17.1. Physical properties of grinding materials

Properties		Alumina $\alpha\text{-Al}_2\text{O}_3$	Silicon carbide $\alpha\text{-SiC}$
Crystal structure	–	hexagonal	hexagonal
Density	g/cm^3	3.99	3.2
Hardness, Mohs	–	9	(9.6)
Hardness, Knoop	(100 g load)	~ 2,000	~ 2,800
Melting point	°C	2,050	> 2,200 °C decomposition
Thermal expansion	K^{-1}	6.10^{-6}	$3.4 \cdot 10^{-6}$

Fillers and Reinforcements

The stability of phenol resin bonded grinding wheels and their flexural strength, the heat resistance and the toughness in particular are improved by the addition of fine grained fillers. Suitable fillers are aluminum oxide, iron oxide, silicates, chalk and asbestos. The addition of cryolite, pyrite, zinc sulfide, lithopone and potassium fluoroborate are effective.

Cure is accelerated by the addition of calcium or magnesium oxide. Sulfur containing additives (pyrite, zinc sulfide, lithopone) prevent the formation of strongly adhering metal oxide layers via a high temperature redox reaction; they delay oxidative degradation of the phenol resin bond and prolong the service life of the wheel. Cryolite melts at high surface temperatures leading to increased porosity and more effective grinding. Concurrently, the melt can also serve as a lubricant. Different reinforcements are used to improve the strength of the wheels; these include glass cloth, textile cloth, non-wovens or kraft paper.

Resins

A combination of liquid and pulverized phenol resin is used for grinding wheel manufacture. The liquid resin functions as wetting agent for the abrasive grain, powder resin and fillers. In addition, furfural, furfuryl alcohol, cresols and anthracene oil are also used in combination with liquid phenol resins. Liquid phenol resins and furfural possess very good wetting properties and good compatibility with pulverized phenol resins. The qualitative requirements of phenol resins are high, especially as far as the uniformity of the batches is concerned[6]. Sodium hydroxide or sodium carbonate are preferred for the production of liquid PF resols (Table 17.2). Low viscosity resins with the highest possible dry solids content are required.

The storage of liquid resol resins in refrigerated rooms is recommended due to their limited shelf life (Table 17.4). At temperatures of 5–10 °C, the increase in viscosity is minimal for several months. The importance of the conditioning of the resins, i.e. a constant temperature and humidity as well as observance of the required storage conditions cannot be underestimated. Powder resins are relatively high MW novolaks with a low free phenol content, high melting range, medium to short flow and high melt viscosity (Table 17.3)

The novolak resin is mixed with HMTA and then finely ground. The HMTA portion may be between 4 and 14%, preferably 9%. Lower HMTA content leads to lower

Table 17.2. Properties of a liquid phenol wetting resin[7]

Dry solids content	70–80%
Free phenol content	4–20%
Water content	1– 5%
Viscosity at 20 °C	2,500–3,500 mPa · s

Table 17.3. Phenol powder resins for grinding wheels[7]

Melting range	95/98–98/100 °C
Flow distance (125 °C)	25–30 mm
Size distribution (0.06 mm screen)	below 1%
HMTA content	ca. 9%
Humidity content	below 1%

Table 17.4. Phenol resol viscosity variation during storage[7]

Storage time days	Viscosity mPa · s	
	Storage at 20 °C	Storage at 40 °C
1	2,600	2,600
10	2,750	5,800
18	3,200	20,000

Table 17.5. Effect of amino silane coupling agent on flexural strength of phenolic resin bonded Al_2O_3 composites[8]

Al_2O_3 grain size	Dry flexural strength (N/mm^2)		% Flexural strength improvement
	No silane	With silane	
12	14.3	21.1	48
20	20.6	28.2	37
36	28.8	36.7	28
60	38.2	44.2	16

Composition: 92 wt% Al_2O_3 and 8 wt% of an 80/20 mixture of powdered and liquid phenolic resin. Silane A-1100 is applied at 0.1 wt% based on grain weight

crosslinking and softer bonds. High HMTA content results in higher crosslink density and thereby higher hardness and heat resistance of the wheel.

The elasticity and toughness of the phenolic bond can be improved by the addition of 10–20% epoxide, phenoxy or polyvinylbutyral resin. The burst resistance of those wheels is improved significantly.

After a storage time of 3 to 6 months the white pulverized novolak becomes yellow but other properties are not changed. In general, pulverized resins should not be stored longer than 1 to 2 months. The storage room must be kept cool and dry to avoid humidity absorption and sinter formation. Prior to use all ingredients should be conditioned to room temperature. Moisture condensation on the resin because of differences in temperature must be avoided.

One of the recent important goals is the development of grinding wheels with an increased resistance against alkaline cooling media. Because of microvoids, cracks, and imperfect adhesion of the resin to the abrasive grain cooling fluids diffuse into the resin-grain interface and deteriorate grinding performance. A significant improvement is achieved by silane coating of the abrasive grains.

Table 17.5 illustrates that a much stronger grinding wheel composite is obtained with amino silane treated Al_2O_3[8]. Thus a greater safety margin for wheels is achieved during frictional heat buildup and under wet grinding conditions.

Caustic-free resins with a favourable MWD and low-HMTA solid resins with low gas formation during cure have been developed. Flow and flexibility are adjusted by addition of epoxides and elastomeric modifiers.

17.1.2 Manufacturing of Grinding Wheels

Two methods are used in the manufacture of grinding wheels: cold molding process and the hot compression molding process. In the cold molding process, green wheels are made first, then cured in an oven. In the hot compression molding process, the wheels are formed and cured in one step. Two mixers are used in most cases to prepare the mixture. The abrasive grain is covered with the wetting resin in the first mixer, the powder resin, additives and fillers are mixed in the second. Then, the wetted abrasive grain is mixed with the mixture in the second mixer.

Preparation and processing of the mixture should be performed in an air-conditioned room at a constant temperature and humidity to obtain reproducible plasticity.

The hot compression molded wheels possess a denser structure, higher strength and longer service life. Cold molded wheels have a more open structure, allow cooler grinding and show, in general, higher cutting efficiency. From economy of production the hot compression molding process is somewhat less favorable than the cold molding process because more molds are necessary due to the long molding times. Therefore, the hot compression molding process will only be applicable to the manufacture of special wheels such as high-density wheels.

Cold Molding Procedure for Non-Reinforced Wheels

A cut-off wheel can be manufactured according to the following formulation:

Table 17.6. Formulation for the manufacturing of cut-off wheels[7]

74 –77	pbw	Abrasive grain NK 24, 30, 36
4 – 7	pbw	Phenol liquid resin
10 –13	pbw	Phenol powder resin
2 – 5	pbw	Pyrite
3 – 5	pbw	Cryolite
0.1– 0.4	pbw	Calcium oxide
0.2– 0.8	pbw	Antimony trisulfide

The final mixture should be homogeneous, free flowing but still plastic enough, so that it can be kneaded by hand with slight pressure.

The weight ratio of liquid to pulverized phenol resins is generally within the range of 1:2 to 1:4. It is dependent upon the size distribution of the abrasive grain, the type and the quantity of the filler and the viscosity of the liquid resin used. The weighed mixture is fed into the mold on a rotating table and evenly distributed.

Mold pressure is within the range of 10–40 N/mm², preferably 15–25 N/mm². The "green" wheels are then cured in an electrically heated drying chamber or tunnel dryer with fresh air supply according to the predetermined program. The preforms must be placed at such distances that adequate air circulation is maintained. Deformations will occur if they are heated too fast and the viscosity of the bonding agent decreases faster than the simultaneous compensating increase by the condensation reaction. Blisters may occur if the surface layer is curing too fast and the volatile components cannot escape.

The green bond changes into a fused mass before the temperature reaches 80 °C; the water contained in the resol and formed during the condensation evaporates as the temperature approaches 100 °C. Resol crosslinking begins in this range. At about 115 °C HMTA decomposition commences and ammonia will escape. Oven temperature is controlled in a way so that 80–85 °C is quickly obtained. Between 80–100 °C a longer holding time is maintained so that a homogeneous melt can form and the volatile components can escape. In case the temperature of 100 °C is approached rapidly or exceeded, gaseous inclusions result in a fine porous structure of the bond with reduced strength. A further holding time is placed in the range of 120–130 °C to enable the ammonia to escape. The hardness and toughness of the bond is influenced by the final temperature. After complete cure, the wheels are slowly cooled to 50–60 °C in the furnace by circulating air. In this way deformation of the wheels and the occurrence of cracks due to thermal stress is prevented.

High-Speed, Reinforced Cut-off Wheels

To increase wheel rotational speed from 45 m/s to 60, 80, and 100 m/s, the impact and flexural strength and burst resistance of the wheels must be improved[9, 10]. The in-

Fig. 17.2. Phenol resin bonded cut-off wheel and glass cloth reinforcement. (Photo: Krebs & Riedel KG Schleifscheibenfabrik, D-3522 Karlshafen)

crease to 60 m/s is achieved by use of epoxy-modified powder resins without changing the manufacturing process. A further improvement in burst resistance is obtained by reinforcement with glass cloth impregnated with thermosetting resins exhibiting strong adhesion to glass fibers, high strength and some flexibility. The cure rate of these additional resins (epoxy or PF resin) must corresponding to the reactivity of the other resins used in wheel construction. A cut-off wheel and the reinforcing glass cloth is shown in Fig. 17.2.

For wheel manufacture the mixture is divided according to the number of prepreg insertions and the mold is charged alternately with the mixture and prepreg layers. The preform is molded at RT and 15–35 N/mm^2 pressure. The cure is completed within 24 h in the oven at a maximum temperature of between 170–180 °C under slight pressure by metal plates.

Snagging Wheels

For heavy surface grinding, for instance cast steel billets, snagging wheels are used which are pressed onto the working piece with automatic pendulum grinding machines at a pressure of 20–80 N/mm^2. The billet moves automatically back and forth under the wheel. Snagging wheels with a diameter of up to 800 mm possess exceptional strength which is obtained by using high-performance components and high compression. While normal grinding wheels have a density of 2.4–2.7 g/cm^3, max. 2.9 g/cm^3, for high-density wheels it is 3.1–3.5 g/cm^3. This high density is only obtained by compression molding. The void volume should not exceed 1%. The preparation is as follows: a preform is made at 15–25 N/mm^2 pressure which is then preheated to 90–130 °C. After 40–60 min, the mass is plastic but still shape retaining and degassed that the preform can then be cured immediately in the hot mold at 20–40 N/mm^2 pressure and 150–170 °C within 30–60 min. Mold cure duration must be sufficient to avoid blister formation during removal and post-curing process in the furnace at 160 °C for 8–12 h. The subsequent cooling must be done very carefully.

Fibrous Laminated Wheels

High-speed grinding wheels with high impact, flexural and burst strength are obtained by reinforcement with kraft paper, non-woven materials or fabrics. The webs are coated with liquid phenol resin by roll coating, scattered with abrasive grain and dried in a tunnel dryer at 86–95 °C. As many rondels as necessary are punched out according to the thickness of the wheel and molded between chrome plated sheets in a mold at a pressure of 10–20 N/mm^2 and at 155–165 °C. The cooling is performed either under pressure in the mold or in clamping devices to prevent warpage.

17.2 Coated Abrasives

Abrasive belts, discs, sheets and drums (Fig. 17.3) are used primarily for polishing or finishing operations where high dimensional accuracy is not required rather than to remove thick sections. The efficiency is relatively low due to the single grain layer. However, by use of synthetic resins instead of animal glue, abrasive belts with higher flexibility and joint strength[11–13] can be made and this material can be used for contour grinding.

Fig. 17.3. Various kinds of coated abrasives. (Photo: Norddeutsche Schleifmittel-Industrie, D-2000 Hamburg)

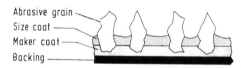

Abrasive grain
Size coat
Maker coat
Backing

Fig. 17.4. Structure of a coated abrasive[12]

In general, the abrasive grain is bonded to the support (backing) by two adhesive coats, the maker coat and the size coat (Fig. 17.4).

17.2.1 Composition of Coated Abrasives

Abrasive Materials

Glass (broken and pulverized), flint, emery (impure corundum mixed with iron oxide), garnet and mainly aluminum oxide and silicon carbide are used for coated abrasives[2, 4, 6, 11]. The quantity applied depends upon the type and size of the abrasive grain used and how densely it is scattered over the backing material, designated as open or closed coat. For open coat, spaces are left between the grains so that wet removal of the polished material is possible. Backings for coated abrasives are paper, fabrics, vulcanized fibers or combinations thereof.

Kraft paper can be used, but in most cases, special papers (70–220 g/m^2) which have already been modified during their production with rubber, acrylic resin and other polymers, are used to improve the flexibility, strength and water resistance. Anisotropic strength is preferred for belts. This can be achieved during the production of the paper by fiber orientation. Specially woven cotton or linen fabrics with high uniformity are used if materials with high flexibility and tensile strength are required. Very strong and tough vulcanized fiber produced from pure cotton paper made by the sulfuric acid- or zinc chloride process, is used particularly for abrasive discs.

Adhesives and Coatings

To an increasing extent animal glue, vegetable oils and natural resins are being displaced by synthetic resins, mainly in areas where water and temperature resistance is required. However, animal glue is still widely used today to bond flint, glass and garnet for dry grinding and when low pressures are applied. Compared to phenolic resins, aqueous animal glue requires only short drying times at low temperatus of 30–50 °C and no curing. Therefore, only relatively simple installations with low energy consumption are necessary. Since animal glue is a natural product, there are differences in quality and its availability is limited. Phenolic resins are superior to all other bonding agents especially in regard to water- and temperature resistance; they lead to higher grinding efficiency and accuracy. Heavy-duty industrial belts are manufactured with synthetic resins only. Alkyd-, polyurethane- or epoxy resins are also used for some applications. In some instances, animal glue and phenolic resin are used in combination.

The use of coated abrasive belts in high-pressure grinding applications is growing as a result of improvement in abrasive grain and backings. Presently both ferrous and non ferrous foundries are benefiting especially in the grinding of investment castings for the aircraft industry. Two major developments such as 40/60 zirconia-alumina grain and polyester backings have facilitated high-pressure grinding applications[14].

17.2.2 Coating Process

The backing material, available in rolls up to 2 m wide, is coated with the liquid resin in a roll coating machine and dried as shown schematically in Fig. 17.4. The quantity for the first coat, which is also called the maker coat, is approximately 100–400 g/m^2 and depends mainly upon the size of the abrasive grain.

Two processes are commonly used to apply the abrasive grain: gravity coating and electrostatic coating. With the older scattering process, the grain descends from a hopper onto the resin loaded web or is expelled against the web by a rotating brush as shown in Fig. 17.4. In the newer electrostatic deposition process, the grains are directed to the web from below against gravity by an electrical field. The grain should be fixed in the resin loaded backing vertical to the longitudinal axis. This is achieved more evenly by electrostatic deposition. Thus, such belts show higher grinding efficiency.

The loaded web is then transported into the festoon dryer where it is dried and only precured in the first drying section. The webs are placed in long loops over bars which move through the different temperature zones by a chain conveyor. By this arrangement it is easier to make adjustments to modify conditions than in horizontal dryers.

The web is then stretched and either fed to the next roll coater or back to the first coater. Here, a second binder coat is applied which is called the size coat, and the abrasive grain finally imbedded and fixed. The final drying and curing is performed in the main festoon dryer according to an exact program; above 80 °C the temperature is increased very slowly in order to prevent too fast curing of the surface and the formation of blisters. The reduction of viscosity at higher temperature is compensated by the increased cross-linking of the resin so that the abrasive grains do not migrate.

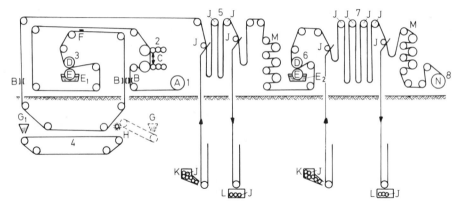

Fig. 17.5. The coating process. (Drawing: Norddeutsche Schleifmittel-Industrie, D-2000 Hamburg) *1* Take-off roll; *2* Printer; *3* Maker coater; *4* Grain coating machine; *5* First dryer; *6* Size; *7* Main festoon dryer; *8* Roll-up machine.
A Backing roll; *B* Thickness measuring instrument; *C* Printing roll; *D* Applicator roll; *E* Pick-up roll; *E1* Resin for the 1st (maker) coat; *E2* Resin for the 2nd (size) coat; *F* Smoothing brush; *G* Grain hopper, mechanical deposition; *G1* Grain hopper, electrostatic deposition; *H* Throwing brush; *J* Rotating poles; *K* Rotating poles supply; *L* Rotating poles deposit; *M* Tension rolls; *N* Product jumbo roll

The maximum temperature should not exceed 120–130 °C; higher temperatures would lead to embrittlement of the web. To adjust the normal moisture content of the web, reconditioning is performed in the last drying zone. Prior to final rolling, the abrasive belt is subjected to a flexing process, which yields the necessary flexibility by a multitude of fine cross and diagonal cracks without damaging the backing. The webs are then made into rolls, blades, sheets, rondels or endless ribbons.

17.2.3 Abrasive Papers

Abrasive papers for dry grinding are bonded with natural glues and synthetic resins. The grain is fixed to the backing with an aqueous animal glue as maker coat. After grain deposition, a large quantity of water is evaporated in about 30 min at 40–50 °C. The second or size coat is made with phenolic resins. They are liquid resins for which the curing rate can be adjusted according to the efficiency of the plant. Resins of medium reactivity and appropriate curing conditions are described in Table 17.7.

The resin for the size coat can contain up to 50% inorganic fillers such as chalk flour ($CaCO_3$). This results in high grinding efficiency, increased bond strength and reduced shrinkage. The fillers thicken the bonding agent so that sagging is prevented. Water and thermal resistance of a pure animal glue bond can be improved by addition of phenolic resins. To increase the shelf life of such mixtures, solvents and stabilizers can be added. The curing temperature is between 90 and 120 °C, depending upon the amount of phenolic resin added.

In high-performance grinding papers, the maker coat as well as the size coat consists of liquid, water miscible phenol resols. Since the liquid resins may deeply penetrate into the paper web and enbrittle it, latex containing papers are used, to which a barrier coat may be applied. The coat can be made of plasticized thermoset resin solutions

Table 17.7. Characterization and curing conditions of phenol resin[7]

Characterization of the resin:	phenol resol	
	solids content	78–79%
	viscosity at 20 °C	2,800–3,200 mPa · s
	gel time at 130 °C (plate)	7–9 min
	gel time at 100 °C	58–68 min
Curing conditions:	heating up time to 75–80 °C	1/2 hour
	holding time at 75–80 °C	1 hour
	holding time at 90–95 °C	2 hours
	holding time at 105 °C	1 hour
	holding time at 120 °C	1 hour
	total time	5 1/2 hours

or thermoplastic polymer dispersions to prevent excessive impregnation. The same or different phenolic resins may be used for the maker and the size coat. In general, the viscosity of the size coat is lower than of the maker coat. For abrasive papers to be used for wet grinding of car body coatings, they must have high flexibility and elasticity besides good water resistance. Latex papers or papers modified with acrylic resins or PVC are often used instead of kraft papers. Alkyd resins are preferred for both coatings which are not always satisfactory as far as strength and grinding efficiency are concerned. Phenolic resins give high hardness and good grinding efficiency, but the papers are quite brittle. Better results are obtained by the use of epoxy resins and polyurethane resins which result in a very hard, elastic and resilient coat. Additional technical effort is required for the development of wet grinding materials, and it appears that alkylphenol resins are yet another alternative.

17.2.4 Abrasive Tissues

Cotton fabrics with an area weight of 350 g/m^2 are used as backing for coarse granulation with high requirements for strength. Lower area weights of between 230–290 g/m^2 are adequate for lower requirements and fine granulation. The grain is fixed with animal glue for dry grinding and after drying, coated with liquid phenolic resins. Fast curing phenol resols are used to shorten the curing cycle and to reduce the thermal stress of the coated web.

A water resistant size is required for fabrics which can be used for wet grinding of glass, stone, synthetic materials, etc. The finish consists of a PVA-dispersion and a low molecular weight, water dilutable phenol resol. The finished fabric should be combined with a barrier coat prior to the application of the maker coat, to prevent resin penetration and embrittlement of the web. Oil-modified alkyd resins or epoxy resins are suitable for the maker and size coat.

17.2.5 Vulcanized Fiber Abrasives

Vulcanized fiber wheels are mainly used for grinding of machine and automobile bodies due to their high grinding efficiency. The wheels usually have a size of 180–

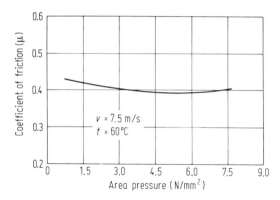

Fig. 18.3. Dependence of coefficient of friction μ with specific pressure[7]

bonded disc pad are shown in the following figures. The reduction of the friction level with increased temperature is called temperature sensitivity. Fading is the increase of pedal pressure during a certain number of consecutive stops, necessary to keep the same deceleration.

Abrasion resistance determines the service life of the lining. If abrasion is high, the brake drum may be affected by the dust debris and leads to erratic braking and often to drum scoring. The abrasion resistance is evaluated by determining the weight loss and thickness of the lining. Another important factor is the water absorption and the influence of moisture on friction performance.

18.1 Formulation of Friction Materials

Generally friction materials consist of a fibrous thermally stable reinforcing material, various metallic or non-metallic powdered or particulate solid fillers, friction modifiers, and a resin to bind the composition. A comprehensive survey concerning brake and clutch linings shows the large variation of material combinations[8]. In most cases the formulation contains up to 20 components (Table 18.1). The exact composition and the process are generally proprietary.

The requirements and composition vary considerably due to the national speed limits and different lining dimensions according to the brake design. In the USA, because of the 55 MPH speed limit, the temperature resistance requirements are not as severe as for the heavy-duty linings in Europe. In general, in the USA the brake linings contain more rubber modified phenolic resin or rubber/phenolic blend than elsewhere.

Fibers

Presently chrysotile asbestos[10] exists as the main fiber component of friction linings. Fibers of different lengths, yarns and fabrics are used. The asbestos fiber provides strength and thermal resistance to the brake combined with relatively low abrasiveness. The physiological hazards[11, 12] and carcinogenic behaviour of asbestos are accelerating the substitution of asbestos in friction elements by other materials.

Table 18.1. Composition of mixtures for friction linings[9]

Fibers	30–50%
Bonding agent	
Phenol- and/or cresol resin	10–15%
Rubber	2–10%
Organic fillers	
Friction dust	0–10%
Inorganic fillers	
Barite	10–20%
Calcium oxide/magnesium oxide	0– 5%
China clay/slate flour	0–10%
Aluminum oxide	0– 2%
Metals	
Powder or shavings	5–25%
Others	
Graphite	1– 5%
Antimony sulfide	0– 5%
Molybdenum disulfide	0– 5%

All asbestos when inhaled in large amounts causes destruction of lung tissue and the development of scars that result in insufficient ventilation of the lungs, decreased oxygenation of blood, pulmonary hypertension and right heart failure. These lung changes comprise the disease "asbestosis". If severe, it is a debilitating and often fatal disease. All four asbestos types – chrysotile, crocidolite, amosite, and anthophyllite – are associated with development of lung cancer (mesothelioma) because of the tendency of the small hollow fibrils (6–13 μm) to attach themselves to lung tissue. OSHA has ruled that not more than two fibers longer than 5 μm per 1 cm^3 of air should be present in the workplace. The ruling is restricted to fibers longer than 5 μm since shorter fibers cannot be accurately counted with a light microscope. A proposed lowering of the present 2 fibers/cm^3 of air to either 0.5 or 0.2 fiber/cm^3 has been suggested by OSHA. The agency estimates that lowering the permissable exposure level will reduce the risk of death for workers exposed to asbestos at least 75%[13].

Mineral fibers, carbon fibers, aramid fibers, metal fibers and combinations thereof are recommended as asbestos replacements in brake linings[14, 15]. Many brake linings manufacturers already sell satisfactory non-asbestos linings in retrofit and original equipment products. In 1983 DuPont commercialized a new aromatic amide short fiber (Kevlar) to replace asbestos for friction applications. Kevlar pulp does not pose the health problems of asbestos and offers the same strength and twice the wear life at a 5% loading as is obtained with a 50% asbestos loading. The semi-metallic[16] non-asbestos linings contain 60–70% metal (Fe), 15–25% graphite as well as 7–10% phenolic resin besides fillers and additives.

Carbon fibers in a carbon matrix (Sect. 10.1) are recommended for the manufacture of aircraft brake linings[17, 18] because of the outstanding performance of the material as a result of low wear of carbon, satisfactory thermal conductivity, high fiber strength and low density.

Fillers

A series of very different inorganic and organic materials are used to improve the friction performance, thermal conductivity, abrasion resistance and strength of the fabricated friction element[8, 19]. Broad compositional variation is shown in Table 18.1. Calcium or magnesium oxide act as cure accelerators for phenolic resins. Metals like iron, nickel, magnesium, copper, brass and zinc in the form of powder or shavings are added to improve the thermal conductivity. Graphite and molybdenum sulfide are used as lubricants and wear modifiers.

The reaction product of cashew nut shell liquid (CNSL, Chap. 2.1.7) with formaldehyde leads to a completely crosslinked polymer designated as "friction dust". It is a common ingredient of friction lining formulations[20, 21]. Friction dust represents 5–15% by weight of the friction element and provides a reduction in frictional wear of the element, moderately uniform coefficient of friction over a broad temperature range and soft pedal action. There are also disadvantages related to friction dust such as increasing cost of the cashew oil, skin irritation and dustiness during the grinding process. The latter sometimes becomes quite hazardous due to smoldering of the dust. Therefore synthetic pathways have been examined to prepare a totally synthetic friction dust prototype. A new phenolic dispersion process has been reported to be a successful route to products whose performance is comparable to CNSL dust.

The two step HMTA catalyzed P/F dispersion process is conducted initially as a resin in water dispersion followed by crosslinking. The product is recovered as discrete particles. The process allows optional introduction of nitrile rubber to the phenolic dispersion at the beginning or upon completion. Thus the rubber phase can be present within the emerging PF condensation phase or as a discrete rubber phase on the surface of the PF resin particle[22]. Friction elements formulated with AN rubber based PF dispersion particles exhibited comparable wear and coefficient of friction. Particulate form friction dust can also be obtained by the HMTA catalyzed condensation of phenol and substituted phenols with formaldehyde in presence of selective dispersing agents[23].

Resins

A wide spectrum of phenolic resins is used as binders for friction materials. Phenol type resins such as novolaks, resols, novolak/resol blends, oil modified novolaks, solubilized oil-phenol types, cresol resins and rubber or thermoplastic modified novolaks.

Oil modification of phenol resins is conducted with CNSL, tung, linseed or soya bean oils, and is reported to yield quieter, softer braking action. These resins are very useful in post forming processes. The latter method requires some flexibility to form the lining without cracking.

Nitrile rubber or other synthetic or natural rubbers are excellent PF resin modifiers introduced as a separate resin phase during PF resin manufacture or as a separate resin component in the formulation (Table 18.1). Rubber modified PF resin bonded disc pads exhibit improved coefficient of friction over an extended temperature range (Fig. 18.2) and pressure (Fig. 18.3). Friction elements comprised of unmodified novolak resins experience "fade" or loss of coefficient of friction at elevated temperature.

Resins of different consistency and reactivity are applied as required for each individual manufacturing process or individual manufacturer. Liquid phenol- and cresol

Table 18.2. Phenol powder resins for friction linings[9]

Melting range	80–100 °C
Gel time at 150 °C (plate)	60–150 s
Flow distance	15– 55 mm

resols, catalyzed with sodium hydroxide or ammonia, may contain small quantities of alcoholic solvents to reduce the viscosity. Phenol novolaks, which can be modified with NBR are dissolved in a mixture of acetone, spirit and toluene. This solution also contains HMTA.

Solid resins in fine, free flowing powder form, blended with HMTA, are used in most instances (Table 18.2). The HMTA level is generally between 8–13%.

Solid cresol resols have in general a lower melting range of 60–70 °C and a gel time of 5–8 min at 130 °C. The production of higher melting resols is not possible because of their high reactivity. Pulverized resol resins sinter on storage; therefore, these resins should be pulverized by the customer shortly before use to ensure free flowing mixtures.

High-temperature phenolics based on naphthalene, phenol, and formaldehyde have been described for friction products[24]. The HMTA cured modified novolak resin exhibited favourable friction properties at temperatures close to 500 °C.

Characterization and Testing Procedures

The macroscopic composition, including grain size, grain distribution and to some extent filler identification is performed by electron microscopy[5] (EMA). Also light micrographs are very useful for these purposes. Both methods are used to study structural changes in the lining during frictional use. Examination of a friction lining after extended use shows a thin carbonized layer (approx. 1 mm) on the surface lining and results during the normal braking operation. This layer provides a favourable change in thermodynamic properties i.e. heat transfer, thermal conductivity, etc. of the friction composite[5]. High-carbon yield phenolics are therefore required (Chap. 10.1).

DSC and TGA studies of frictional composites at temperatures between 600 and 650 °C indicate chrysotile asbestos is transformed into an amorphous structure with a weight loss of 13%[25]. Phenolic binder decomposition and weight loss attributable to carbonization[26, 27] is about 5% at temperatures above 550 °C.

The wear of a friction lining therefore consists of two distinct phenomena, abrasion and thermocarbonization. Other physical properties which are important for the thermo-mechanical performance[28, 29] are modulus of elasticity (via compressive strength), coefficient of thermal expansion, thermal conductivity, heat capacity and specific gravity. These properties should be evaluated over a wide temperature range.

Laboratory tests for evaluation of various components within a friction composition are conducted by constant input and constant output testing method SAE J-661 "Brake Lining Quality Control Testing Procedure (1979)", utilizing a dynamometer or a standard Chase type friction testing machine. These tests provide coefficient of friction under normal and elevated temperature conditions as well as sample wear.

Table 18.3. Molding conditions for production of friction elements

Process	Impregnation	Wet mix	Dry mix
Temperature °C	150–175	150–175	160–180
Molding time, min/mm thickness	0.5–1	0.5–1	0.4–0.6
Molding pressure, N/mm^2	15–20	15–30	15–50

Actual use test involves the evaluation of friction materials in a vehicle as described in Federal Motor Vehicle Safety Standard No. 105-75 or FMVSS-105-75. Testing in the heating mode is known as the "Fade" test. The cooling mode is known as the "Recovery" test. Wear is also measured as the difference in thickness of the friction element from the start to the conclusion of the test.

Manufacturing of Brake- and Clutch Linings

The following processes are the most important amongst those used for manufacturing of resin bonded friction materials[2, 4, 8].
1. Impregnation process for fabric- or yarn linings for drum brakes and clutch facings;
2. Wet mix "dough" process for brake linings;
3. Dry mix process for brake linings.

These processes differ from each other by the nature of the bonding agent which may be added as a solution or in solid form. The curing conditions are shown in Table 18.3, provided the cure takes place by pressing.

18.2 References

1. Lancaster, J. K.: Plastics and Polymers, Dec. 1973, 297
2. Jacko, M. G., Rhee, S. K.: "Brake Linings and Clutch Facings". In: Kirk-Othmer: Encyclopedia of Chemical Technology, Vol. 4, P. 202, J. Wiley & Sons, N.Y. 1978
3. Hodge, J. C.: Plastics and Polymers, Feb. 1974, 27
4. Barth, B. P.: "Phenolic Resin Adhesives-Friction", Chap. 23 in: Handbook of Adhesives, I. Skeist, ed. van Norstand, N.Y. 1977
5. Cordes, H.: „Produktentwicklung am Beispiel von Bremsbelägen", Dissertation TH Aachen, 1983
6. Tanaka, K., Veda, S., Nogodu, N.: "Fundamental Studies on the Brake Friction of Resin-Based Friction Materials", Wear, 23, 349 (1973)
7. Textar GmbH: Technical Bulletin
8. Bohmhammel, H.: „Entwicklung von Reibbelägen für Kupplungen und Bremsen", Gummi, Asbest u. Kunststoffe: 11, 924 (1973); 12, 1063 (1973); 1, 34 (1974); 3, 183 (1974); 5, 370 (1974); 7, 524 (1974); 9, 738 (1974); 11, 926 (1974)
9. Bakelite GmbH: Technical Bulletin
10. Noll, W.: Asbest. In: Ullmanns Encyclopädie d. techn. Chem., Vol. 8, 4th Ed., Weinheim: Verlag Chemie 1976
11. Heidermanns, C., Kuhnen, C., Schutz, A., Prochaska, R.: Staub-Reinhaltung Luft 35, 433 (1975)
12. Gross, P., Braun, D. C.: Chem. Tech. 436 (1980)

13. N. N.: Chemical and Engineering News, April 1984, P. 9
14. Rütgerswerke AG: US-PS 4,373,038 (1983); US-PS 4,384,054 (1983)
15. Ferodo Ltd.: US-PS 4,273,699 (1981)
16. Liu, T., Rhee, S. K.: Wear 76, 213 (1978)
17. Ho, T. L., Kennedy, F. E., Peterson, M. B.: "Evaluation of Aircraft Brake Materials", American Society of Lubrication Engineers 22, 71 (1979)
18. Bill, R. C.: "Friction and Wear of Carbon-Graphite Materials for High Energy Brakes", American Society of Lubrication Engineers 21, 268 (1978)
19. Manhattan Rubber: US-PS 2,025,951
    Ferodo Ltd.: FR-PS 722209
    Union Carbide Corp.: US-PS 1,416,492
20. BP Chemicals: Cellobond Phenolic Resins, Technical Bulletin
21. Weintraub, M. H., Anderson, A. E., Gealev, R. L.: Polymer Science Technology 5 B, 623 (1974)
22. Union Carbide Corp.: US-PS 4,316,827 (1982)
23. Union Carbide Corp.: US-PS 4,420,571 (1983)
24. 3M: US-PS 4,395,498 (1983)
25. Waldrom, G. W. J.: Institution of Mechanical Engineers, Conference Publication 1979 − 11, P. 125
26. Back, L. S., Moran, D., Percival, S. J.: "Polymer Changes During Friction Material Performance", Wear 41, 309 (1977)
27. Jacko, M. G.: "Physical and Chemical Changes of Organic Disc Pads in Service", Wear 46, 163 (1978)
28. Dow, T. A.: "Thermoelastic Effects in Brakes", Wear 59, 213 (1980)
29. Lagedrost, J. F., Eldrige, E. A.: "Thermal Property Measurements in Brake Shoe Materials", Institution of Mechanical Engineers, Conference Publications, 1979 − 11, P. 111

19 Phenolic Resins in Rubbers and Adhesives

Generally, phenolic resins exhibit adhesive characteristics in virtually every type of application[1, 2]. They are always used in conjunction with fillers or reinforcing fibers. Their adhesion to most materials is very good due to the marked polarity of the phenolic structure. Disadvantages, however, are their brittleness in non-modified composition combined with high cure temperature and pressure requirements.

19.1 Mechanisms of Rubber Vulcanization with Phenolic Resins

The remarkable improvement in chemical and physical properties of rubber resulting from sulfur vulcanization[3] was discovered jointly by Goodyear and Hancock in the early 1840's. About a hundred years later, Wildschut found that natural and synthetic rubbers could be vulcanized with selective phenolic resins without any additives[4]. The advantage of phenolics in comparison to sulfur vulcanization is higher strength, heat aging resistance, low compression set and improved chemical resistance. However, the economic significance of phenolic resin vulcanization is marginal. Butyl rubber, which is difficult to vulcanize conventionally because of its low unsaturation and easy reversion, can be favorably crosslinked with phenolic resins.

Resol resins (general formula 19.1) are required for crosslinking. The *para* position must be substituted (R = tert-butyl-, octyl- or nonyl group), otherwise the resin condensation reaction would be much faster than the desired resin-rubber interaction. Appropriate resins with MP's between 55–65 °C, added in the range of between 5–12% depending on rubber type and crosslink density desired, are mostly based on *p*-octylphenol. Activators are used to enhance the vulcanization rate, e.g. pyromellitic anhydride, phthalic anhydride or fumaric acid for NBR, toluene sulfonic acid or chloroacetic acid for SBR, $Sn\ Cl_2 \cdot 2\ H_2O$ or ZnO in combination with chlorine ions containing compounds like Hypalon or Neoprene for BR. A second variety of resins is halomethylated ($-CH_2X$ instead of $-CH_2OH$; 1–2% chlorine or bromine content) and requires therefore no additional activation.

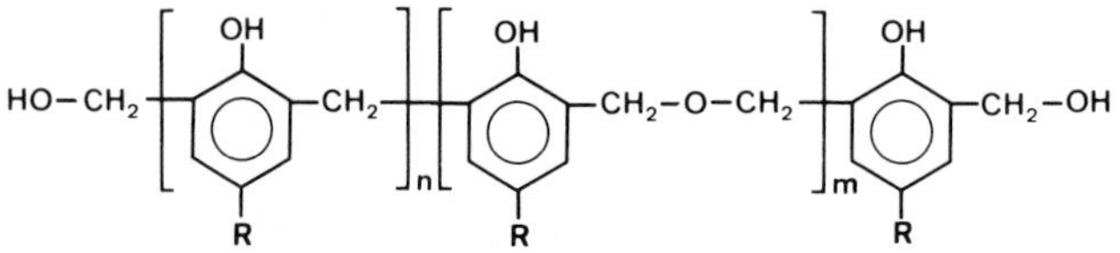

$$HO-CH_2 \left[\begin{array}{c} OH \\ \end{array} -CH_2 \right]_n \left[HO \begin{array}{c} OH \\ \end{array} -CH_2-O-CH_2 \right]_m \begin{array}{c} OH \\ \end{array} -CH_2-OH \qquad (19.1)$$

There are different interpretations regarding the reaction mechanism of rubber crosslinking with phenolics. Hultzsch[5] suggested that the reaction may occur (analogous to drying oils) through quinone methides as intermediates followed by

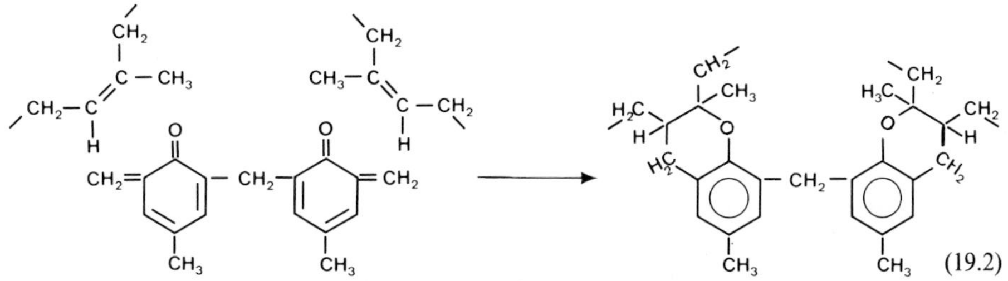

$$(19.2)$$

chroman ring formation (19.2) (see Sect. 3.3.5). As a result of the proposed mechanism dinuclear resols are required. However, the fact that crosslinking can be accomplished with mononuclear phenol compounds casts some doubt on this reaction mechanism.

Van der Meer[6] also proposed quinone methides as intermediates with crosslinking occuring through hydrogen abstraction from the methylene group in the allyl-

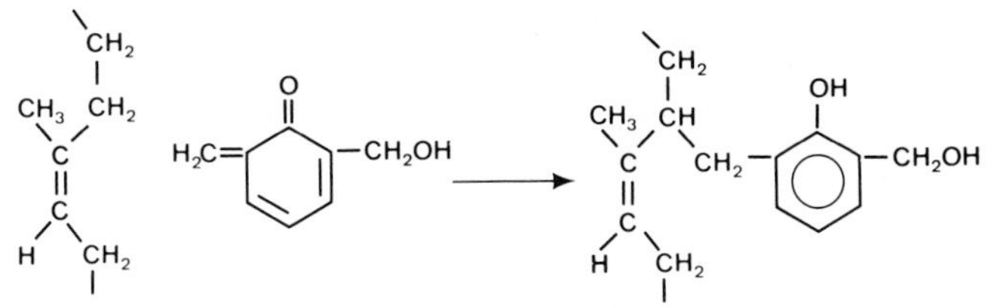

$$(19.3)$$

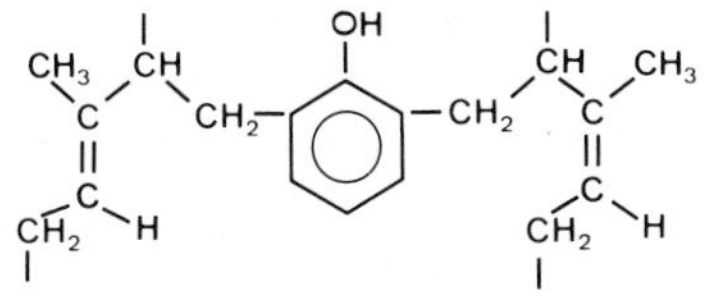

$$(19.4)$$

position of the rubber (19.3). After rearrangement to the aromatic structure, a second quinone methide is formed leading to the mononuclear crosslink (19.4).

The fact, that the rate of vulcanization is greatly increased by addition of acidic catalysts is interpreted by Giller[7] as favouring an ionic chain reaction mechanism. Metal halides are formed, even in small quantities, by the interaction of basic

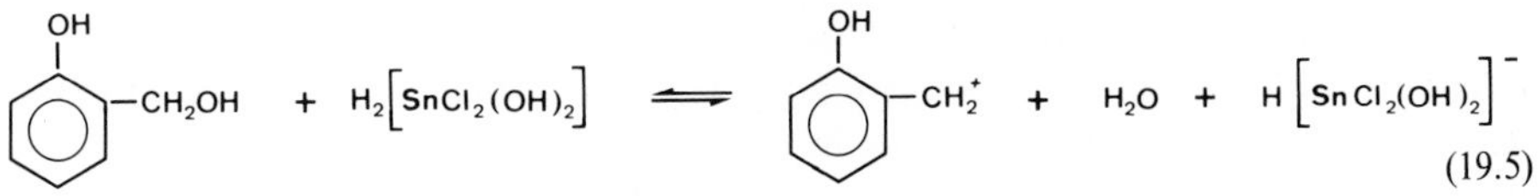

$$(19.5)$$

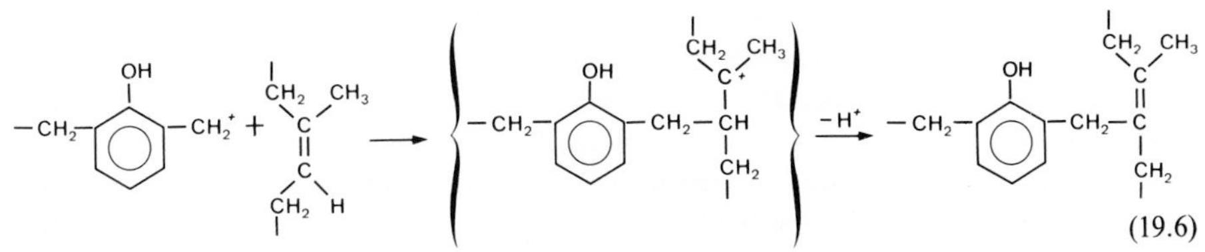

$$(19.6)$$

oxides and halogenated polymers, which act as Lewis acid-type catalysts, via carbonium ion formation (19.5). The dibenzyl ether bridge is also cleaved by the action of acidic catalysts. Based on existing facts, the phenolic vulcanization is best explained by the ionic chain mechanism as shown in equation (19.6). Additional crosslinking reactions may occur with synthetic rubbers depending on their functionality.

19.2 Thermosetting Alloy Adhesives

Phenolic resins are combined with thermoplastic polymers in structural adhesives with the phenolic resin providing metal/metal and rubber adhesion. The characteristic properties of these combinations, commonly referred to as alloys, can be utilized and specific requirements met by the ratio of the resins. As the soft morphological phase, polyvinylacetals, NBR, polyamides and polyacrylates are preferred. The elongation, elasticity and resiliency of the phenolic resin is improved considerably, especially at low temperatures. The toughness of such polymer blends is attributed to chemical reaction, but even more to the morphology of the cured system. Because of the limited solubility of the thermoplastic or elastomeric component, a discrete, well dispersed, discontinuous phase develops in the cured phenolic matrix. The crack propagation is considerably minimized by the two-phase system and the impact strength therefore improved (see Chap. 10.8).

19.2.1 Vinyl-Phenolic Structural Adhesives

Polyvinylacetal-phenolic resin blends are quite frequently used as structural metal adhesives[8]. The polyvinylacetal serves as a coreactant, flexibilizing agent and flow control component.

They are used in the aircraft industry, as copper adhesives for printed circuits, to bind brake lining to brake shoes, for various honeycomb constructions, curtain walls, and ski manufacturing.

High temperature shear strength is higher for polyvinylformal (PVF)/phenol blends than for the polyvinylbutyral (PVB)/phenol combination, but the latter system yields higher peel strength[9]. Cure conditions require a temperature between 140 and 180 °C and pressure of 0.3 to 1.0 N/mm². By addition of accelerators, like resorcinol-formaldehyde resins, the temperature can be reduced to approximately 115 °C; however, some loss in strength will occur.

The ratio of phenol/acetal resin can range from 0.3:1 to 2:1, depending upon the desired elastic modulus, tensile strength, creep and temperature resistance. High molecular weight is preferred for the thermoplastic component; however, sufficient fusion and wetting during the curing reaction must be developed to ensure high adhesive and cohesive strength.

For phenol/PVB alloys alcohols may be used as solvents, while the PVF component can be dissolved only in special solvent combinations, such as toluene/ethanol, ethylene dichloride/MEK/ethanol or diacetone alcohol/ethylene dichloride/cellosolve acetate. Copper adhesives based on phenolic resins/PVB or acrylic resin blends are used in the manufacture of phenolic-paper laminates for printed circuits. High peel strength, dielectric properties, and solvent- and blister resistance when in contact with tin solder at 260 °C are the critical requirements for this application. The laminates according to NEMA FR-2 and XXX-PC are cured and concurrently bonded with 35 μm copper foil at 160 °C and 10 N/mm² pressure for approximately one hour. The adhesive coat consists of between 25 and 40 g/m² (see Sect. 14.1).

In aircraft applications, high temperature shear strength, fatigue resistance, creep, oil and gasoline resistance are most important. Polyvinylacetal PF adhesives are now qualified under tests of MIL-A-25463A Type II (150 °C).

The high viscosity of polyacetal/phenol resin solutions and the necessity of volatilizing large amounts of solvents is very inconvenient in some cases. If so, either the metal substrate is coated first with the liquid phenol resol and then sprinkled with powdered acetal resin after airing. Prefabricated adhesive foils or tapes can be used. The use of adhesive foils offers a series of advantages. Taking a low weight carrier (25–65 g/m²) of fabric or non-woven, a uniform joint thickness is obtained. Here too, the carrier is first impregnated with the resol solution and then the powdered acetal resin is sprinkled on the tacky prepreg.

The phenolic resins for this application are resols catalyzed by sodium hydroxide. The hydroxymethyl group can react with the hydroxyl group of the polyvinylacetal resin. Consequently, the hydroxyl and acetal content are important factors in the selection of the polyvinylacetal component. Furthermore, their particle size distribution is important with regard to the manufacture and application of the adhesive.

19.2.2 Nitrile-Phenolic Structural Adhesives

NBR[10] in conjunction with its high compatibility, reacts more readily with phenolic resins than polymers which are only unsaturated in the polymer chain (19.7). Acrylonitrile (AN) content can vary from 15 to 40% AN; higher AN content favours better phenolic compatibility.

$$\left[\begin{matrix} H & H & H & H \\ | & | & | & | \\ -C & -C & =C & -C- \\ | & & & | \\ H & & & H \end{matrix}\right]_n \left[\begin{matrix} H & H \\ | & | \\ -C & -C- \\ | & | \\ H & C\equiv N \end{matrix}\right]_m \qquad (19.7)$$

Customary trade names for such elastomers are: Perbunan (Bayer AG), Hycar (Goodrich Chemical Co.) and Breon (British Geon Ltd.). Some types contain methacrylic acid as comonomer.

Adhesive bonding of metal and plastic parts by NBR/phenolic alloys has been widely used. A high temperature liquid phenolic resin/NBR adhesive formulation is given in Table 19.1.

The use of 1% silane based on combined weight of NBR/novolak significantly improves Al to Al lap shear strength[11].

The strong reinforcement of NBR and possibly of carboxylated NBR-terpolymers suggests that further reactions between the nitrile and carboxyl group and the methylol group may occur (19.8) and (19.9). NBR-phenolic cements are used for metal-to-metal and rubber-to-metal bonding. For some applications a substitute primer with chlorinated rubber or polyurethane is necessary. Zinc oxide, carbon black, iron oxide, and sulfur are common additives. Magnesium oxide is preferred and is a more active

Table 19.1. NBR/phenolic structural adhesive composition[10]

NBR (>30%, AN)	100
Novolak resin	75 −250
ZnO	5
Sulfur	1 − 3
Accelerator	0.5− 1
Aging inhibitor	0 − 5
Stearic acid	0 − 1
Carbon black	0 − 20
Filler	0 −100
Plasticizer	0 − 10

curing agent for carboxylated NBR than zinc oxide. Compared to vinylphenolic

$$\text{HC}-\text{C}\equiv\text{N} + \text{HOCH}_2-\bigcirc-\text{CH}_2\text{OH} \longrightarrow \text{HC}-\underset{\text{NH}}{\overset{||}{\text{C}}}-\text{O}-\text{CH}_2-\bigcirc-\text{CH}_2\text{OH} \qquad (19.8)$$

$$\text{HC}-\text{C}\overset{O}{\underset{\text{OH}}{\diagup}} + \text{HOCH}_2-\bigcirc-\text{CH}_2\text{OH} \longrightarrow \text{HC}-\overset{O}{\overset{||}{\text{C}}}-\text{O}-\text{CH}_2-\bigcirc-\text{CH}_2\text{OH} \qquad (19.9)$$

adhesives, NBR-phenolic blends exhibit higher peel strength and impact resistance. Humidity, oil, and salt spray resistance is very good; the poor fileting characteristic[9] is a disadvantage.

19.3 Phenolic Resins in Contact Adhesives

Contact adhesives are adhesives which are applied to two substrates, dried and combined under sufficient pressure to result in good contact. The bond is immediate and sufficiently strong to hold components together without further clamping, pressing or air drying.

Adhesives based on polychloroprene[12] (neoprene) and NBR[10] exhibit high green strength and excellent adhesion to various substrates. The strength, temperature and creep resistance can be improved and adhesive cost reduced by the addition of phenolic resins. They are used in shoe manufacturing (to bind leather, cloth, plastic and rubber), in the automotive- (interior upholstery), furniture- and construction industries. Chloroprene adhesives yield high peel strength and have outstanding green strength properties. Nitrile adhesives exhibit excellent oil and grease resistance. Heat and metal oxide reactive phenolic resins according to formula (19.1) are used (see Sect. 3.4.3). p-tert-butyl phenol resins offer the best hot cohesive strength and form a single phase system with CR.

19.3.1 Chloroprene-Phenolic Contact Adhesives

The polymerization of 2-chlorobutadiene may occur via 1,4- as well as 1,2-addition [Formula (19.10)]. The allyl chlorine in the case of 1,2-addition is far more reactive and can easily be displaced. Commercially available CR's are Neoprene (Du Pont) and Baypren (Bayer). Some CR-types exhibit crystallization behaviour because of stereochemical reasons depending also on the ratio of *cis*- and *trans*-configuration. A *trans*-1,4-configuration increases crystallization whereas the other configurations *cis*-1,4, 1,2, and 3,4 decrease crystallization. Crystallization is a reversible process, which can be reverted by increased temperature or dynamic stress.

$$
\begin{array}{ccc}
\underset{\text{Cl}}{\overset{\text{Cl}}{-CH_2-C=CH-CH_2-}} & \underset{\underset{\text{CH}=CH_2}{|}}{\overset{\text{Cl}}{-CH_2-C-}} & \underset{\underset{\underset{\text{CH}_2}{||}}{\overset{|}{C-Cl}}}{-CH_2-CH-}
\end{array}
\tag{19.10}
$$

1,4-polychlorobutadiene   1,2-polychlorobutadiene   3,4-polychlorobutadiene
(cis and trans)

Thin films, prepared from such elastomer solutions, show a high contact adhesive and a high cohesive strength due to rapid crystallization. This property makes them attractive for use as contact adhesives[13]. The crystallization, however, is lost at temperatures above 60–70 °C. Temperature and solvent resistance can be considerably improved by additional crosslinking with phenol resins, which also improve tack and adhesion. Thermal resistance is enhanced further by the addition of calcium-, zinc- and cadmium oxide. Experience has shown that magnesium oxide is most convenient and provides maximum heat resistance, and contributes favorably to the stability of the formulation by acting as hydrochloric acid acceptors[14]. Heat aging resistance is enhanced through addition of common antioxidants.

By increasing the amount of resin the heat resistance is improved, however, the elongation will be reduced with corresponding brittle adhesive layer, as well as tack, and flexibility. A 40–45% resin level seems to be optimum.

CR-types with a medium or high crystallization behaviour are preferred. CR is soluble in aromatic chlorinated hydrocarbons, in some esters, ketones, and also in mixtures of certain solvents. The metal oxide-prereacted phenolic resin is also soluble in these solvents. An important feature of a contact adhesive is the relationship between open time and green strength. For an ideal system a rapid build-up of green strength and long tack retention is desired. Normally, the green strength increases with open time to a maximum value and then begins to decrease. The dried bond strength is greater at shorter open times. The presence of the phenolic resin influences the open time and peel strength to a great extent. Methylol groups and dimethylene ether bridges are the important constituents which must be contrasted with the crystallization behaviour of the CR. The higher the proportion of hydroxyl- and dimethylene ether groups, the shorter the open time; peel strength and elevated temperature resistance will be increased considerably. This relationship occurs if the rubber shows a medium crystallization rate. If the crystallization rate is high, non- or less base reactive phenolic resins are preferred[15].

These room temperature curing adhesives (Table 19.2) are formulated in two parts. Neoprene AC is masticated on a two-roll mill to improve solubility; the aging inhibitor is introduced, and finally zinc oxide is added. The temperature of the rubber should not exceed 60–80 °C. The homogenized rolled sheet is then cut in sections and dissolved in the solvent mixture. For neoprene contact adhesives solvent systems have

Table 19.2. Guide formulation for a neoprene-
phenolic contact adhesive[17]

Polychloroprene (Neoprene AC)	100
Phenolic resin (Bakelite KA 773)	40
Magnesium oxide	5
Zinc oxide	4
Antioxidant (Neozone A)	2
Toluene	150
Ethyl acetate	150
White spirit	150

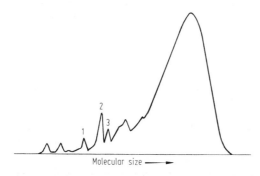

Molecular size ⟶

Fig. 19.1. Molecular weight
distribution (GPC) of a p-tert-
butylphenol resin for contact
adhesives.
$1 = p$-tert-butylphenol
$2 =$ hydroxymethyl p-tert-
butylphenol
$3 =$ bis(hydroxymethyl)p-tert-butyl-
phenol

been developed by considering solubility parameters, hydrogen bonding index and
plotted to identify best solvent combination. For the second portion the alkyl phenol
resin is prereacted with magnesium oxide in toluene by addition of minor quantities
of water (2%). Maintaining a certain temperature range (25–30 °C) and reaction time
(5–15 h) are important for the adhesive quality. A slurry process can be used for those
compounders without milling equipment. The solubility of the rubber can be improved by adding a part of the solid resin during the mastication. Occasionally floc-
culation or settling of the metal oxides and filler particles and phase separation may
occur depending on the type of neoprene, resin and solvent system. The MWD of the
phenolic resins affects adhesive stability; low molecular weight portions are detrimental[16] and cause phase separation. Evidence[12] has been presented that dialcohols of
t-butyl phenol are responsible for phase separation. MWD of a satisfactory p-tert-
butylphenol resol resin is shown in the Fig. 19.1.

The adhesive application method for bonding materials used for shoe manufacturing is described briefly. The strips to be adhered are first buffed with emery
paper. The adhesive coating is applied with a brush to both sides, and the solvents
volatilized. Bonding can be affected by the presence of small amounts of residual
solvents or by activation with hot air or by IR radiation. The activation temperature is
approximately 90 °C. The strips are then combined under pressure of 0.3–0.7 N/mm²
for 30 s. Several days are required for the development of maximum bond
strength.

Water-borne neoprene adhesives have been commercialized to avoid air pollution,
fire explosion and toxicity hazards of solvent systems[12]. Major uses consist of
packaging, carpet installation, construction mastics and general purpose industrial
and consumer contact adhesives.

Phenolic resins based on tert-butylphenol can be emulsified and blended with neoprene latex[18, 19]. Neoprene latex has a higher molecular weight than bulk polymers which must be handled in solution; therefore, better strength retention at elevated temperatures can be expected, and the use of metal oxides can be omitted in some cases. MgO cannot be used in aqueous dispersions because it leads to coagulation. Newer water-borne contact adhesives can be formulated with either neoprene or acrylic latex and dispersion phenolic resin[20] (Chap. 5). The resulting water based contact adhesives/neoprene or acrylic phenolic blend exhibit a good balance of contact properties on canvas, metal, wood and high pressure laminates.

The use of CR-adhesives in the shoe industry[21] has levelled although this industry still represents the largest market. The bonding of oil treated uppers and especially of PVC materials is troublesome. Further disadvantages are the poor resistance in the presence of plasticisers (PVC), poor light resistance and the very complicated adhesive formulation. They are being replaced by polyurethanes to some extent and therefore, the market for alkyl phenol resins in adhesives is decreasing.

19.3.2 Nitrile-Phenolic Contact Adhesives

NBR[10] adhesives lack some of the disadvantages which limit the application of CR adhesives and are considered as alternatives to urethanes. They offer good adhesion to PVC and good resistance against plasticisers. The oil and white spirit resistance is clearly higher depending on the nitrile content. The open time and the adhesion to sole rubber is, however, not always adequate. Raw material costs are higher than for neoprene/phenolic cements. The phenolic resin addition improves tack, solvent release and hot strength.

The adhesive formulation is much simpler (Table 19.3) as compared to a neoprene type adhesive.

Table 19.3. NBR-phenolic contact adhesive formulation[17]

NBR	(Perbunan 3810)	100
Phenol resol	(Bakelite KP 785)	40
Magnesium oxide		5
Methylethyl ketone		450

19.4 Phenolic Resins in Pressure-Sensitive Adhesives

Pressure-sensitive adhesives are permanent tacky materials which upon light pressure develop adhesive characteristics when in contact with a second substrate. In contrast, the tack of contact adhesives is limited in time. Generally, an adhesive tape consists of a backing, tie coat, pressure-sensitive adhesive and in some cases a release liner. Backings of polyester or polyethylene foils possess poor adhesive properties so that the application of a coupling layer is critical. For soft PVC plasticized sheets the use of primer is necessary as a barrier to prevent migration of the plasticiser, because natural and neoprene rubber are susceptible to them. In general, the adhesive consists of a base resin, tackifiers, plasticisers, fillers and stabilisers. It can be applied as an emulsion or latex, solution or hot melt. While adhesives containing solvents are prevalent, hot melts offer favorable economic advantages, i.e. ease of application, safety, labour and time saving features. Disadvantages are limited temperature range and low strength[22].

In most cases, the base resin is rubber (SBR, BR, IR or EPDM), but different thermoplastic polymers, for example, vinyl polymers, acrylates or urethanes, can also be used. For optimum adhesive function, the key properties of adhesion and tack, must be balanced. Incremental addition of a tackifier to the base resin increases tack to a maximum and then with further addition the tack decreases. In this first stage a continuous phase of the base resin saturated with the tackifying resin is formed. By further addition, a second discrete phase with appropriate distribution of both polymers among each other is formed. Finally, a coat of tackifying resin forms at the surface.

Resol, novolak or substituted phenolic resins may be used to build tack within hydrocarbon resins, rosin and cumarone resins[23]. The prominence and presence of the phenolic group is necessary for satisfactory adhesion; adhesion diminishes when the hydroxyl group is etherified. Resins with medium to high MW are needed in minor quantities to obtain optimum tackiness rather than those with lower MW; however, a broad MWD is necessary. The octyl group is a more effective substituent than the tert-butyl group; the methyl group shows only a weak action.

Terpene-phenolic resins are reaction products of phenols with terpenes, e.g. limonene[24]. The alkylation is performed in acidic medium whereby sulfuric acid, p-toluene sulfonic acid, boron trifluoride or acidic ion exchange resins are preferred as catalysts. These thermoplastic resins have also found application in coatings. Also phenol-acetylene resins, made from p-tert-butylphenol and acetylene using mercury salts or organic bases as catalysts, are appropriate tackifiers for rubber (Koresin, BASF).

19.5 Rubber-Reinforcing Resins

Phenol novolaks in combination with fine particle HMTA can easily be incorporated into rubber blends[25] and represent one of the earliest interpenetrating polymer networks (IPN, Chap. 10.8). During processing they function as a processing aid and reduce compound viscosity. During vulcanization, the crosslinking reaction of the novolak resin with HMTA leads to a second solid phase. No chemical reaction with rubber occurs; the cured phenolic forms a threedimensional reinforcing network (IPN). PF-resins with a P/F molar ratio of approximately 1 : (0.75–0.85) are used. The compatibility with rubber can be improved by resin modification with cashew nut shell liquid[26] or tall oil. Reinforcing resins are recommended for SBR and NBR; the resin addition can amount to 50% or more for NBR. Homogeneous (up to 100% addition) and transparent nitrile rubber compounds with high hardness and strength as well as oil resistance are obtained[27]. Phenolic reinforcing resins find application in shoe soles and heel top lift compounds and solid and pneumatic tires. In general, they impart increased hardness and stiffness, which is maintained even at elevated temperature, increased tear resistance and decreased elongation[28]. Growing use is expected in tire construction, particularly in the non dynamic areas of radial tire applications[27].

19.6 Resorcinol-Formaldehyde Latex Systems

Resorcinol-formaldehyde prepolymers[29, 30] have been used in combination with vinyl pyridine copolymer latex for tire cord adhesion ever since rayon replaced cotton in the early 1940s. They find application in bonding almost all currently used tire reinforcing materials, including rayon, nylon, polyester, glass, aramide fibers and steel wire. Common bonding systems include silica/resorcinol prepolymer/HMTA, resorcinol prepolymer/HMTA and resorcinol resin/methoxymethyl melamine resin combinations. A blocked, latent source of formaldehyde leads to single component resins which are stable at ambient working conditions. The curing rate is influenced by temperature and pH as shown in Fig. 11.8, Chap. 11. The alkalinity of most rubber compounds is sufficient to affect complete resin cure at normal vulcanization times and temperatures of at least 145 °C[27].

Most fiber reinforcing materials develop sufficient adhesion on treatment with the resorcinol-formaldehyde latex (RFL) system. Considerable difficulty is experienced with polyester fibers. A double dip system is required for satisfactory adhesion. A comprehensive study of the surface characteristics of polyester fibers before and after treatment with RFL by a variety of techniques such as FTIR, ESCA; laser microscope was not as effective as transmission electron microscopy[31]. The TEM microphotographs clearly demonstrated the capillary flow of the RFL solution from outer fiber surfaces to fiber interstices; hence the double dip process facilitates adhesion between the polyester and rubber phase[32].

19.7 References

1. Flick, E. W.: Adhesive and Sealant Compound Formulations, Ind. Edition Noyes Publication, Park Ridge, N. J. 1984
2. Flick, E. W.: Handbook of Adhesive Raw Material, Noyes Publication, Park Ridge, N. J. 1982
3. Coleman, M. M., Shelton, J. R., Koening, J. L.: Ind. Engng. Chem. Prod. Res. Develop., Vol. 13, 154 (1974)
4. Wildschut, A. J.: Rec. Trav. Chim. 61, 898 (1942)
5. Hultzsch, K.: Chemie der Phenolharze: Berlin, Göttingen, Heidelberg: Springer 1950
6. van der Meer, S.: Rec. Trav. Chim. 63, 147 (1944)
7. Giller, A.: Kautschuk u. Gummi-Kunststoffe 19, 188 (1966)
8. Lavins, E., Snelgrove, J. A.: "Polyvinyl Acetal Adhesives" In: Handbook of Adhesives, 2nd ed., Chapter 31, Skeist Ed., Van Nostrand and Reinhold, N. Y. (1977)
9. Eby, L. T., Brown, H. P.: Thermosetting Adhesives. In: R. L. Patrick (ed.): Treatise on Adhesion and Adhesives. Vol. II. New York: Marcel Dekker 1969
10. Morrill, J. P.: "Nitrile Rubber Adhesives", In: Handbook of Adhesives, 2nd ed., Chapter 17, Skeist Ed., Van Nostrand and Reinhold, N. Y. (1977)
11. Barth, B. P.: "Phenolic Resin Adhesives", In: Handbook of Adhesives, 2nd ed., Chapter 23, Skeist Ed., Van Nostrand and Reinhold, N. Y. (1977)
12. Steinfink, M.: "Neoprene Adhesives", In: Handbook of Adhesives, 2nd ed., Chapter 21, Skeist Ed., Van Nostrand and Reinhold, N. Y. (1977)
13. Bayer AG: Baypren Chloroprenkautschuk. Technical Bulletin
14. Schunck, E.: Kunstharz-Nachrichten Hoechst 15, 28 (1979)
15. Schunck, E.: Kunstharz-Nachrichten Hoechst 4, 33 (1973)
16. Garret, R. R., Lawrence, R. D.: Adhesives Age, October 1966, 22 US-PS 3,394,099

17. Bakelite GmbH: Alkylphenolharze für Polychloropren-Klebstoffe, Technical Bulletin
18. Azrak, R. G., Barth, B. P.: Adhesives Age, June 1975, 23
19. Union Carbide Corp.: DE-AS 2 420 684
20. Fry, J. S., Stregowski, R. J.: ACS Polymer Preprints 24 (2), 203 (1983)
21. Meuser, H.: Adhäsion 12, 332 (1975)
22. Mooncai, W. W.: Adhesives Age, October 1968, 28
23. Giller, A.: Gummi, Asbest, Kunststoffe 29, 766 (1976)
24. Schmelzer, H.: Adhäsion 5, 174 (1973)
25. Hoechst AG: DE-AS 1 769 456
26. Fries, H., Esch, E., Kempermann, Th.: Kautschuk Gummi Kunststoffe 32, 860 (1979)
27. Hooker Chemicals: Durez Resins for the Rubber Industry, Technical Bulletin
28. Yurcick, P. A.: Rubber World, October 1976, 47
29. Gils, G. E.: I & EC Product Research and Development, Vol. 7/2 151 (1968)
30. The General Tire & Rubber Co.: DE-AS 24 22 769 (1974)
31. Gillberg, G., Kemp, D.: J. Appl. Polym. Sci., 26, 2023 (1981)
32. Gillberg, G., Sawyer, L. C., Promislow, A. L.: J. Appl. Polym. Sci., 28, 3723 (1983)

20 Phenolic Antioxidants

Most plastics are subjected to thermal or oxidative stresses during processing or on exposure. All polyolefins require stabilization against thermal, UV or metal-catalyzed oxidation. The combined oxygen and UV-radiation has an even greater destructive effect. The oxidation of polymers by molecular oxygen is in most instances a free radical chain reaction, with hydroperoxide as the primary product. Stabilizers can be effective as peroxide scavengers, metal deactivators, chain terminators and UV absorbers.

In contrast to amine antioxidants, phenolic compounds cause only minor discoloration of rubber or thermoplastics[1-3].

Two peroxy radicals can be inactivated by one phenolic molecule yielding a 2,4-peroxycyclohexadienone which can be isolated easily at low temperatures[4].

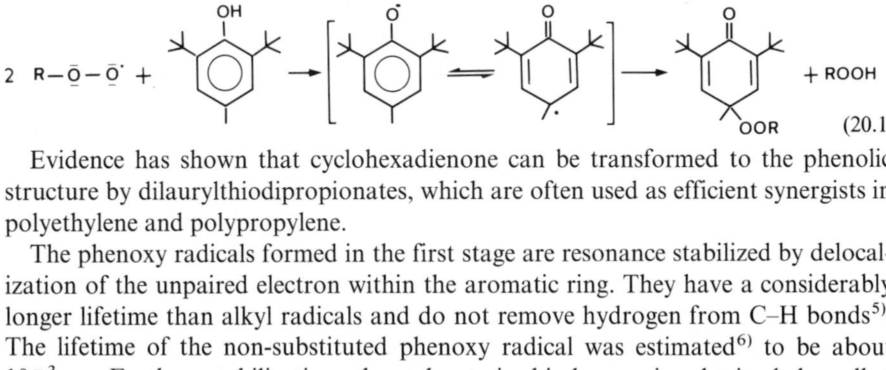

$$(20.1)$$

Evidence has shown that cyclohexadienone can be transformed to the phenolic structure by dilaurylthiodipropionates, which are often used as efficient synergists in polyethylene and polypropylene.

The phenoxy radicals formed in the first stage are resonance stabilized by delocalization of the unpaired electron within the aromatic ring. They have a considerably longer lifetime than alkyl radicals and do not remove hydrogen from C–H bonds[5]. The lifetime of the non-substituted phenoxy radical was estimated[6] to be about 10^{-3} s. Further stabilization through steric hindrance is obtained by alkyl substitution. Some of these substituted aryloxy radicals are stable over periods of hours or even days. They are paramagnetic and can be highly colored in solution. In the solid state they can exist as colorless dimeric quinol ethers[5]. The relative rate of hydrogen abstraction of mononuclear phenols by peroxy radicals provides a good correlation with Hammer's σ-values, but steric effects also have considerable influence on antioxidant (AO) efficiency.

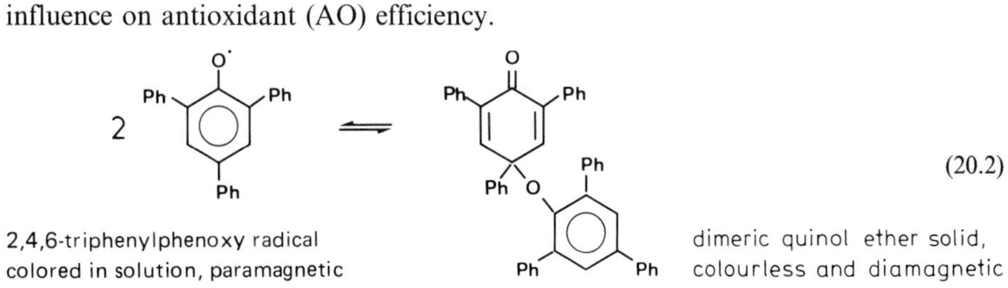

$$(20.2)$$

2,4,6-triphenylphenoxy radical
colored in solution, paramagnetic

dimeric quinol ether solid,
colourless and diamagnetic

The value of flash photolysis as a probe for studying antioxidant activity of a number of phenolic antioxidants was recently reported[7]. The kinetic microsecond flash photolysis technique provides useful information on the relationship between AO structure and efficiency of phenoxy radical production.

Factors to be considered in the selection of an antioxidant for a specific use are:

– FDA (non-toxic);
– colorless, odorless;
– thermally, UV stable;
– low volatility, low extractibility;
– low color development.

For alkyl substituted phenols (20.3) maximum efficiency was found for:

R_1 = methyl or tert-butyl,
R_2 = tert-butyl and
R_3 = methyl

(20.3)

Bulky group in the *para* position decrease AO-activity.

This indicates that although steric hindrance of the phenolic hydroxyl group is required, too much hindrance renders the compound less effective[8, 9, 10]. Partially hindered phenols provide better stability at the expense of greater color development[11]. Electron releasing groups (methyl, tert-butyl, methoxy) in general increase the AO activity. Electron attracting substituents, such as nitro, carboxyl and halogen, have the opposite effect.

It has been shown that mononuclear phenolic compounds are effective antioxidants. They have, however, certain disadvantages. Their vapour pressure is relatively high so that during the extrusion or molding operation at temperatures between 180–220 °C considerable quantities may be lost due to volatility. Many plastics are used as sheets and fibers and further loss may arise due to migration and volatilization of low molecular weight AO's at even lower temperatures. Di- and poly-nuclear phenols, however, have in addition to lower vapour pressure additional advantages. They are able to bind metals (polyolefin catalyst residues) under complex formation. Many metal ions, for example copper and manganese, catalyze the decomposition of hydroperoxides and accelerate oxidative degradation. This catalytic activity of metals with variable valency is attributed to the formation of a coordinate complex with hydroperoxide followed by electron transfer between the hydroperoxide and the metal ion. Through competitive interaction with stronger chelating agents, the catalytic activity of the metal can be suppressed. Instead of formaldehyde, diolefins can be used for alkylation of phenols, for example dicyclopentadiene or limonene, in the presence of Lewis acid catalysts.

Of the mononuclear phenols, 2,6-di-tert-butyl-4-methylphenol, also called BHT, Ionol or Topanol O, is the best known compound[8]. It is obtained by reaction of isobutylene and *p*-cresol as a colorless crystalline powder with a MP of 70 °C. BHT is approved for use in the food sector. Styrenated phenols, valuable rubber AO's, are obtained by arylation of phenol or *p*-cresol with styrene (20.4).

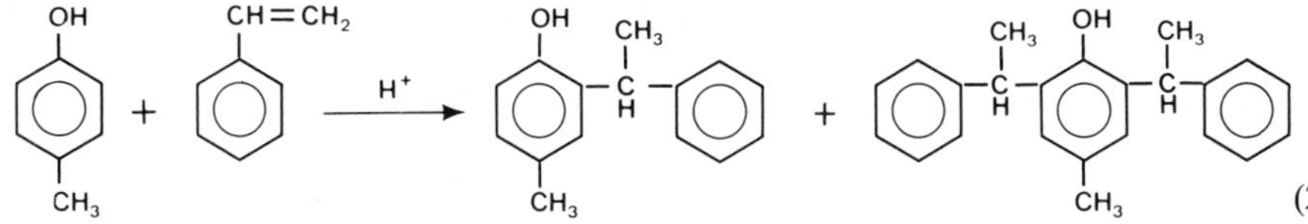

$$(20.4)$$

Styrenated phenols are liquid, this applies also to alkylated or terpene substituted phenols. Bisphenols or higher molecular weight compounds, on the other hand, are crystalline or grindable solids, have lower vapor pressure and higher migration stability.

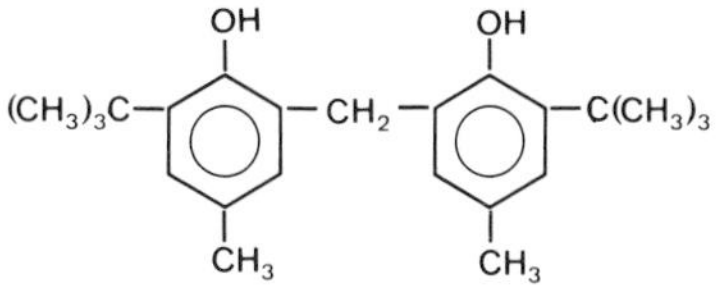

$$(CH_3)_3C \quad CH_2 \quad C(CH_3)_3$$

2,2'-methylene-bis-(4-methyl-6-tert-butylphenol)
Orthophene OMB (Bakelite)
Anti-aging agent BKF (Bayer)   (20.5)
Anti-oxidant 2246 (Cyanamid)

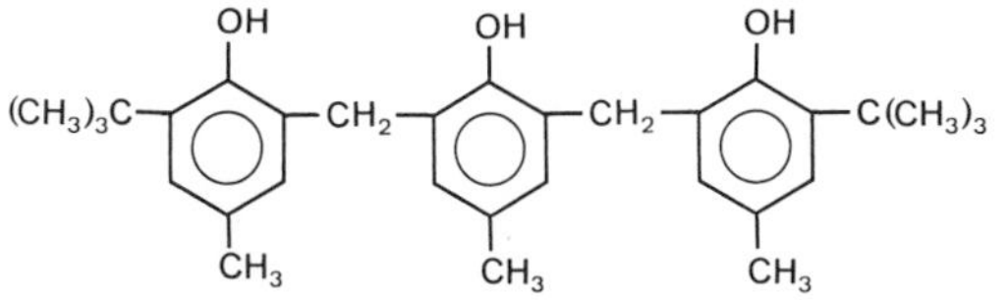

$$(CH_3)_3C \quad CH_2 \quad CH_2 \quad C(CH_3)_3$$

trisphenol   (20.6)

$$HO \quad S \quad OH$$

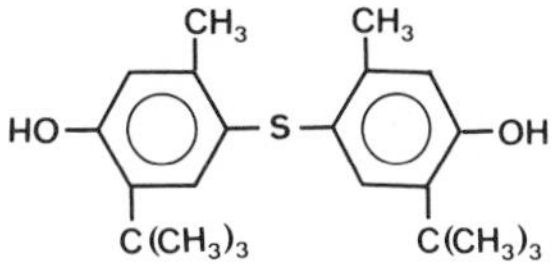

4,4'-thio-bis-(3-methyl-6-tert butylphenol)
Rütenol (Chem. Fabrik Weyl)   (20.7)

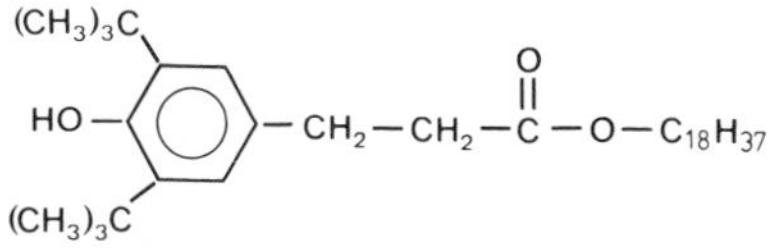

$$HO - CH_2-CH_2-\overset{O}{\overset{\|}{C}}-O-C_{18}H_{37}$$

β-(3,5-di-tert-butyl-4-hydroxyphenyl)-propionic acid-n-octadecylester, Irganox 1076 (Ciba-Geigy)

$$(20.8)$$

Amines, e.g. N,N'-dialkylphenylenediamines, are effective stabilizers in rubber compounds containing carbon black. Conversely, hindered phenols are more effective than aromatic amines in rubber formulations without carbon black reinforcement. As a rule, phenolic AO's are used in polyolefins. Bisphenols and multivalent phenols are preferred because of their increased effectiveness[9, 10, 12]. The use of synergistic mixtures of phenols with organic sulfides especially β,β'-thiodipropionic acid esters, is wide spread.

20.1 References

1. Lloyd, D. G.: Kautschuk u. Gummi-Kunststoffe 27, 477 (1974)
2. D'Ianni, J. D., Widmer, H.: Gummi, Asbest, Kunststoffe 9, 718 (1973)
3. Voigt, J.: Stabilisierung der Kunststoffe gegen Licht und Wärme. Berlin, Heidelberg, New York: Springer 1966

4. Rampley, D. N., Hasnip, J. A.: J. Oil Col. Chem. Assoc. 59, 356 (1976)
5. Mihajlovič, M. L., Cekovič, Z.: Oxidation and Reduction of Phenols. In: S. Patai, (ed.): The Chemistry of the Hydroxyl Group. New York: Interscience 1971
6. Stone, T. J., Waters, W. A.: Chem. Soc. 213 (1964)
7. Allen, N. S., Parkinson, A., Loffelman, F. F., Susi, P. V.: Polym. Degrad. Stab. 5 (6) 403 (1983)
8. Kurze, W.: Antioxidantien. In: Ullmanns Encyclopädie d. techn. Chem. 4. Ed. Weinheim: Verlag Chemie 1976
9. Ohkatsu, Y., Haruna, T., Osa, T.: J. Macromol. Sci.-Chem. A11, 1975 (1977)
10. Roginskii, V. A.: "Sterically Hindered Phenols – Antioxidants for Polyolefines. Relation of Antioxidative Activity to Structure. Review": Polymer Sci. USSR, 24, 2063–2088 (1982), Transl. Vysokomol, soyed. A 24, No. 9, 1808 (1982)
11. Bartz, K. W.: Polymer Preprints 25 (1) 74 (1984)
12. Roginskii, V. A. et al.: Eur. Polym. J. 13, 1043 (1977)

21 Other Applications

21.1 Refractory Linings and Taphole Mixes

In recent years, service conditions in the steelmaking processes, e.g. BOF-process, have become more severe because of increased operating temperatures caused by an increase in the continuous casting ratio and production of high grade steel. Improved lining performance[1a] is required in converters and other steelmaking equipment. Oxidized magnesia-dolomite bricks are the main lining material in LD-converters. A substantial increase in thermal shock resistance, slag resistance, and abrasion resistance of magnesium oxide-dolomite bricks is achieved by using carbon[1, 2] as additive and binder (MgO-C brick).

A premix is formulated from magnesia clinker, carbon, and a solid novolak resin with HMTA and a glycol solvent. After mixing, kneading and shaping, the preform is heated to 1,300 °C and the phenolic resin is pyrolyzed to carbon. Thus the high-temperature resistant bond developes as carbon formation begins (see Chaps. 8.1 and 10.1).

Generally blast furnace taphole mixes consist of finely granulated Al_2O_3, SiO_2, SiC and coke and a coal tar or pitch binder[3]. Since blast furnaces have become larger and are operated under high top pressure, reliable taphole mix performance has become increasingly important. Conventional tar bonded mixes are relatively weak in resistance to heat and erosion, emit black smoke, and require lengthy hardening times. As a means of avoiding these disadvantages, phenolic resin bonded mixes were developed[4, 5]. The early green strength occurs due to the resin bond. A relatively strong, abrasion resistant vitreous carbon bond is formed at higher temperatures due to the carbonization reaction. Since these materials develop a high strength in a short period, the stuffing machine can be moved back after about 10 min; 60 min are necessary with coal tar based binders. The material possesses high hot-flexural and compressive strength[4], erosion resistance, low porosity and low smoke density. A phenolic resin bonded material[5] contains 15–25% clay, 25–35% chamotte, 15–30% silicon carbide, 15–25% coke and approximately 12–18% binder which consists of a phenol novolak (MW 400–800), HMTA and a plasticizing solvent containing hydroxyl groups (diethylene glycol, glycerin).

21.2 Phenolics for Chemical Equipment

Acid resistant molding materials based on asbestos and phenolic resins for the manufacture of chemical equipment have been available since the 20's and are known as

Haveg. These materials are recognized for their outstanding chemical resistance to non-oxidizing inorganic or organic acids at elevated temperature[6]. Alkali resistance is improved by partial etherification[7] of the hydroxyl group. The use of 2% MgO accelerates cure rate with improved properties. Cresol resols (high *m*-cresol content with some etherification) furfuryl alcohol and/or resins[8] provide excellent, alkali resistance[9]. Some formulations still contain anthophyllite and crocidolite asbestos. Carbon and glass fibers combined with mica are being recommended as asbestos replacements due to physiological hazards and carcinogenesis of asbestos. Multilayer coating conditions are used with intermediate curing for each layer for an optimum coating thickness of 15–35 mm. Conditions for small items such as pipes to large equipment parts including reactor vessels are provided in several company brochures[10].

21.3 Socket Putties

Phenol resins are used as socket putties for light bulbs, radio valves, and the like because of their high temperature resistance and strong adhesion. The putties consist of a powdery mixture of fast curing phenol novolak/HMTA resin and mineral fillers. Resin composition is 12–15%. They are mixed to a paste with 10 pbw of a solvent like ethanol or isopropanol. This paste is transferred into the hopper of a socket machine and injected into the metal jacket by a metering device. The glass bulb is then placed on the jacket under spring pressure and the putty hardened in a tunnel oven or carousel at temperatures of 180–200 °C or above.

21.4 Brush Putties

The brush manufacturing industry was transformed to phenolic resins to fasten the hairs and bristles because of resin water and solvent resistance. High-solids (80%) phenol resols, differing in viscosity and reactivity, are used in combination with inorganic acids as hardening agents. The acids are diluted with alcohols or glycol. The formulation of the putty includes silica flour. Cure is effected either at 90–95 °C in an oven within 6–8 h or at room temperature within 2 days.

21.5 Synthanes

Synthanes or synthetic organic tannins are divided into pre-, post-, exchange-, alum-, bleach-, shrink-, and combination synthanes according to their application. Depending upon their chemical composition, they have a somewhat strong (synthetic replacement tannins) or no self-tanning effect (auxiliary tannins)[11].

Synthetic aromatic tannins are condensation products of aryl-, mono- or polynuclear phenol-sulfonic acids or carboxylic acids with formaldehyde or other carbonyl compounds, whereby further aromatic compounds can be incorporated (for instance amines, amides, polynuclear aromatic compounds, lignin-sulfonic acids).

The tanning effect of phenols and PF-resins is due to the ability of the phenolic hydroxyl group to form strong hydrogen bonds with the polypeptide groups and thereby have a crosslinking effect on the collagen micellae[12]. Reactive polymers polymerize in the cutaneous structure and, after the displacement of water, lead to the isolation of the collagen fibrils and to the irreversible change of the skin[13]. Condensation products of sulfonated phenols and formaldehyde were suggested initially as tanning agents by Stiasny[14]. The sulfonation (and neutralization) is necessary in order to obtain the desired water solubility for commercial tanning processes. The sulfomethylation of phenols with sodium sulfite and formaldehyde (21.1) or later sulfonation of novolaks (21.2) are additional reactions resulting in synthanes[15, 16]. Naphthols and resorcinol are also recommended as phenolic compounds[17].

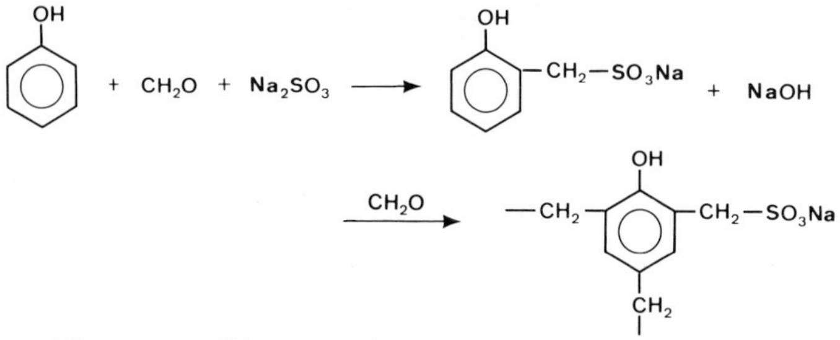

$$\text{OH} + CH_2O + Na_2SO_3 \longrightarrow \text{OH}-CH_2-SO_3Na + NaOH$$

$$\xrightarrow{CH_2O} \quad -CH_2-\text{OH}-CH_2-SO_3Na$$

(21.1)

$$\text{OH}-CH_2-\text{OH}-CH_2-\text{OH}$$
$$SO_3H \quad SO_3H \quad SO_3H$$

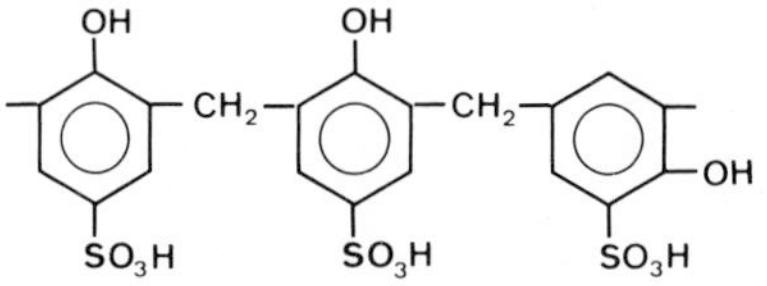

(21.2)

Resins similar in structure are also used for ion-exchange resins (Chap. 10.7) and concrete additives.

21.6 Concrete Additives

A variety of additives are applied in concrete technology to improve distinct properties. Concrete fluidizers improve the fluidity and plasticity of the concrete mixture, thus permitting a lower water/cement ratio without deterioration of other properties. Commercial concrete fluidizers are predominantly lignin sulfonate solutions, aqueous sulfonated MF-oligomers and sulfonated naphthalene formaldehyde condensates. Sulfonated PF-resins[18-23] and combination with previously mentioned materials are reported in the patent literature. Resorcinol phenol formaldehyde resin added to Portland cement was reported[24] to produce a material with superior ductility making it suitable for use in rapid repair systems for deteriorated concrete bridge decks and highways.

21.7 Casting Resins

Casting-type resins which are used for the production of decorative articles, bowling balls and the like are prepared using a high formaldehyde concentration (P/F ratio 1:2–3) with sodium hydroxide or alkaline earth hydroxide as catalysts[25]. After the reaction is completed, water is removed completely and diluents, e.g. ethylene glycol, polyethylene glycols or glycerol, are added to reduce the viscosity and enhance the flexibility of the casting. The pH is usually adjusted to about 5–7 with weak organic acids. Fillers and pigments may be added. The resin is poured or cast into open molds and maintained below the boiling point of water. Usually a period of 70–200 h is required at temperatures between 75–95 °C.

The curing temperature can be reduced to ambient temperature by the addition of organic acids, e.g. lactic acid, phenoxyacetic or sulfamic acid. However, post curing temperatures between 40 and 50 °C are necessary. The light transparency depends upon the choice of the acidifying agent[26]. Lactic acid for example gives opaque castings while phenoxyacetic acid gives transparent products.

21.8 References

1a. Koltermann M., Sperl H.: Stahl u. Eisen 105, 43 (1985)
1. Abe, A.: Jpn. Kokai Tokkyo Koko 79, 107, 952 (1979) C.A. 92: 23602
2. Watanabe, A., Takeutschi, Y.: Taikabatsu Overseas 1, 40 (1981)
3. Rütgerswerke AG: DE-PS 2021469, DE-OS 1960461, DE-OS 1960462
4. Ochiai, T., et al.: Sprechsaal 10 (1979) P. 755
5. Nippon Steel Corp.: DE-OS 2624288; C.A.: 100, 125864z (1984)
6. Kessler, G., Schacht, E.: Rohre und Apparaturen aus Phenol-Formaldehyd-Harzmassen. In: Haus der Technik-Vortragsveranstaltungen 287, 31 (1972)
7. Campbell, N. R.: Werkstoffe-Korrosion 22, 219 (1971)
8. Röbe-Oltmanns, G.: DE-OS 1965165 (1969)
9. Hefele, J.: DE-OS 1494099 (1962)
10. Bakelite GmbH: Harze für den Apparate- und Behälterbau. Technical Bulletin
11. Scheer, W.: Zur Nomenklatur der Textilhilfsmittel, Leder und Pelzhilfsmittel, Papierhilfsmittel, Gerbstoffe und Waschrohstoffe. Verband der Gerbstoffindustrie e. V., Frankfurt (1970), Limburger Vereinsdruckerei GmbH
12. Faber, K., Komarek, E.: Leder und Gerbung. In: Ullmanns Encyclopädie d. techn. Chem., Vol. 11, 3. Ed., P. 595, München: Urban und Schwarzenberg
13. Stather, F.: Gerbereichemie und Gerbereitechnologie. Berlin: Akademie Verlag 1967
14. Stiasny, E.: DE-PS 262558 (1911)
15. Küntzel, A., Plapper, J.: Leder 6, 176 (1955); 7, 60 (1956)
16. Noerr, H., Mauthe, G.: DE-PS 693923
17. Wegler, R., Herlinger, H.: Polyaddition u. -kondensation v. Carbonyl- u. Thiocarbonylverbindungen. In Houben-Weyl: Methoden d. Org. Chem., Vol. XIV/2, Thieme: Stuttgart 1963
18. Bayer AG: DE-OS 2935719
19. Holmen GmbH: DE-PS 2811098, DE-PS 2857306
20. Chemische Werke Wildhausen: DE-OS 2653801
21. Showa Denko: DE-OS 2405437
22. Rütgerswerke AG: DE-AS 2033015
23. Sika AG: DE-OS 3143257
24. Sugama, T., Kuckacka, L. E., Horn, W.: J. Appl. Polym. Sci. 26, 1453 (1981)
25. Sandler, S. R., Karo, W.: Polymer Syntheses, Vol. II, P. 57, New York: Academic Press 1977
26. Harris, T. G., Neville, H. A.: J. Polym. Sci. 4, 673 (1953)

Subject Index

Abietic acid   175, 176, 252
Ablation   150, 159, 160
Abrasives   3, 269 f., 276
Acetaldehyde   14, 22, 47, 51, 69, 98
Acetylene   22
– resins   296
Acid curing   43
– reactivity   220
Acidity   74
– of phenols   26
Acoustical insulation   213
Acrolein   22
Activated carbon process   112
Activation energies   131
Adipic acid   6
Air treatment processes   108
Aldehydes   14 f.
– physical properties   15
Aldimines   36
Alkali extraction of phenols   11
Alkylphenols   10, 11, 252
Alkylphenol resins   51, 248
Alloy adhesives   290
Allyl chloride   148, 254
Aluminum oxide   270, 271
– sulfate   191
α-Aminoalkylation   35
4-Aminoantipyrine method   119
Aminolysis   79
p-Aminophenol   150
Ammonia   34
Ammonium polyphosphate   182
Analytical methods   118 f.
Aniline-formaldehyde resins   152
Animal glue   277
Antifouling paints   250
Antioxidants   10, 299
Aralkylethers   153
Arrhenius energy of activation   48
Arylphenol resins   251
Asbestos   199, 282, 304
– toxicology of   283
Automotive coatings   245
Azomethine group   34
Azomethines   53

Barium hydroxide   33, 130, 217
Basalt   215
Base reactivity of phenols   51
BASF-process   15
Battery separators   242
Benzaldehyde   69
Benzoxazine   34, 53, 54
Benzylamines   53
Bisphenol-A   5, 6, 14, 40, 47, 169, 248
Blowing agents   220
Bond angles   81, 85
– distances   81
Boric acid   33, 149
Boron carbide   270
Boron-modified resins   149
Branching   48, 62, 123, 133
Bromination   72, 77, 78
Brush putties   304
B-stage time determination   136
Bulk weight determination   135
Burning rating code   113
Butanol   148
Butylated hydroxy toluene   10
p-tert-Butylcalix[4]arene   80
p-tert-Butylphenol   5, 40, 106, 251, 295, 296
p-tert-Butylphenol resin   294
Butyl rubber   288
Butyraldehyde   15, 22

Calcium hydroxide   217
– resinate   251
Calixarenes   67, 68, 70, 73, 76, 80
– crowned   77
Can coatings   248
Cannizzaro reaction   17, 30
ε-Caprolactame   6
Carbazol   153
Carbon, artificial   157
– composites   158, 160
– fibers   159, 166, 203, 283
– foams   159
– glassy   158
– materials   156
– polymeric   158
– yield   156, 160, 285

Carbonless paper   168, 169
Carborundum   270
Cashew nut shell liquid   13, 284, 296
Casting resins   305
Catalysts   31, 33, 40, 181, 220
– latent   165
Cation transport   77
Cellulose fibers   201, 202
Char yield   115
Charge distributions   28
Chelating ability   169
China wood oil   45, 235, 284
Chloromethanol   47, 48
Chloromethylation   47
Chlorophenols   106
Chloroprene adhesives   292
Chroman   45
Chrysotile   283
Claisen rearrangement   72
Clay   256
Clean Air Act   103, 244
Clean Water Act   103
Clutch linings   282, 286
Coal tar   11, 303
Coatings   3, 244 f.
– for metal containers   247
Coil coating   248
Cold-box process   265 f.
Cold-runner molding   209
Colophonium   191, 252, 253
– modified resin   251, 252
Colorimetric determination of phenols   118
Combustion   111
– enthalpies   91
– of phenolics   113
Complexation of cations   76
Compression molding   207
Computer simulation   48, 58
Concrete additives   305
Configuration in solution   25
Conformational analysis   85, 86
– inversion   87
Conformations   28, 77, 80, 82
Coniferyl alcohol   175
Contact adhesives   292, 293, 295
Container coatings   247
Corundum   270
Cotton fabric laminates   238
Cresols   5, 10, 11, 12, 40, 106, 237, 252
– preparation methods   9
Cresol resols   234, 238, 304
Critical oxygen index   225
Croning process   260
Crotonaldehyde   22
Crystal structures   80
– – of phenol   25
– – of linear oligomers   81

Cumarone resins   296
Cumene   8
– process   7, 9
p-Cumylphenol   40
Cyclic compounds   62, 67
– formals   19
Cyclohexadienone   299

Decorative laminates   3, 238 f.
– – production of   239
– – standards for   240
Degradation, biochemical   109
– microbial   108
Dehalogenation   64
Diastereomers   66
Diazabicyclooctane   54
Diazidodiphenylsulfone   167
Diazonaphthoquinone   166
Dibenzylethers   45
1,1-Dichlorodimethyl ether   93
Dicyandiamide   152, 218
Dicyclopentadiene   300
Diethylene glycol   303
Differential scanning calorimetry   53, 130, 285
Differential thermal analysis   53, 130
Dihydroxydibenzyl ether   41
Dihydroxydiphenyl methanes   40, 47
– – reactivity of   50
4,4-Dihydroxydiphenyl sulfone   169
Dihydroxymethylphenols   31
Diisopropyl ether   112
Dimethylene ether group   69, 293
Dimethylethylamine   265
Dimethylphenols   11, 40
Dioxolanes   19
Diphenolic acid   246
Diphenylcresylphosphate   235
Dipole moment   87, 133
– – of phenols   27
Directing effects   26, 32
Dispersions   91, 98, 99
Disposal of waste material   107
Dissociation constant   76
Drying oils   251
Dust explosions   20, 100
Dynamic mechanical analysis   131
Dynamic nuclear polarization   88

Efficiency of resins   217
Elastic constants   132
Electrical laminates   3, 230 f.
– – production of   235
– – standards   231
Electrodeposition   245, 246, 247
Electron density   27, 29, 37
– microscopy   285
– spectroscopy   123

Enhanced oil recovery   172
Entropy of activation   39, 48
Environmental protection   103 f.
Environmental Protection Agency   103
Enzyme immobilization   171
Epichlorohydrine   55
Epoxy-phenolic compounds   201
Epoxy resins   154, 201, 247, 248, 249, 250,
   253, 272, 274
– – reaction with   54
Electron spectroscopy for chemical analysis
   297
Esterification   71, 149
Etherification   148
Explosion hazards   100
Exposure levels   105
Extraction processes   111

Factory Mutual   226
Fading   282
Felt bonding   3
Ferric chloride test   118
Fiber boards   190 f.
– – classification of   190
– – medium density   190
– – production of   191
Fiber composites   160 f.
Fibrous insulation materials   3, 213 f., 218
Filters   3, 241
Fire retardancy   163
Fish toxicity   106
Flame retardants   116, 182
Flammability   113, 213
Flavan   45
Flexibilization   148, 234
Floral foam   219, 222
Flow distance, determination of   136
Flow tests   206
Foaming equipment   221
Foam properties   220
Formaldehyde   14 f., 105
– alcohol equilibria   29
– carcinogeneity   105
– conversion   40
– determination   119
– emission   177, 179, 185, 219
– production processes   16
– properties   16
– specification   18
– toxicology   104
– use   17, 18
Formals   30
Formic acid   18
Foundry resins   3, 256 f.
– processes   257
– sands   257

Fourier transform infrared spectroscopy
   120, 297
Free phenol, determination of   95
Friction dust   13, 284
Friction materials   3, 281 f.
– – formulation of   282
– – production of   281, 286
– – testing procedures   285
Friedel-Crafts-acylation   72
Fructose   21
Functionality   24, 58
Fungicides   182
Furan resins   21, 253, 257, 263, 266
Furfural   14, 15, 20, 21, 271
Furfuryl alcohol   21, 220, 257, 261, 263, 266,
   271, 304

Gas chromatography   128 f.
Gas scrubbing processes   112
Gelation   25, 53, 54, 193
Gel permeation chromatography   33, 48, 49,
   58, 97, 130, 181, 294
Gel point   25
Gel time, determination of   136
Glass fibers   213, 215
– – production process   216
Glass transition temperature   162
Glucose   21
Glyoxal   14, 15
Gold lacquers   248
Graphite materials   156
Green sand molds   256
Grinding materials   271
– – coated   275 f.
– wheels   269 f.
– – manufacturing of   273
Guanidine   17

Hardox process   266
Haveg   304
Heat curing   43
– deflection temperature   196
– insulation   214
Hemiformal resol resins   52
Hemiformals   30, 32, 52, 98, 165
Hexamethylenetetramine   18, 20, 35, 52, 53,
   54, 101, 134, 261, 271
– formation   35
– hydrolysis   36
– novolak reaction   53
– physical properties   20
– use   20
High-energy radiation, resistance to   144
High-ortho resins   50, 66, 70, 97, 98, 165
High-performance liquid chromatography
   124 f., 128
High-pressure laminates   238

Hock process   14
Hot-box process   261 f.
Hot cone technique   208
Hot tops   267
Hydrocarbon Research process   12
Hydrocarbon resins   296
Hydrochloric acid   49, 93
Hydrogenation   73
Hydrogen bonds   74, 77, 81, 85, 87
– – isodromic   81
Hydrogen peroxide   109
Hydroquinone   66, 73
Hydroxybenzyl amine   79
2-Hydroxyethylacrylate   245
Hydroxylamine hydrochloride   119
Hydroxymethylation   36
–, ortho   50, 66, 70
Hydroxymethylphenols   31
4-Hydroxystyrene   154
Hyperacidity   51, 54, 70, 74
Hyperthermal runner system   208

Incineration, catalytic   110
– thermal   110
Industrial laminates   230 f.
Infrared spectroscopy   50, 119
Injection molding   208
Internal bond   184
Interpenetrating polymer networks   171, 296
Ion exchange resins   169 f.
Ionol   300
Isobutylene   300
Isobutyraldehyde   15
Isocyanates   55, 56, 180, 181, 265
–, blocked   56, 247
4,4′-Isocyanatodiphenylmethane   56, 265
Isopropylbenzene   8

Karl Fischer method   134
Kevlar fiber   283
K factor   223
Kinetic studies   35, 77
Kjeldahl method   134
Kriewitz-Prins reaction   17
Koppeschaar method   118

Labeling of resin compositions   107
Laminated tubes   237
Laminated wood   189
Lederer-Manasse reaction   31
Light fastness   148
Lignin   12, 175
Lignosulfonates   181, 218
Limiting oxygen index   114
Limonene   296, 300
Linear oligomers   62, 63, 65, 73
Lurgi process   11

Magnesium oxide   33, 51, 203, 271, 284, 291, 293
MAK-value   6, 108
Maleic acid   93
– – anhydride   253
Mannich reaction   35, 53
Marine paints   249
Market segments of phenolics   3
Mark-Houwink-Sakurada expression   133
Mass spectrometry   124
Melamine   17, 56, 152, 218
– phenol compounds   199
– reaction with   56
– resins   187, 234, 239
Melting point   73, 74
– – determination of   135
Membranes   76
Metal-modified resins   151
Metasolvan process   11
Methane oxidation   15
Methanol   15
Methanolysis   79
Methylene bridge formation   41, 73
Methylene glycol   29, 30, 32
Methyl ethyl ketone peroxide   266
Methylstyrene   8
Mica   203
Michael addition   44
Mineral fibers   213 f.
– – production of   216
Mineral fillers, properties of   203
– flour   202
Mold defects   261
Molding compounds   3, 196 f.
– – composition of   200
– – production of   204
– – properties of   209
– – standardization of   199, 200
Molecular shielding   58
– structure   25
– weight distribution   33, 48, 49, 58, 97, 130, 181, 294
Multilayer circuit boards   154

Naphthalene   153, 285
Naphthols   305
Neoprene rubber   292
– latex   295
Nitration   73
Nitrile-butadiene rubber   159, 291, 296
Nitrile-phenolic adhesives   291
Nitrogen-modified resins   152
Nitrophenol   73, 75
No-bake process   263
p-Nonylphenol   5
Novolak   172
Novolak resins   24

– chiral   66
– continuous process   94
– crosslinking   52
– epoxidized   54
– high-*ortho*   33, 46, 49, 50, 97, 124, 201,
   265
– production   93 f.
– specification   201
Nuclear magnetic resonance spectroscopy
   30, 42, 43, 48, 50, 52, 54, 56, 57, 100, 120 f.

Occupational Safety and Health
   Administration   103
p-Octylphenol   5, 40, 251
Oil-modified phenolic paints   251
"Oil-reactive" resols   252
Oiticica oil   235
Oriented strand board   185, 186
Ortho-linked oligomers   66, 70
Oxalic acid   46, 49, 93, 201
Oxidation   73
– of phenols   143
Oxygen index   115
Ozone   110

Paper chromatography   128
Paper impregnation   230 f.
Paraffin   181, 182, 191
Paraformaldehyde   17, 18, 19, 39, 51, 52, 92,
   98, 149, 192, 193, 267
– properties of   19
Particle boards   177 f.
– formaldehyde emission of   185
– production of   182
– properties of   183, 184
– water resistance of   184
Peel back   261
Pentaerythrol   18
Perforator test   177, 179, 218
Phenol   5, 6, 40, 106
– capacities   6
– health risks of   104
– hemiformal   52
– physical properties of   5
– production processes   7
– recovery of   111
– removal   109
– toxicology   104
Phenol-furfural resins   21
Phenol sulfonic acid   43, 165, 169, 220, 264,
   304
Phenolic composites   163
– compounds   196 f.
– fibers   159, 165, 166
– foam   115, 159, 214, 219 f.
– – properties   222, 224
– – production   222

Phenolic oligomers   74
– thermospheres   99
Phenols from coal   11
– from petroleum   11
Phenoraffin process   11
Phenosolvan process   11, 112
Phenoxide ions   31
Phenoxy radicals   143, 299
Phenoxy resin   272
p-Phenylphenol   40, 169, 251
Phenylurethanes   55
Phosphoric acid   149, 201, 220, 250, 264
Phosphorus-modified resins   151
– oxychlorides   151
– oxyhalides   149
Photo resists   154
Plywood   186 f.
– glue formulation   188
– production of   188
Polyacetal resins   18
Polyacetylene   242
Polyamide fibers   203, 283
Polybutadiene resins   246
Polycarbonate   14
Polychlorobutadiene   293
Polychloroprene   292
Polyesters   165
Polyesterimides   171
Polyhydroxystyrene   154
Polyimides   171
Polyimide precursors   57, 171
Polymethylene glycol   16
Polyoxymethylene   30
Polystyrene foam   213, 214
Polysulfone   14, 52
Polyurethanes   263
Polyurethane foam   213, 214
Polyvinylacetal   290, 291
Polyvinylbutyral   248, 249, 250, 253, 272
Polyvinylformal   290
Poly-4-vinylphenol   154, 167
Poly-p-xylylene   147, 153
Positional reactivities   39
Post-forming process   240, 241
Pott-Hilgenstock process   11
Precoated shell sand   260
Prepreg   163, 164
Pressure-sensitive adhesives   295
"Pres-tock"-process   190
Prins reaction   253
Printed circuit boards   166, 230
Printing inks   251
Propionaldehyde   15
Protective groups   63, 64
Pyrolysis   140

Quartz   258
Quinol ethers   299
Quinone methides   42, 43, 44, 45, 46, 51, 65, 66, 79, 289

Radiation resistance   144
Radicals   299
Raschig-Hooker process   7
Rate constants   38
Reaction enthalpy   91
– injection molding   164
– kinetics   36, 48
– mechanisms   24 f.
– *ortho*-specific   65
– rates   37
Reactivity of phenols   25, 77
Reactors   92
Refractive index, determination of   135
Refractory linings   303
Resinates   252
Resin content, determination of   135
– efficiency   217
Resin production   3, 91 f.
Resists   168
Resol resins   25
– high-*ortho*   52, 98, 165
– production of   95
– high-solids   98
– solid   99
Resorcinol   5, 13, 14, 19, 69, 106, 181, 192, 290, 305
– adhesives   192
– rate of reaction   192
– resins   187, 189, 192, 220, 297
Resorcinol-formaldehyde latex   297
Rock wool   214
Rosin   176, 250, 296
– acids   176
– modified resins   252
Rotational barriers   29
Rubber reinforcing resins   296
– vulcanization with phenolics   288

Safety data of phenol   6, 108
Salicylic acid   246, 261
Saligenin   31, 50
Sasol   11
Scanning electron microscopy   223
Scrap recycling   209
Sensitizers   167
Shellac   269, 270
Shell molding process   259 f.
Shell sand   260
Shop primer   249, 250
Sieve analysis   136
Silanes   151, 218, 263, 266, 272, 273, 291

Silica sands   257, 258
Silicon carbide   270, 271
Silicones   150
Silicon-modified resins   150
Siloxanes   150
Silver catalyst processes   15
Smoke density   113, 114, 164
– emissions of foam materials   225
– generation   160, 163, 213
SO_2-process   266
Socket putties   304
Sodium carbonate   33
– hypochlorite   110
– silicate   256, 257
Solution properties   25, 85, 133
Specific heat capacity   132
Specific gravity, determination of   135
Spray dried resins   101
Stamp volume, determination of   135
Starches   21
Statistical approach   58
Stereoisomers   66
Styrenated phenols   300
Substitution reactions   32, 72
Sugar   21
Sulfonamides   152
Sulfonation process   7, 10
Sulfur   153
– modified resins   152
Surfactants   221
Synthanes   304

Tall oil   296
Taphole mixes   303
Temperature resistance of phenolics   150
Terpene-phenolic resins   296
Textile felts   226 f.
Thermal degradation of phenolics   140 f.
– expansion   211
– insulation   3, 213
– resistance   147 f., 161, 209
Thermodynamic properties   132
Thermogravimetric analysis   53, 130, 140, 141, 285
Thermoset flow   205
Thickness swelling   184
Thin layer chromatography   128
Thioglycolic acid   47
Threshold odor concentration   106
– taste concentration   106
Throwing power   246
Titanium-modified resins   151
p-Toluene sulfonic acid   12, 43, 46, 163, 220, 264, 296
Torsion angles   81
– braid analysis   53, 131
Toxicity of gaseous products   113, 164

Toxicology   103 f.
Toxic Substances Control Act   103, 244
Transalkylation   71
Transfer molding   207
Transmission electron microscopy   297
Triethylamine   33, 265
Trihydroxymethylphenol   31, 83, 85
Trimethylsilyl phenols   130
Trioxane   19, 52, 149
Tung oil   45, 235, 284

Underwriters Laboratories   226
UOP process   8
Urea   17, 56, 181, 217, 218
– reaction with   56
Urea-formaldehyde resins   152, 177, 187,
  261, 263, 266
Urea-melamine resins   187
Urea-melamine-phenol resins   180
UV-spectra   75

Veneers   188, 189
Veratrole   69
Vinsol   261
Vinyl-phenolic adhesives   290
Vinyl pyridine latex   297
Viscosity, determination of   135
Vulcanization   288, 296, 297
Vulcanized fiber abrasives   279

Wafer board   185, 186
Wash primer   250
Waste water treatment   108
Water-borne coatings   246
Water-borne adhesives   294
Water content   95
Water dilutability   135
Water reducible systems   249
Wax   191
Weight distribution functions   58
Wood   175
– adhesives   176
– chips   179
– fibers   190
– flour   201, 202
– gluing   192
– materials   175 f.
Wood working industry   3

X-ray diffraction analysis   80, 132
Xylenols   5, 10, 106, 252
–, production methods   9
Xylenol resols   234
Xylok resin   153

Zinc acetate   49, 50
– oxide   248, 291, 293
– silicate   249
– sulfide   271
Zircon   258